D0079138

Analog signal processing
and instrumentation

Analog signal processing and instrumentation

ARIE F. ARBEL

Associate Professor, Department of Electrical Engineering
Technion, Israel Institute of Technology, Haifa

CAMBRIDGE UNIVERSITY PRESS

Cambridge
London   New York   New Rochelle
Melbourne   Sydney

Published by the Press Syndicate of the University of Cambridge
The Pitt Building, Trumpington Street, Cambridge CB2 1RP
32 East 57th Street, New York, NY 10022, USA
296 Beaconsfield Parade, Middle Park, Melbourne 3206, Australia

© Cambridge University Press 1980

First published 1980

Printed in Great Britain by
J. W. Arrowsmith Ltd., Bristol

Library of Congress cataloging in publication data
Arbel, Arie F
Analog signal processing and instrumentation.
Includes bibliographical references and index.
1. Electronic instruments.   2. Signal processing.
I. Title.
TK7878.4.A72   621.38′043   79-13461
ISBN 0 521 22469 1

To Miriam

Contents

Preface

This book is intended for the electronic system designer who utilizes readily available integrated circuit modules. Active circuits are treated as black boxes, whose properties are specified by the manufacturer. Such an approach does not detract from the importance of a thorough understanding of the circuits employed, but this aspect has not been included here.

Most of the material grew out of a two-semester course taught by the author at the Technion – Israel Institute of Technology. Much of it has been taken from his designer's notebook. In order to make the book self-contained without duplicating previous efforts, the author has preferred to adapt some review articles rather than recapitulate topics already well-covered in literature. Also included are excerpts from application notes published by manufacturers – a type of material which often includes excellent expositions of design principles but is generally not available in libraries. Quotations are duly referenced, and permission for using them is gratefully acknowledged.

Chapter 1 deals in detail with the properties of amplifiers, the most widely employed building blocks in analog signal processing.

Chapter 2 classifies transducers with regard to the signal parameter they produce, and deals with aspects arising from their interconnection with amplifiers.

Linear signal processing is discussed in Chapter 3, with emphasis on precision, noise performance and interference suppression. In the section on non-linear amplification and shaping, the comparator and baseline restorer have been singled out for detailed treatment – the first because of its importance as a basic building block, and the latter because outside the field of nuclear electronics it is only rarely employed, in gross disproportion to the variety of potential applications in signal processing.

The properties of an analog signal may be measured either in the frequency or in the time domain. However, because high-speed A/D conversion and the fast Fourier transform are widely available, direct analog signal measurement in the frequency domain is only rarely employed. Hence, Chapter 4 deals specifically with analog signal measurement in the time domain.

The subject of A/D and D/A converters is well documented in the professional literature. Some of this material has been adapted, information scattered throughout various publications has been collected and the subject matter has been brought up-to-date in Chapters 5, 6 and 7.

Chapter 8 considers the design of data acquisition systems, which employ a combination of building blocks described in the preceding chapters.

Signal enhancement utilizes the statistical properties of waveforms and events, respectively, in order to extract signals buried in noise and to improve the accuracy of digital time measurements. Chapters 9 and 10 describe the techniques involved and present practical applications.

One problem faced by the design engineer is the question about where to draw the line between calculations by hand and by a computer. Computer programs for system design are available in cases where an algebraic solution is impracticable, but the complexity of the computations involved often prevents a qualitative understanding of the basic relationships, causing the designer not to see the wood for the trees. It is one objective of this book to present simplified models and equations, which can be used to obtain first-order quantitative results through hand calculations. Such models are defined as having been stripped of all factors obscuring their dominant behavior, which enables the designer to identify their desirable properties and also their shortcomings, and to modify the design so as to enhance the former and suppress the latter. These preliminary computations provide guidelines for possible modifications of the design, and the results thus obtained can be further refined by using them as a starting point for an accurate computer aided analysis or for the construction of a prototype, whose performance may be optimized by trial and error. Chapter 11 provides a practical example for the interplay between hand and machine computations involved in the design of a particular signal processing system.

The appendixes deal with computational methods employed throughout the book. Appendix B.4 describes the derivation of the loop transmission of feedback amplifiers from the Kirchhoff equations, on the basis of the circuit topology employed. This analysis may be extended to multivariable systems employing more than a single amplifier. It also provides the conditions for applying mismatch to feedback amplifiers as described in Appendix B.5, and is simpler and more revealing than the (inaccurate) one employed in most textbooks.

Particular emphasis is laid on computations based on a simplified system as referred to above. One kind of simplification relates to the response of feedback amplifiers under conditions of infinite loop transmission at dc. The simplified loop transmission may be derived from the actual one by employing the gain-bandwidth theorem, which is defined in Appendix C.3. This simplification is useful in evaluating stability conditions, transient response, settling time, sensitivity, and noise performance.

Another kind of simplification deals with the design for short settling time. Initial data may be obtained from a design of the loop transmission, for which the dominant pair among the closed-loop singularities forms a double pole on the real axis, before branching out into the complex frequency plane. The effect of the various design parameters – and that of signal delay within the loop – on this limiting condition are readily identifiable from relatively simple computations. This method is based in part on the simplified loop transmission introduced above, and is dealt with in Appendixes C.1 and C.2.

A further kind of simplification deals with the characterization of a system response in the frequency and time domains by its time moments. Appendix D deals with this aspect. The above simplifications provide insight even into comparatively involved systems and frequently yield, by inspection, results which would otherwise entail rather lengthy calculations.

Normalization, frequently employed throughout the book, is another important aspect. Representation of parameters such as performance characteristics, figure of merit or response in normalized form facilitates comparison between related systems with widely different parameters, and aids the practical engineer in comprehending their basic properties.

Relevant problems are scattered through the book and formulated so as to bring out finer points of the subject matter covered in the corresponding chapter; they add important information to the text and should be treated as an essential part of it. Many problems have been chosen as illustrations of design problems which are thought-provoking and promote the process of learning.

I am indebted to the 'nuclear electronics community': numerous discussions with those involved in nuclear electronics have left their imprint on the material presented in this book. It is a pleasure to acknowledge fruitful discussions with Professor Israel Bar-David and

Amnon Adin. I wish to thank the many students whose questions and suggestions have contributed to the clearness of presentation. Thanks are also due to Jetti Tibor for her meticulous typing of the often barely legible manuscript. Financial support by the Technion Vice President's research fund is gratefully acknowledged.

Arie F. Arbel
Haifa, Israel   March 1979

Introduction

Electronic instrumentation serves to amplify, shape and convert an analog signal into a form suitable for measurement, permanent recording and/or further processing by a digital computer. In many cases, the analog signal is derived from a transducer which converts the physical quantity to be measured into an electrical signal.

The following diagram shows a typical example of a signal processing system, in which an arbitrary analog signal is simultaneously displayed, converted by an A/D (*analog-to-digital*) converter into a digital format, and recorded in analog and digital form. In certain cases, such as the automatic control of an industrial process, the processed signal will be analysed by a computer, which can then control the original process.

In this book we shall mainly deal with analog instrumentation employed between transducer and digital computer, with a short chapter

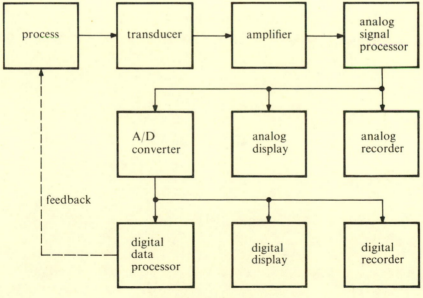

Signal processing system.

on transducers themselves. Let us briefly discuss the purpose of the various blocks in the diagram.

In most cases, the transducer signal must be amplified to a level suitable for further processing. Frequently, special care must be taken to minimize the unavoidable deterioration of the SNR (*signal-to-noise ratio*) available from the transducer due to the noise contribution of the preamplifier employed.

The signal processor prepares the signal for the extraction of the desired information. In its broadest definition, this stage may include a filter which optimizes the SNR, and the means of obtaining amplitude and/or timing information from the signal.

Finally, the desired information is delivered to one or more output stages, where it is displayed, recorded or converted into digital form.

Signal processing can also be performed digitally; this is slower but preserves accuracy better, and is considerably more flexible. At any point, the digital signal can be reconverted into analog form by a D/A (*digital-to-analog*) converter for analog recording, display and/or further processing. The choice of the most appropriate arrangement must be taken by the system designer.

1

Basic and feedback amplifiers

Amplifiers are an essential part of most analog signal processing systems. Clearly the most significant parameter of an amplifier is its gain, which may be specified as power, voltage or current gain. Strictly speaking, gain is a dimensionless factor, but transimpedance or transadmittance certainly also defines the 'gain' of corresponding amplifiers.

In most cases of signal processing for instrumentation systems, either voltage or current is the signal parameter of interest. The other parameter is invariably undesirable and, although always present, should be minimized by proper design. This is the basis of the principle of *mismatch*, which is widely employed in the design of broadband signal processing systems.

We shall distinguish between basic amplifiers and feedback amplifiers, the multitude of commercially available IC (*integrated circuit*) amplifiers being representative of the former, and the latter consisting of a basic amplifier connected to a passive feedback network which stabilizes the overall transfer function. To become familiar with basic amplifiers, we shall discuss: (*a*) their idealized characteristics; (*b*) deviations from their idealized characteristics; and (*c*) the use of feedback in the stabilization of transfer functions.

1.1 Classification of basic amplifiers

The *transfer function* of an amplifier is defined as the ratio between the output and the input parameters. With a choice between voltage and current as two possible information-bearing parameters, we have a total of four combinations between output and input parameters. This leads to four basic amplifiers, whose ideal characteristics are tabulated in Table 1.1 and whose equivalent diagrams are shown in Fig. 1.1: the voltage controlled voltage source (VCVS) or voltage amplifier, the current controlled current source (CCCS) or current amplifier, the voltage controlled current source (VCCS) or transadmittance amplifier, and finally the current controlled voltage source (CCVS) or transimpedance amplifier.

3

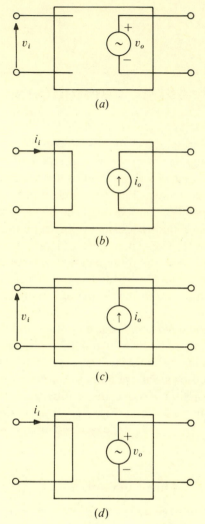

Fig. 1.1. The four idealized basic amplifiers: (*a*) VCVS; (*b*) CCCS; (*c*) VCCS; (*d*) CCVS.

The idealized characteristics of a particular kind of amplifier are easily arrived at, if we consider its insertion in a signal processing chain. For example, consider the arrangement shown in Fig. 1.2, where the information-bearing signal parameter is current throughout. An ideal amplifier will maximize the signal transfer, which in this case requires an ideal current sink at its input ($z_i = 0$), an ideal current source at its output ($z_o = \infty$), and finally an infinite and frequency-independent current gain

Table 1.1   *Ideal characteristics of basic amplifiers*

	Symbol	Impedance		Parameter	
		Input	Output	Input	Output
VCVS	A_v	∞	0	v	v
CCCS	A_i	0	∞	i	i
VCCS	Y_T	∞	∞	v	i
CCVS	Z_T	0	0	i	v

A_i. Identical considerations lead to the values of z_i and z_o given in Table 1.1 for the remaining kinds of amplifiers.

Both the input and output of an amplifier may be either single-ended or differential. An amplifier with *single-ended input* has one input and one reference terminal, which is usually grounded. Using the reference terminal as input and the input terminal as reference adversely affects the specified input impedance and often many other parameters. With a *differential input*, the two input terminals are equivalent except for polarity, and the input impedance remains unchanged if either is employed as reference. Most voltage amplifiers use a differential input stage.

A *differential output* delivers two signals of equal amplitude but opposite polarity from two separate output terminals. Most amplifiers, however, have a single output terminal only – a *single-ended output*.

1.2   Differential voltage amplifiers

Of the four kinds of basic amplifiers, the differential input single-ended output voltage amplifier is most extensively used. A model for its transfer function is shown in Fig. 1.3, with

$$v_o = A_b v_b - A_a v_a. \tag{1.1}$$

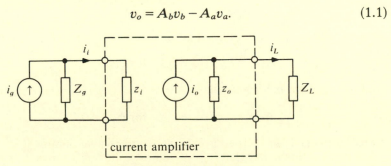

Fig. 1.2. Current amplifier; $i_o = A_i i_i$.

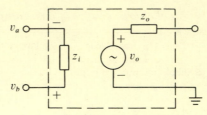

Fig. 1.3. Differential amplifier.

For an ideal differential amplifier, $A_a = A_b = A$. Hence an ideal differential amplifier responds only to the voltage difference between v_b and v_a. Its response to a common mode voltage (for which the difference between v_b and v_a remains constant) is zero.

The following sections deal with specifications commonly listed in the manufacturers' data-sheets. All specifications should be checked carefully against design requirements and environmental conditions, before a particular device is chosen. Watch the small print on a data-sheet!

1.2.1 Common mode rejection ratio

In practice, because the gains A_a and A_b are finite and slightly different, common mode input voltages are not entirely subtracted at the output. In order to evaluate the performance of such an amplifier, we consider the input signal to consist of two components: $v_{DM} = (v_b - v_a)/2$, the *differential mode input* signal, which usually conveys useful information, and $v_{CM} = (v_b + v_a)/2$, the *common mode input* signal, which is an undesirable by-product.

Since we consider the small-signal response of the amplifier to be linear, we may employ superposition in order to obtain v_o as a function of v_{CM} and v_{DM}:

$$v_o = v_{DM}A_{DM} + v_{CM}A_{CM}, \tag{1.2}$$

with the *differential gain* defined as

$$A_{DM} = 2v_o/(v_b - v_a) \quad \text{for } v_{CM} = 0, \tag{1.3}$$

and the *common mode gain* as

$$A_{CM} = 2v_o/(v_b + v_a) \quad \text{for } v_{DM} = 0. \tag{1.4}$$

Substituting (1.1) in (1.3) and (1.4), respectively, we obtain

$$A_{DM} = A_b + A_a \quad \text{and} \quad A_{CM} = A_b - A_a, \tag{1.5}$$

from which it follows that for an ideal differential amplifier $A_{CM} = 0$, since $A_b = A_a$.

This is in agreement with the consideration that the input circuit of an ideal differential amplifier is floating so that its output voltage is not affected by the voltage between the two input terminals and ground (v_{CM}), as long as the voltage difference between them (v_{DM}) remains unchanged.

The *common mode rejection ratio* (CMRR) is defined as

$$\text{CMRR} = A_{DM}/A_{CM}. \tag{1.6}$$

It serves as a useful figure of merit for differential amplifiers, and its value in decibels is usually included in the manufacturer's data-sheet together with A_{DM}, the 'differential gain' or simply 'voltage gain'. The parameters A_a and A_b are not normally specified, although the difference between them is the physical cause for a finite CMRR.

Note that, because we define $v_{DM} = (v_b - b_a)/2$, the value of A_{DM} is twice that of what is commonly defined as the gain of a differential amplifier. The factor of $\frac{1}{2}$ in v_{DM} has been introduced for reasons of mathematical convenience.

Another interpretation of the CMRR is obtained by substituting (1.3) and (1.4) in (1.6):

$$\text{CMRR} = \frac{(v_b + v_a)/v_o}{(v_b - v_a)/v_o}. \tag{1.7}$$

Accordingly, the CMRR is defined as the ratio between the amplitudes of v_{CM} and v_{DM} producing the same output voltage. Finally we may substitute (1.5) in (1.6) to obtain a third definition:

$$\text{CMRR} = (A_b + A_a)/(A_b - A_a). \tag{1.8}$$

Practical values of CMRR for commercial IC amplifiers lie between 60 and 100 dB.

Specifying CMRR precisely[1] is complicated by the fact that it can be a highly non-linear function of v_{CM} and also varies with temperature. Usually, published CMRR specifications apply only to dc or very low-frequency input signals (although this may not be stated). Its value decreases beyond the half-power frequency ω_h.

1.2.2 Frequency dependence of amplifier gain

The gain of practical amplifiers is not only finite but also decreases beyond a certain frequency. The Bode plot of a typical 'uncompensated

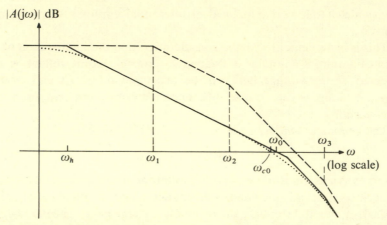

Fig. 1.4. Typical Bode plot of uncompensated (dashed line) and compensated (solid line) second stage of IC amplifier.

amplifier' gain function is shown by the dashed line in Fig. 1.4, indicating that the amplifier introduces two corner frequencies ω_1 and ω_2 within the passband of the amplifier. As a consequence, disregarding the pole at ω_3, the asymptotic frequency response of the complete amplifier falls off at high frequencies with 40 dB/decade, corresponding to a phaseshift of 180° at $\omega = \infty$. Since such an amplifier creates stability problems if used in a feedback circuit, it is customary to 'compensate' it by an internal capacitor.

The effect of this *pole-splitting* capacitor[2] is demonstrated by the solid line in Fig. 1.4, showing that the poles at ω_1 and ω_2 separate into a dominant one at ω_h, the *half-power frequency*, and another beyond ω_0, the *extrapolated gain-bandwidth product*. For any transfer function exhibiting a dominant pole, the gain-bandwidth product is equal to that frequency at which the extrapolated gain of the region over which the gain drops by 20 dB/decade, falls to unity.

A simplified block diagram typical of most compensated IC amplifiers is shown in Fig. 1.5. The transconductance gm is reasonably well stabilized, and the pole-splitting capacitor C connected between the terminals c and d determines the transimpedance of the amplifier A. Hence, the transfer function is

$$\lim_{|A(j\omega)|\to\infty} v_o/(v_b - v_a) \approx gm/j\omega C. \tag{1.9}$$

This is true to a very good approximation over the frequency range $\omega_h < \omega < \omega_0 = gm/C$.

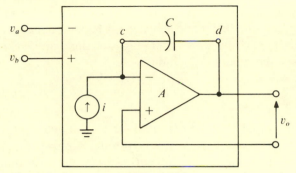

Fig. 1.5. Compensated IC amplifier; $i = \mathrm{gm}\,(v_a - v_b)$.

Although C is an integrated capacitor, in externally compensated amplifiers another capacitor may be connected in parallel with C, terminals c and d being accessible through connector pins. This enables the designer to control ω_0.

The solid line in Fig. 1.4 shows the asymptotic Bode plot of a practical compensated amplifier, and the dotted line shows the actual response

$$A(\mathrm{j}\omega) = A(0) / \Big[(1 + \mathrm{j}\omega/\omega_h) \prod^{n} (1 + \mathrm{j}\omega/\omega_i) \Big], \qquad (1.10)$$

with $\omega_0 = A(0)\omega_h$.

The parameter ω_0 is reasonably well controlled, whereas the production spread of $A(0)$ and hence of ω_h may be of the order of one decade. This is fortunate, since the stability of the amplifier employed in a feedback circuit is primarily a function of ω_0 and not of ω_h (see Section 1.7.3).

The frequency ω_0 is to be distinguished from ω_{c0}, the *unity-gain frequency*. As shown in Fig. 1.4, $\omega_{c0} < \omega_0$ because of the parasitic poles. These poles are due to the ω_i in (1.10) and are always present. Two of them are shown in Fig. 1.4. With most internally compensated amplifiers, they are located sufficiently far beyond ω_0 to make the total phaseshift at ω_{c0} less than 180°. This ensures the amplifier's unconditional stability when used as a feedback amplifier if no additional phase lag is introduced by the feedback network.

The small-signal frequency response of commercial amplifiers is generally specified by the gain-bandwidth product $f_0 = \omega_0/2\pi$, where 'small-signal' indicates that, in general, it is not possible to obtain a FS (*full-scale*) voltage swing at high frequencies due to slew rate limiting. For most differential amplifiers, f_0 is valid for both the inverting and the non-inverting input. However, some wideband amplifiers with feedforward design (a circuit technique in which the high-frequency components of the signal bypass the slow input stages of the amplifier) have a fast response only on the inverting input.

1.2.3 Offset

Ideally, the output of a differential voltage amplifier whose input terminals are each connected to ground through arbitrary resistors should equal zero. Any output voltage appearing under these conditions will be due to undesirable internal voltage and current sources, whose total effect may be lumped into an independent series voltage source termed the *offset voltage* and an independent parallel current source termed the *offset current*, both being connected to the amplifier's input. This procedure can be shown formally by considering the amplifier as an active two-port (Fig. 1.6(*a*)), describable by the chain equations[3]

$$\begin{bmatrix} v_1 \\ i_1 \end{bmatrix} = \begin{bmatrix} a & b \\ c & d \end{bmatrix} \begin{bmatrix} v_2 \\ i_2 \end{bmatrix}. \tag{1.11}$$

The internal independent sources causing the offset can be extracted from the amplifier, as shown in Fig. 1.6(*b*), by rewriting (1.11) as

$$\left. \begin{aligned} v_1 &= \bar{a}v_2 + \bar{b}i_2 + V \\ i_1 &= \bar{c}v_2 + \bar{d}i_2 + I \end{aligned} \right\}, \tag{1.12}$$

where V and I are the independent offset sources referred to the amplifier's input, and $\bar{a}$, $\bar{b}$, $\bar{c}$ and $\bar{d}$ are the chain matrix coefficients of the same amplifier but without offset. The same technique will be used in Section 1.2.4 for referring the noise sources of an amplifier to its input.[4]

The preceding analysis relates to a single-ended input circuit. The offset sources of an amplifier with a differential input stage are modelled

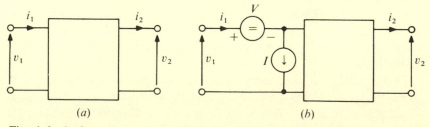

(a)   (b)

Fig. 1.6. Active two-port, described by chain equations. Offset sources (*a*) included, (*b*) extracted.

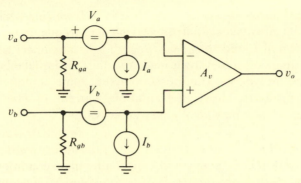

Fig. 1.7. Differential amplifier with dc offset sources.

in Fig. 1.7, in which the amplifier input terminals are connected to source resistors R_{ga} and R_{gb} in order to demonstrate computation of the output offset voltage for practical conditions. In this computation it is good engineering practice to disregard the difference in the gain factors A_a and A_b, which can be shown to have a secondary effect. Hence, $A_a = A_b = A_v$, and

$$v_o = A_v(V_a - V_b + I_a R_{ga} - I_b R_{gb}). \qquad (1.13)$$

Equation (1.13) shows that (*a*) V_a and V_b can be lumped into a single offset voltage source $\Delta V = V_a - V_b$, and (*b*) if $R_{ga} = R_{gb} = R_g$, then we may also employ a single differential bias current offset source $\Delta I = I_a - I_b$ as shown in Fig. 1.8:

$$V_o = A_v(\Delta V + 2\Delta I R_g). \qquad (1.14)$$

The condition of equal resistance seen by the input terminals of a difference amplifier is common design practice for dc amplifiers, since it

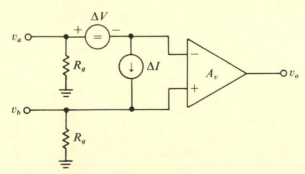

Fig. 1.8. Differential amplifier with simplified dc offset sources.

restricts the current offset to the effect of ΔI which is, with ICs, a controlled quantity whose value is often smaller by one order of magnitude than that of I_a and I_b.

In most amplifiers, provision is made to adjust the initial offset voltage to zero with an external trim potentiometer. Some amplifiers employ internal compensation to reduce initial bias current, in which case the residual bias current may be of either polarity.

Temperature drift[1]

Temperature drift is by far the most important source of errors in most dc applications. The *temperature coefficients* of ΔV and ΔI are defined as the average values of the derivatives over a specified temperature range. In general, however, drift is a non-linear function of temperature and the derivative values are greater at extremes of temperature than around normal ambient (25 °C).

Because of this non-linearity, a definitive specification of temperature coefficients is only meaningful if the specification method is known. Ambiguity on this point can lead to some startling conclusions. For example, one popular method is to subtract arithmetically the measured offset values at the upper and lower temperature extremes and then divide this difference by the temperature excursion. This can yield an extremely misleading result, particularly where offset drifts in the same direction at the two extremes. For example, consider Fig. 1.9. With ΔV_H and ΔV_L denoting the offset voltage at high and low temperatures, respectively, and $\mathcal{T}_H$, $\mathcal{T}_L$ and $\mathcal{T}_0$ the high, low and reference temperatures, the temperature coefficient as defined above becomes

$$\frac{\mathrm{d}(\Delta V)}{\mathrm{d}\mathcal{T}} = \frac{\Delta V_H - \Delta V_L}{\mathcal{T}_H - \mathcal{T}_L} = 0 \ \mu\mathrm{V}/°\mathrm{C}!$$

Analog Devices employs the 'butterfly' characteristic for drift specifications, which overcomes the above deficiencies. Referring to Fig. 1.10, the 'butterfly' characteristic insures that if the amplifier offset is adjusted to zero at $\mathcal{T}_0$, the offset at any temperature would, in no case,

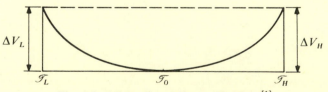

Fig. 1.9. Misleading drift specification.[1]

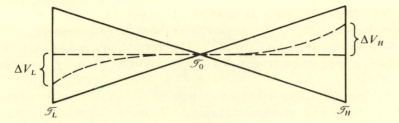

Fig. 1.10. 'Butterfly' characteristic specification of temperature drift.[1]

exceed the value predicted by multiplying the specified drift rate (in $\mu V/^\circ C$) by the temperature excursion.

The actual specified drift is

$$.\Delta V/\Delta \mathcal{T} \leqslant |\Delta H_H|/(\mathcal{T}_H - \mathcal{T}_0)$$

or

$$\Delta V/\Delta \mathcal{T} \leqslant |\Delta V_L|/(\mathcal{T}_0 - \mathcal{T}_L),$$

whichever is larger.

Drift in the value of the offset sources may also occur because of variation in the supply voltage. As components age, drift occurs as a function of time and thus will be specified for high-precision amplifiers.

1.2.4 Noise[5]

The process of amplification invariably introduces noise. It therefore leads to a deterioration in the SNR available at the source, or to a noise figure F greater than 0 dB (see Appendix A.5).

Design philosophy
The technique of noise figure computation is widely used in the design of communication receivers. The design target is to minimize the noise figure of a cascade of n amplifying devices, derived in Appendix A.5 as

$$F = F_1 + (F_2 - 1)/P_{a1} + \cdots + (F_n - 1)\bigg/ \prod^{n-1} P_{ai}.$$

The relatively narrow frequency band involved in communication makes it possible to employ matching networks in order to modify the noise figure and power gain of the individual stages in such a way as to optimize the overall noise figure F.

The design philosophy for instrumentation is in most cases different from that employed in communication systems for two reasons.

(*a*) The signal involved covers a wide range of frequencies (frequently including dc), which eliminates the possibility of using matching networks.

(*b*) The information-bearing parameter is in most cases either voltage or current, leading to mismatch oriented design throughout the system. Hence, as shown in Appendix A.5, we shall express the noise figure in terms of voltage or current ratios rather than power ratios.

Amplifier model

Evaluation of the noise figure requires an amplifier model with the internal noise sources referred to its input, which involves the same reasoning as with offset sources. The resulting model for a differential amplifier is shown in Fig. 1.11(*b*), with the spectral densities given as

$$S_{va} = 4k\mathcal{T}R_{sa}, \qquad S_{vb} = 4k\mathcal{T}R_{sb}, \qquad S_{ia} = 4k\mathcal{T}/R_{pa}, \qquad S_{ib} = 4k\mathcal{T}/R_{pb},$$
$$(1.15a)$$

where k is Boltzmann's constant, $\mathcal{T}$ the temperature in Kelvin, and R_s and R_p are the equivalent series and parallel noise resistances of the amplifier, referred to its corresponding input terminals a and b (see Appendix A.3).

For convenience, the total noise of an amplifier or gain element is often modelled in terms of the *equivalent series noise resistance*

$$R_{eq} = R_s + |Z_g(j\omega)|^2/R_p, \qquad (1.15b)$$

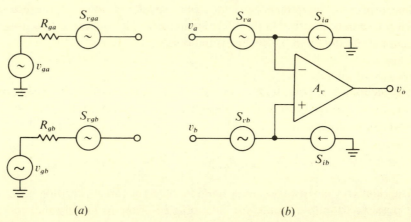

(*a*)   (*b*)

Fig. 1.11. Equivalent noise diagram of (*a*) signal sources, and (*b*) differential amplifier.

whose second term is obtained by transforming the parallel mean squared noise current density across the source impedance Z_g into its equivalent Thevenin voltage source.

Transistor noise parameters

The noise performance of a well-designed amplifier is dominated by the input stage, which consists in most practical cases of a single transistor or two transistors connected as a differential pair. Because of their importance, we shall define the noise parameters of transistors (without proof).

The equivalent series noise resistance of a bipolar transistor is

$$R_{eq} = r_x + (r_e/2) + |Z_g(j\omega)| + r_e + r_x|^2 |A(j\omega)|^2 / 2h_{FE}r_e \quad (1.16a)$$

where $r_e = k\mathcal{T}/qI_E$ is the dynamic resistance of the base-emitter diode, r_x the *base-spreading resistance* and h_{FE} the CE (*common emitter*) dc current gain. The factor $|A(j\omega)|^2$ models the frequency dependence of the noise due to the random partition of charge carriers between the base and the collector:

$$|A(\omega)|^2 = 1 + f^2/f_c^2, \qquad f_c = (f_\alpha f_\beta)^{1/2} \quad (1.16b)$$

where f_c is the *noise corner frequency* of the transistor, and f_α, f_β are the half-power frequencies of the transistor's CB (*common base*) and CE current gains.

The noise parameters of a junction FET (*field effect transistor*) are

$$R_{eq} = R_s + |Z_g(j\omega)|^2 / R_p \quad (1.17)$$

where $R_s \approx 0.7/gm$, with gm the transconductance of the FET; $R_p = 2V_T/I_l$, where $V_T = k\mathcal{T}/q$, the *voltage equivalent of the absolute temperature*, and I_l is the leakage current of the gate. Note, however, that in the noise model of an FET the series noise squared voltage source $4k\mathcal{T}R_s$ is connected internally to the gate, and the gate source capacitance appears directly in parallel with Z_g, whereas in the noise model of a bipolar transistor, the series noise squared voltage density $4k\mathcal{T}(r_x + r_e/2)$ is connected externally between the base emitter impedance and Z_g. Hence, the term in the output, due to series noise, is a function of $Z_g(j\omega)$ in the case of a bipolar transistor, but not in the case of an FET.

These relationships should be taken into account in the complete evaluation of an amplifier's noise performance.

Although there is a certain degree of correlation between parallel and series noise sources of transistors and amplifiers, this correlation is negligible for most practical purposes and will therefore be disregarded throughout this book. We shall furthermore restrict our considerations to the frequency range over which the spectral density of the noise sources of an amplifier may be considered as being independent of frequency. Thus, low-frequency noise (such as flicker noise and popcorn noise) will be ignored. Similarly, we shall disregard the increase in amplifier noise referred to input beyond the noise corner frequency, due to a reduction in the gain of the amplifying devices at high frequencies. Such a simplified treatment is sufficient to explain the basic noise properties of amplifiers.

Specifications for high-performance ICs generally include noise data. As an example, Analog Devices specifies low-frequency noise in a passband of 0.01 Hz to 1 Hz as peak-to-peak with a 3.3σ uncertainty, implying that 99.9% of the observed peak-to-peak excursions fall within the specified limits. Wideband noise in a passband of 5 Hz to 50 kHz is specified as rms.

Information about the noise performance of high-performance low-noise amplifiers is available from plots of source resistance versus frequency, with the noise figure or the equivalent noise temperature (see (1.22)) as parameter.

Computation of noise figure

The noise figure F is a measure of an amplifier's noise contribution if it is connected to a signal source of known SNR. Consider Fig. 1.11 with one signal source connected to each input terminal of the differential amplifier. First, the total noise voltage at the output must be found, which is then referred to the desired input terminal and modelled by an equivalent noise voltage source in series with the signal voltage source:

$$\overline{v_{no}^2} = 4k\mathcal{T}(R_{ga} + R_{gb} + R_{sa} + R_{sb} + R_{ga}^2/R_{pa} + R_{gb}^2/R_{pb}) \int_0^\infty |A_v(j\omega)|^2 \, df.$$

$$(1.18a)$$

If we assume that

$$R_{sa} = R_{sb} = R_s, \qquad R_{pa} = R_{pb} = R_p, \qquad R_{ga} = R_{gb} = R_g,$$

we obtain

$$\overline{v_{ng}^2} = 8k\mathcal{T}R_g \, \Delta f \quad \text{and} \quad v_{ni}^2 = 8k\mathcal{T}(R_s + R_g^2/R_p) \, \Delta f, \quad (1.18b)$$

where $\overline{v_{no}^2}$ is referred to either input and Δf denotes the ENB (*equivalent noise bandwidth*). Hence

$$F = 1 + v_{ni}^2/v_{ng}^2 = 1 + R_s/R_g + R_g/R_p. \qquad (1.19a)$$

F attains a minimum value for the optimum source resistance

$$R_{g\,\text{opt}} = (R_s R_p)^{1/2},\tag{1.20}$$

which gives

$$F_{\min} = F(R_{g\,\text{opt}}) = 1 + 2(R_s/R_p)^{1/2}.\tag{1.19b}$$

In applications such as computerized optimization, it is convenient to express the noise figure as the sum of $F_{\min}$ and a function of the difference between $R_{g\,\text{opt}}$ and the actual source resistance:[4,6]

$$F = F_{\min} + (R_{g\,\text{opt}} - R_g)^2/R_g R_p = F_{\min} + (1 - R_g/R_{g\,\text{opt}})^2 R_s/R_g.\tag{1.21}$$

Equation (1.21) can be extended to the general case of complex source impedance and correlation between the two amplifier noise sources.

The equivalent noise temperature
Equations (1.19a) and (1.21) imply that source and amplifier are at the same temperature. But noise due to thermal fluctuations and leakage may be reduced by cooling. Hence, it is customary in certain cases to cool the transducer and/or the amplifier. As a result, the amplifier may be at a temperature different to that of the source. In that case, these equations are not applicable, and the noise performance of the amplifier should be specified by its *equivalent noise temperature*.

The equivalent noise temperature $\mathcal{T}_{\text{eq}}$ is derived from the total noise voltage referred to the amplifier's input, in accordance with (1.18b):

$$\overline{v_{ng}^2} + \overline{v_{ni}^2} = 4k(\mathcal{T}_g R_g + \mathcal{T}_a R_{\text{eq}})\,\Delta f = 4k R_g(\mathcal{T}_g + \mathcal{T}_{\text{eq}})\,\Delta f,$$
$$R_{\text{eq}} = R_s + R_g^2/R_p \quad \text{and} \quad \mathcal{T}_{\text{eq}} = \mathcal{T}_a R_{\text{eq}}/R_g.\tag{1.22}$$

Hence, $\mathcal{T}_{\text{eq}}$ is defined as the increase in source resistance temperature required to produce the observed available noise power at the output of the amplifier, the amplifier being noiseless. Unlike the noise figure defined by (1.19a), $\mathcal{T}_{\text{eq}}$ is a function of $\mathcal{T}_a$, the temperature of the amplifier, and finally

$$F(\mathcal{T}) = 1 + \mathcal{T}_{\text{eq}}/\mathcal{T}_g.$$

It is customary to specify the equivalent noise temperature of an amplifier by a plot of source resistance versus frequency, with $\mathcal{T}_{\text{eq}}$ as parameter.[5(c)] Usually, these plots are taken for $\mathcal{T}_a = 290$ K.

Noise matching
The technique of minimizing F has been termed *noise matching*, which is not to be confused with the concept of a matched filter. Practically, noise

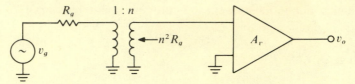

Fig. 1.12. Transformer matching between source and amplifier.

matching can be achieved by modifying éither the source impedance seen by the amplifier or the noise parameters of the amplifier, as will be shown in the sequel.

If we assume the noise parameters of the amplifier are invariant, (1.20) defines the optimum source resistance needed to achieve the minimum possible noise figure. This conclusion is sometimes claimed to be misleading, since it seems to suggest that the noise figure could be improved simply by connecting a resistance in series with R_g when the latter is less than $R_{g\,\text{opt}}$. However, as is well known, the correct way of modifying R_g is to use a lossless transformer.

This conflict is resolved if we recognize that the dubious improvement in F is achieved by *increasing* the source noise, which causes a reduction in the SNR of the source alone. Although this reduces the relative noise contribution of the amplifier, it also results in a deterioration of the SNR measured at the output of the amplifier, clearly an undesirable result.

Impedance matching by a lossless transformer, as shown in Fig. 1.12, modifies the source resistance seen by the amplifier without affecting the

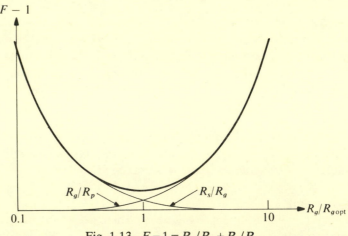

Fig. 1.13. $F - 1 = R_g/R_p + R_s/R_g$.

SNR of the source. This is therefore the correct way of noise matching if the noise parameters of the amplifier are invariant.

The physical significance of (1.19a) and (1.21) is illustrated by Fig. 1.13, which shows that series noise is dominant if $R_g < R_{g\,opt}$, whereas parallel noise is dominant if $R_g > R_{g\,opt}$. If $R_g = R_{g\,opt}$, parallel and series noise contribute equally, and this is the condition for the amplifier's noise contribution to be a minimum.

As already mentioned, noise matching by transformer is mainly employed with narrowband high-frequency amplifiers, due to the restricted frequency response of matching networks or transformers. A different technique, applicable also to broadband amplification including dc, modifies the amplifier's noise parameters. This may be accomplished by parallel connection of n identical gain elements as shown in Fig. 1.14, for which the combined equivalent series and parallel noise resistances are R_s/n and R_p/n, respectively. Hence, the optimal number of gain

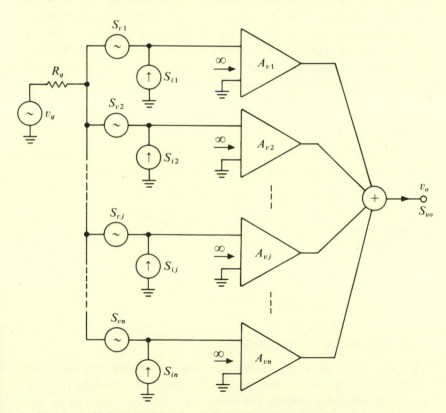

Fig. 1.14. Noise matching through parallel connection of gain elements.

elements is

$$n_{\mathrm{opt}} = (R_s R_p)^{1/2}/R_g. \tag{1.23}$$

We notice that parallel connection of gain elements reduces the series but increases the parallel noise contribution. So, if the series noise contribution of a single element connected to the same source turns out to be dominant, connecting gain elements in parallel will improve the noise figure. If the opposite is true, parallel connection leads to a deterioration of the noise figure and should not be employed. With FETs, whose parallel noise is extremely low and can even be further reduced by cooling, parallel connection is frequently employed with low-noise pre-amplifiers designed for capacitive detectors (see Section 1.7.6) in order to obtain optimal noise matching. In principle it is also possible to use several complete amplifiers in parallel, but this is hardly ever done.

With bipolar transistors it can be shown that an inversely proportional non-linear relationship exists between $R_{g\,\mathrm{opt}}$ and the emitter current. This may be utilized for noise matching those ICs which provide an external adjustment of the quiescent current of the bipolar differential input stage. Note, however, that this relationship is restricted for low values of collector current by a reduction in the current gain, and for high current values by the series noise due to the base spreading resistance r_x, which limits any further significant reduction of the series noise.

Problems

1.1. Verify (1.23) and (1.21).

1.2. Fig. 1.15 shows an amplifier with single-ended input, with

$$A_v(\mathrm{j}\omega) = A(0)/(1 + \mathrm{j}\omega\tau).$$

All externally connected resistors contribute noise. The reference terminal v_b is grounded through a small resistor R_1.
   (a) Compute the noise figure $F(f)$.
   (b) What value of r_o makes F nearly independent of A_v and R_L? (r_o is the dynamic output resistance of the amplifier, which does *not* contribute noise.)
   (c) Compute the rms noise across R_L if $r_o = 0$.

1.2.5   Non-linearity and distortion

Throughout the preceding sections we have considered only small-signal transfer functions, which require a definition of the operating point. The region in which the amplifier operates, in this case, is said to be quasi-linear. Large-signal analysis should take into account the deviation of the

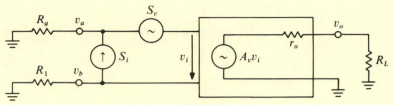

Fig. 1.15. Circuit for problem 1.2; $z_i = \infty$.

amplifier response from this linear relationship, which may be specified as non-linearity of gain in the time domain, or as distortion in the frequency domain.

Time domain non-linearity is derived from observing the input–output relationship, with a linear ramp being applied to the input. This definition is useful in considering a system including an A/D converter (see Section 6.4). Non-linear amplification distorts the ratio between signal levels and therefore affects the linearity of conversion. Hence, the performance of a perfect A/D converter is wasted if it is fed by a non-linear amplifier.

Specification of distortion in the frequency domain is customary in applications such as high-fidelity sound reproduction systems or high-frequency power amplification.

Definition of linearity in the time domain
Linearity may be specified as integral or differential linearity. *Integral linearity* is a measure of the maximum deviation from a straight line of a plot of the ideal input–output characteristic of an amplifier over its operating range, expressed as a percentage of FS.

Differential linearity is a considerably more sensitive measure than integral linearity. It measures the deviation of the small-signal gain from its average value over the full output range of the amplifier. Its value is typically worse by one order of magnitude than that obtained for integral linearity.

Distortion in the frequency domain
The transfer function of an amplifier is, in practice, a non-linear function of the output voltage v_o similar to the one shown in Fig. 1.16:

$$v_o = f(v_g). \tag{1.24}$$

This can be expanded into a Maclaurin series:

$$v_o = V_o(0) + \frac{\partial v_o}{\partial v_g} v_g + \sum_2^n \frac{1}{i!} \frac{\partial^i (v_o)}{\partial (v_g)^i} (v_g)^i = V_o(0) + A_1 v_g + \sum_2^n A_i (v_g)^i, \tag{1.25}$$

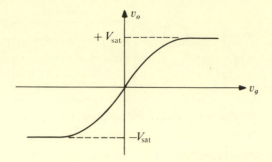

Fig. 1.16. Non-linear amplifier gain characteristics.

where the first term defines the quiescent dc operating point, the second the small-signal linear gain, and the remainder the non-linearity of the transfer function.

Computation of the first three terms for $v_g = V_g \cos \omega t$ yields

$$v_o = V_o(0) + \tfrac{1}{2}A_2 V_g^2 + A_1 V_g \cos \omega t + \tfrac{1}{2}A_2 V_g^2 \cos 2\omega t \qquad (1.26)$$

in which we recognize $\tfrac{1}{2}A_2 V_g^2$ as the dc shift due to second harmonic distortion and $\tfrac{1}{2}A_2 V_g^2 \cos 2\omega t$ as the second harmonic distortion term. Defining $\sum_2^n V_{Si}$ as the total dc shift due to all harmonics and D_i, the *i*th *harmonic distortion factor*, as the ratio between the amplitude of the *i*th harmonic and the fundamental, (1.25) can be written as

$$v_o = V_o(0) + \sum_2^n V_{Si} + A_1 V_g \left(\cos \omega t + \sum_2^n D_i \cos i\omega t \right). \qquad (1.27)$$

The distortion terms in (1.27) can be modelled as shown in Fig. 1.17 as independent sources in series with the amplifier output. This model will be employed in the section dealing with reduction of distortion by feedback.

Another kind of distortion – slew induced distortion – occurs at those frequencies at which the slope of the undistorted output signal would exceed the slewing capability of the amplifier[7] (see Section 1.2.7).

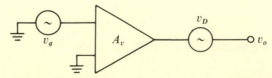

Fig. 1.17. Modelling of distortion; $v_g = V_g \cos \omega t$, $v_D = V_g A_v \sum_2^n D_i \cos i\omega t$.

1.2.6 Rated output[1]

Rated output voltage is the minimum peak output voltage which is guaranteed at rated output current before clipping or excessive non-linearity occurs. *Rated output current* is the minimum guaranteed value of current supplied at the rated output voltage.

1.2.7 Slewing rate and large-signal response

The *slewing rate S* of an amplifier defines the maximum rate of change of the output voltage for a large, overloading input step. It is usually expressed in $V/\mu s$ and may have a different value for positive and negative polarity.

This effect can be explained by again considering Fig. 1.5, which models a typical compensated IC. We have $v_o \simeq (v_b - v_a)\mathrm{gm}/\mathrm{j}\omega C$, C being the compensating capacitor which determines the gain-bandwidth product $f_0 = \mathrm{gm}/2\pi C$. The slewing rate is determined by the maximum current I_{max} which can be supplied by the transconductance gm:

$$S = I_{max}/C \text{ V/s}.$$

The large- and small-signal response characteristics of amplifiers differ substantially. An amplifier will not respond to large-signal changes as fast as the small-signal bandwidth characteristics would predict, primarily because of slew rate limitations. Large-signal response is specified at two frequencies:

(*a*) *Maximum unity-gain response*, which is obtained from Fig. 1.5 as

$$V_{max}(f_{c0}) < S/\omega_{c0} \simeq S/\omega_0 = I_{max}/\mathrm{gm}. \tag{1.28}$$

(*b*) *Full-power response*, which is defined as the rated maximum amplitude sinusoidal output at rated load, without exceeding a predetermined distortion level. Again, due to slewing rate limitations, there will be a maximum frequency at which full-power response can be attained.

1.2.8 Overload recovery[1]

Overload recovery defines the time required for the output voltage to recover to the rated output voltage from a saturated condition, the latter being defined by a given percentage overdrive. In some amplifiers, overload recovery increases for large impedances in the input circuit.

1.2.9   Absolute maximum ratings[1]

Under this heading fall all ratings which, if exceeded, may lead to permanent damage of the device. Some of them are a function of temperature. Most of them are self-explanatory.

supply voltage

power dissipation, specified for case temperature (typically 125 °C). Should be derated for ambient temperatures above (typically) +75 °C

lead temperature soldering, 60 s duration (typically)

differential input voltage

input voltage

output short-circuit duration (may be infinite)

operating temperature range

storage temperature range

1.3   Transimpedance amplifiers

A transimpedance amplifier exhibits, by definition, low input and output impedances. The low input impedance is obtained in the circuit of Fig. 1.18 by using a CB connected input transistor, which serves as an impedance converter. An IC may be employed as gain element, preferably compensated by a capacitor C_f as shown. The controlled transimpedance is then $1/sC_f$, and the input impedance equals the parameter h_{ib} of the input transistor. The transistor is connected to the positive and negative supply voltages through biasing resistors. Stable dc conditions

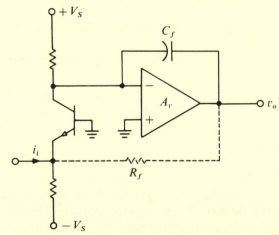

Fig. 1.18. Basic transimpedance amplifier.

are, however, not assured unless v_o is connected to the emitter through a feedback resistor R_f (shown by the dashed line in Fig. 1.18).

The LM 3900 – a current-differencing amplifier

The LM 3900 – termed a Norton amplifier by National Semiconductor Corp., Inc. – contains four independent amplifiers in a single package. Unlike differential input voltage amplifiers, a Norton amplifier does not need a negative current source supplying the differential input stage and therefore requires only a single positive power supply, a useful feature in certain designs.

The basic circuit diagram of this amplifier is shown in Fig. 1.19(a), where the non-inverting input terminal is seen to be connected to a current mirror[8] consisting of a diode in parallel with the input of a CE connected transistor, Q_1, which injects the input current with opposite polarity into the inverting terminal. Another CE connected transistor, Q_2, serves as gain element. This transistor is classified, in spite of its high input impedance ($\sim 1\,M\Omega$), as the input stage of a transimpedance amplifier, since its base current is ($i_a - i_b$). Hence, its base serves as a

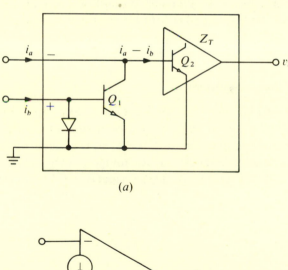

(a)

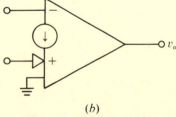

(b)

Fig. 1.19. The LM 3900; (a) basic diagram, (b) symbol.

current-summing node. The amplifier is internally compensated for a roll-off of 20 dB/decade over its useful frequency range.

The dc level of both input terminals is one junction voltage above ground. Hence its performance as a dc amplifier is not as precise as that of a standard IC amplifier operating with split supplies, but it is adequate for many less critical situations. Input V_{BE} match between non-inverting and inverting inputs occurs for a mirror current of about 10 μA. Simplicity of design in conventional applications and some novel ideas are demonstrated in reference [9].

1.4 Transadmittance amplifiers

Transadmittance IC amplifiers are available as CA 3060, 3080 and 3094, manufactured by RCA. They feature a differential voltage input and a push-pull current output capable of supplying a true dc bipolar current signal to the output load. Their main properties are shown symbolically in Fig. 1.20. The transadmittance is controlled by a current I fed into a separate terminal, which allows the output not only to be modulated linearly, but also to be switched on and off. This facility enhances the device's flexibility, enabling it to be used in a broad spectrum of applications.

1.5 Current amplifiers

A basic current amplifier should exhibit low input and high output impedances. It can be realized by cascading a transimpedance amplifier with a transadmittance output stage, as shown schematically in Fig. 1.21. Current amplifiers offer certain advantages, but at the same time present particular design problems. These aspects will be discussed in Section 1.6.4.

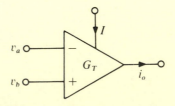

Fig. 1.20. Basic transadmittance amplifier; $i_o = \mathrm{gm}(v_b - v_a)$, $\mathrm{gm} = 19.2I$ for the RCA CA3080 amplifier.

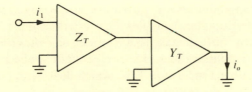

Fig. 1.21. Basic current amplifier.

1.6 Feedback amplifiers, various configurations

Basic amplifiers are ill suited for accurate signal processing. Transfer functions with narrow tolerances and modified impedances can only be realized by employing negative feedback. Appendixes B and C provide the basis for the feedback analysis technique employed throughout this section and will be referred to in the corresponding passages in the text.

We shall describe the transfer function of a feedback amplifier by

$$H(s) = \frac{1}{B(s)} \frac{[-\mathrm{LT}(s)]}{[1 - \mathrm{LT}(s)]}. \tag{1.29}$$

A definition of $B(s)$ and $\mathrm{LT}(s)$ follows. $H(s)$ is the product of two factors: the first, $1/B(s) = \lim_{|\mathrm{LT}| \to \infty} H(s)$, is a function of the feedback network only and describes the ideal behavior of $H(s)$. The identification of $1/B(s)$ in terms of the circuit parameters is discussed in Appendix B.1. Its nature can be A_v, A_i, Y_T or Z_T irrespective of the kind of basic amplifier employed. It is determined solely by the kind of feedback applied to the output and input of the basic amplifier as shown in Table 1.2.

Although the nature of $1/B$ is not affected by the choice of the basic amplifier, that choice has important implications on the overall performance of the feedback amplifier: making this choice in accordance with the kind of feedback applied (i.e. a transimpedance amplifier for

Table 1.2

1/B	Kind of negative feedback applied to		Resulting impedance level at the	
	output	input	output	input
$v_2/i_1 = Z_T$	shunt	shunt	low	low
$i_2/v_1 = Y_T$	series	series	high	high
$v_2/v_1 = A_v$	shunt	series	low	high
$i_2/i_1 = A_i$	series	shunt	high	low

shunt-shunt, transadmittance for series-series, voltage for shunt-series and current for series-shunt) enables the properties of the basic amplifier to be enhanced by the feedback. This is termed an *enhancing combination* between a basic amplifier and the kind of feedback applied to it. More will be said on this subject in Section 1.6.2 which deals with transimpedance amplifiers.

The second factor in (1.29), $[-LT(s)]/[1-LT(s)]$, is a function of the LT (*loop transmission*) and describes the deviation of $H(s)$ from ideal behavior. The LT is given by

$$LT(s) = A(s)B'(s) \tag{1.30}$$

which is the topologically exact expression if $A(s)$ and $B'(s)$ are derived from the coefficients of the Kirchhoff matrix as shown in Appendix B.4 and repeated here for reference:

$$A(s) \triangleq -A/I_{22}, \qquad B'(s) \triangleq -B/I_{11}. \tag{1.31}$$

In general, $B'(s) \neq B(s)$, and the frequently employed description $H(s) = A(s)/[1 - A(s)B(s)]$ is only correct in the special case, in which $B(s) = B'(s)$. Identification of LT by *source splitting* is discussed in Appendix B.2.

In the following analyses of feedback amplifiers we shall assume *mismatch conditions* to be satisfied between the impedances of the feedback network and those of the amplifier. These conditions are obtained in Appendix B.5 from the topologically exact expression for LT. They are defined by simple inequalities, on whose basis the input and output impedances of the basic amplifier may be assumed to exhibit ideal values: infinite input and zero output impedances for a voltage amplifier, infinite input and output impedances for a transconductance amplifier, etc. Mismatch conditions also reduce direct signal transmission through the feedback network from input to output, permitting the corresponding factor to be disregarded.

Applying these simplifications, the expression for LT obtained by the source splitting method becomes the same as the 'exact' one obtained from the Kirchhoff equations. The ensuing equations are simple and provide intuitive insight into the effect of the various parameters on the overall design.

Frequency dependence of the controlled amplifier sources will be retained wherever its effect on the performance is investigated. Otherwise, the gain of the basic amplifier will be assumed to be infinite.

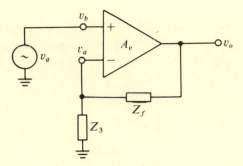

Fig. 1.22. Feedback-stabilized voltage amplifier.

1.6.1 Voltage amplifiers

Fig. 1.22 shows a basic voltage amplifier connected to a feedback network. The latter applies series feedback to its input and shunt feedack to its output, which is commonly called a shunt-series feedback connection (output first). Thus, in accordance with Table 1.2, $1/B$ describes the ideal voltage gain, which is derived in Appendix B.1 as

$$1/B = \lim_{A_v \to \infty} v_o/v_g = (Z_f + Z_3)/Z_3. \tag{1.32}$$

The LT is identified in Appendix B.2 from Fig. B.4 by splitting the controlled source into two. Fig. B.4 is redrawn here as Fig. 1.23, applying mismatch conditions (i.e. $z_i = \infty$, $z_o = 0$) to the basic voltage amplifier. Hence,

$$LT_s = LT = -A_v Z_3/(Z_f + Z_3). \tag{1.33}$$

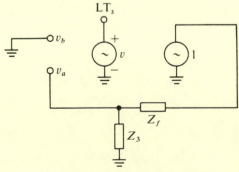

Fig. 1.23. Source splitting for computation of LT; $v = A_v(v_b - v_a)$.

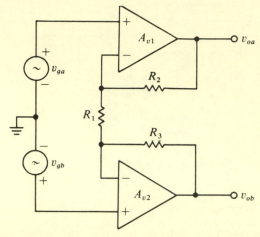

Fig. 1.24. Feedback-stabilized differential input, differential output voltage amplifier.

A feedback-stabilized differential input differential output voltage amplifier is shown in Fig. 1.24. Its ideal transfer function is recognized as

$$\lim_{A_{v1}=A_{v2}\to\infty} (v_{oa} - v_{ob})/(v_{ga} - v_{gb}) = [1+(R_2+R_3)/R_1] \qquad (1.34)$$

for the differential mode, and

$$\lim_{A_{v1}=A_{v2}\to\infty} (v_{oa} + v_{ob})/(v_{ga} + v_{gb}) = [1+(R_2-R_3)/R_1] \qquad (1.35)$$

for the common mode.

From (1.34) and (1.35) it follows that the CMRR will be greatest if $R_2 = R_3$, which is, therefore, a desirable design condition for this amplifier.

A rigorous evaluation of LT involves the analysis of a multivariable feedback circuit discussed in Appendix B.6. However, the practical

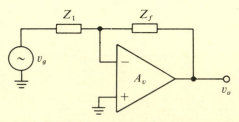

Fig. 1.25. Feedback-stabilized inverting amplifier.

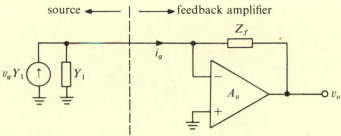

Fig. 1.26. Inverting feedback amplifier redrawn as current driven transimpedance amplifier.

approach of analysing one amplifier at a time while assuming the other to exhibit infinite gain completely satisfies in this case the requirements of the design engineer, who is concerned with conditions for stability and a reasonably accurate prediction of the transient response. With this simplification, the LT of each amplifier is

$$LT_1 \simeq -A_{V1}R_1/(R_1+R_2) \quad \text{and} \quad LT_2 \simeq -A_{V2}R_1/(R_1+R_3).$$

$$(1.36)$$

1.6.2 Transimpedance amplifiers

Fig. 1.25 shows one of the most widely used circuits, a voltage amplifier in an inverting feedback connection. Recognizing Z_f as the feedback element applying shunt feedback to both the input and the output, the nature imposed by Z_f on the transfer function will be, in accordance with Table 1.2, that of a transfer impedance driven by a current source. The circuit is accordingly redrawn in Fig. 1.26, for which we identify

$$1/B = \lim_{A_v \to \infty} v_o/v_g Y_1 = -Z_f. \qquad (1.37)$$

Assuming mismatch conditions to be satisfied,

$$LT = -A_v Z_1/(Z_1 + Z_f), \qquad (1.38)$$

also derived as (B.18) in Appendix B.5. We note that with respect to feedback Z_1 constitutes the source impedance, although functionally it belongs to the feedback network.

A modification of the transimpedance amplifier, shown in Fig. 1.27, is frequently employed as a differential input single-ended output voltage amplifier. Its idealized transfer function is

$$\lim_{A_v \to \infty} v_o = -v_{ga}Z_2/Z_1 + (v_{gb} + V_5 Z_3/Z_4)[Z_4/(Z_3+Z_4)][(Z_1+Z_2)/Z_1].$$

$$(1.39)$$

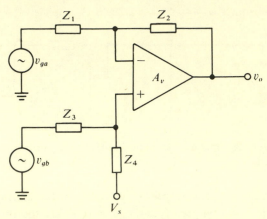

Fig. 1.27. Differential input, single-ended output feedback amplifier.

Note that v_o is referenced to a voltage V_s, commonly called the *sense voltage* (see Section 3.1.7 on referencing differential amplifiers, p. 168). Equation (1.39) for $Z_4/Z_3 = Z_2/Z_1$ becomes

$$v_o - V_s = (v_{gb} - V_{ga})Z_2/Z_1. \qquad (1.40)$$

Assuming infinite input impedance of the basic amplifier, the LT is identical to the one given by (1.38).

Note that different impedances are seen by the signal sources v_{ga} and v_{gb}. Hence, in a practical design, the output impedance of the two sources should either be negligible or incorporated in Z_1 and Z_2 respectively. The alternative of making $Z_3 + Z_4 = Z_1$ is usually impractical, since it leads to awkward resistor values unless the gain equals unity.

Enhancing versus non-enhancing combinations
Fig. 1.26 represents a *non-enhancing combination* between a basic amplifier and shunt–shunt feedback. This configuration is widely employed in operational amplifiers, where dc stability is important. The *enhancing combination*, on the other hand, postulates the use of a basic transimpedance amplifier, whose impedances will accordingly be enhanced by the shunt–shunt feedback. The input impedance obtained with an enhancing combination is lower by at least two orders of magnitude than that obtained with a non-enhancing one, for the same value of LT. Furthermore, the source impedance is practically short-circuited to ground by the low input impedance of the basic amplifier and therefore does not affect the LT. This has important implications for active differentiators (see Section 3.1.3).

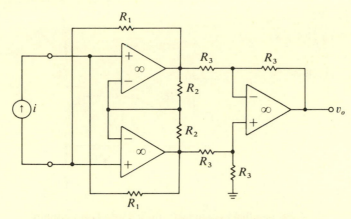

Fig. 1.28. Floating input transimpedance amplifier.

However, when an enhancing combination with shunt–shunt feedback is used, the excellent dc stabilizing properties of the basic voltage amplifier are lost. Fortunately, these properties become less important just in the case where the enhancing combination yields the greatest benefit (i.e. in differentiation) – a good example of the guidance provided by nature to the intuitive.

Problem
1.3. (*a*) In the circuit shown in Fig. 1.25, replace the voltage amplifier by a transimpedance amplifier, whose transfer function is $v_o/i_i = Z_T$. Find the appropriate mismatch conditions and show that in this case $\text{LT} \simeq -Z_T/Z_f$.
(*b*) Fig. 1.28 shows a transimpedance amplifier with floating input.[10] Show that its transfer function equals $v_o/i = -2R_1$.

1.6.3 Transadmittance amplifiers

Unipolar output Y_T amplifier
Fig. 1.29 shows a series–series feedback applied to a basic transadmittance amplifier, which consists of a voltage amplifier connected in cascade with an FET. Because the FET is a VCCS, it is ideal for applying series feedback to the output. The circuit of Fig. 1.29 is usually employed as a constant current source, in which case $v_g = 0$. It may in principle also serve as a small-signal transadmittance amplifier, provided that the small signal v_g is superimposed on the voltage bias V. Its drawbacks are that the polarity of i_o can only be as indicated in Fig. 1.29 and that there must be a voltage difference between load and source to provide the drain voltage for the FET.

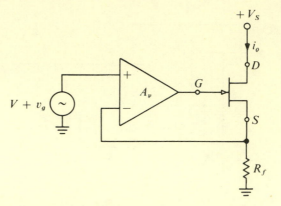

Fig. 1.29. Feedback-stabilized transadmittance amplifier.

We identify

$$1/B = \lim_{A_v \to \infty} i_o/(V + v_g) = 1/R_f. \tag{1.41}$$

From Fig. 1.30, the LT is identified by opening the loop, for mismatch conditions, as

$$LT_s = LT = -gmR_fA_v/(1 + gmR_f). \tag{1.42}$$

Bipolar output Y_T amplifiers
Another kind of transadmittance amplifier is shown in Fig. 1.31 where a differential input voltage amplifier is fed from a floating power supply.[11] The ideal transfer function is found to be

$$1/B = \lim_{A_v \to \infty} i_o/vg = 1/Z_f, \tag{1.43}$$

with $Z_f = Rf/(1 + sR_fC)$.

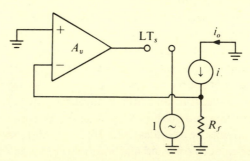

Fig. 1.30. Source splitting applied to Fig. 1.29; $i = 1gm/(1 + gmR_f)$.

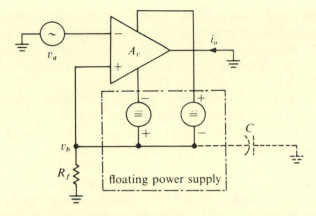

Fig. 1.31. Transadmittance amplifier providing bipolar output.

The floating power supply ensures that the current i_o flows through Z_f and is not shunted to ground, as would happen if the power supply *were* grounded. The only limitation to this ideal condition is the unavoidable stray capacitance C between power supply and ground, which shunts R_f at high frequencies.

In Fig. 1.32, the dependent source of the amplifier is split into two, which yields the LT:

$$LT_s = LT = -A_v Z_f / (Z_f + r_o). \tag{1.44}$$

Unlike the circuit of Fig. 1.29, this circuit is able to supply a bipolar current to the output load – a definite advantage. On the other hand, the design of the floating power supply poses practical problems.

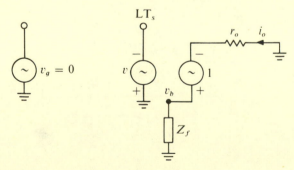

Fig. 1.32. Source splitting applied to Fig. 1.31; $v = A_v(v_b - v_g)$, $Z_f = R_f/(1 + sR_f C)$.

The transadmittance amplifier shown in Fig. 1.31 may be converted into a differential one by connecting the originally grounded terminal of R_f to the output of a voltage follower. Disregarding C, i_o is then proportional to the difference between v_g and the voltage follower input, divided by R_f.

The desirable high output impedance of a transadmittance amplifier can be obtained not only by applying negative series feedback, but also by applying positive parallel feedback to the output of a basic amplifier. A possible configuration is the Howland current source[12] shown in Fig. 1.33. With $v_{ga} = v_{gb} = 0$, $A_v = \infty$, $R_2/R_1 = R_4/R_3$ and Z_L disconnected, it is easy to show that the current i flowing into the output terminal, due to voltage source v, equals zero. It follows that the output impedance of the amplifier is infinitive. Evaluation of the idealized transfer function with Z_L connected shows that this is a differential input transadmittance amplifier:

$$\lim_{A_v \to \infty} i_L = (v_{gb} - v_{ga})/R_3. \tag{1.45}$$

The absence of Z_L in (1.45) is another indication of the infinite output impedance of this circuit, but this condition depends on accurate matching of the resistor ratios $R_2/R_1 = R_4/R_3$.

Another, less critical circuit also providing bipolar output current into a grounded load is shown in Fig. 1.34.[13]

This circuit will be balanced if $(V_{E3} + V_{E4})/2 = V_g$, in which case $i_L = 2V_g/R_1$. The voltage swing capability of the amplifier must be greater by a factor of $(R_2 + R_3)/R_3$, than that required at the gates of Q_1

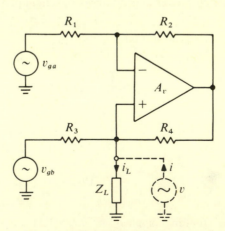

Fig. 1.33. Howland current source.

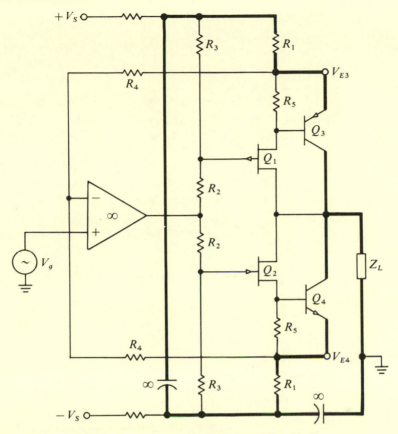

Fig. 1.34. Transadmittance amplifier providing bipolar output.

and Q_2 to drive the output. Transistors Q_3 and Q_4 boost the output current capability without reducing the output impedance. The enhanced lines in the diagram indicate the path taken by the small-signal component of i_L, which should be kept as short as possible if a reasonably fast response is desired. As in all low-impedance current loops, response will be severely limited by undesirable inductances in this path. The capacitors whose values are shown in the diagram to be infinite should have values as large as possible.

Problems

1.4. Consider the circuit shown in Fig. 1.35, whose transfer function is i_o/v_g.

   (*a*) Define the kind of feedback applied to the output and the input.

   (*b*) Find the transfer function of the amplifier.

   (*c*) Find the output admittance $-i_o/v_o$ of the circuit.

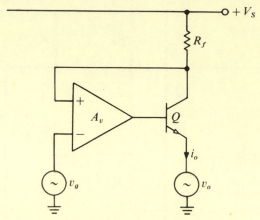

Fig. 1.35. Problem 1.4.

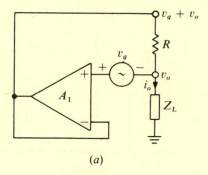

(a)

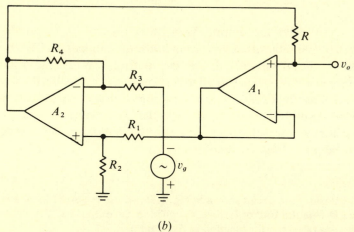

(b)

Fig. 1.36. Problem 1.6.

1.5. Consider the Howland current source shown in Fig. 1.33.

(a) Show that for $R_3 = R_1$, $R_4 = R_2$, the output resistance 'seen' by Z_L equals

$$R_o = (R_1 \| R_2'),$$

where

$$R_2' = R_2[1 + A_v R_1/(R_1 + R_2)].$$

(b) Show that for $\lim_{A_v \to \infty}$ the output conductance

$$1/R_o = G_3 + G_4 - G_4(G_1 + G_2)/G_2$$

(c) Find the LT of this circuit.

1.6. Fig. 1.36(a) shows a transadmittance amplifier, whose transfer function is $i_o = v_g/R$. Fig. 1.36(b) shows its practical implementation employing a voltage follower and a difference amplifier. The resistors R_1, R_2, R_3 and R_4 are nominally equal to R_0, and their deviation from the nominal value is given as $R_j = R_0(1 + \delta_j)$ for $j = 1, 2, 3, 4$; δ_1, $\delta_4 > 0$; and δ_2, $\delta_3 < 0$.

(a) Assume A_1, $A_2 \gg 1$ and independent of frequency, and $\delta_j = 0$. Find the output resistance R_{out}.

(b) Assuming $A_k \simeq 10^5$ ($k = 1, 2$), $|\delta_j| = 10^{-3}$, find the dominant factor reducing the output resistance to below infinity.

(c) Assume $A_k(s) = A_k(0)/[1 + sA_k(0)\tau_{0k}]$, $\delta_j = 0$. Show that for

$$\omega \ll \left[\frac{2}{A_2(0)} + \frac{1}{A_1(0)}\right]/(\tau_{01} + 2\tau_{02}),$$

the output impedance Z_{out} is resistive.

(d) Show that the frequency range over which Z_{out} is resistive increases if $\delta_j \neq 0$.

1.6.4 Current amplifiers

Fig. 1.37 shows series-shunt feedback applied to a basic current amplifier. A bipolar transistor is employed in cascade with a transimpedance amplifier, serving as a series feedback element in the output stage of the basic current amplifier. dc conditions are disregarded in the circuit shown.

Converting the input circuit into a Norton source, in accordance with the shunt feedback applied to the input, we obtain

$$1/B = \lim_{A_v \to \infty} \alpha_Q i_o/v_g Y_1 = \alpha_Q(1 + Z_f/Z_3). \tag{1.46}$$

Note that the parameter stabilized by feedback is i_o and not $i_o' = \alpha_Q i_o$, so that $1/B$ includes the CB current gain α_Q. If an FET is employed as Q, then $\alpha_Q = 1$.

The LT is obtained as

$$LT = \frac{-Z_T \beta}{[\beta + (Y_f + Y_3)(z_o + h_{ie})]Z_f}, \tag{1.47}$$

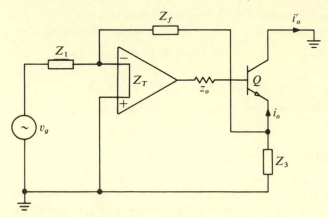

Fig. 1.37. Feedback-stabilized current amplifier, small-signal diagram; $i'_o = \alpha_Q i_o$.

where z_o is the output impedance of Z_T, and β, h_{ie} are the CE current gain and input impedance of the transistor Q, respectively.

Problems

1.7. A line driver employing the LM 3900 differential Norton amplifier is shown in Fig. 1.38. $I_{CQ} = 10$ mA, $I^+ = 10$ μA, $R_3 = 470$ Ω, $R_1 = R_2 = 1$ MΩ, $R_L = 100$ Ω, and $R_5 = R_4$.

(a) Find a suitable value for $R_4 + R_5$, so that V_{CQ} (the collector voltage of Q) becomes +20 V.

(b) What is the function of C_3?

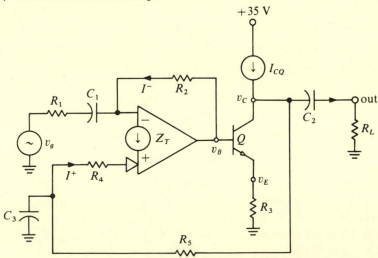

Fig. 1.38. Line driver employing LM 3900; $C_1, C_2 \rightarrow \infty$. Problem 1.7.

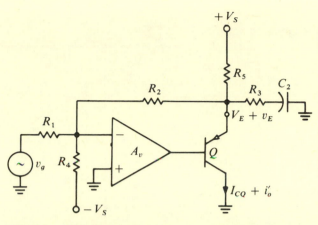

Fig. 1.39. Line driver employing IC voltage amplifier. Problem 1.8.

(c) Find $1/B$ and LT of the circuit. Make the simplifying assumptions that the small-signal voltages $v_B = v_E$, and that the CB current gain α of the transistor equals unity.

(d) In practice, $v_B \neq v_E$. Is this effect reduced by feedback or does it enter the ideal response $1/B$?

(e) Repeat part (d) for $\alpha \neq 1$.

1.8. Fig. 1.39 shows a line driver employing an IC voltage amplifier. $V_S = 15$ V, $R_1 = 1$ kΩ, $R_2 = 10$ kΩ, $R_3 = 22$ Ω, $C_2 = 10$ μF, $I_{CQ} = 5$ mA, and $\alpha_Q = \alpha_0/(1 + s\tau_\alpha)$.

(a) Find R_4 and R_5 to make $V_E = 10$ V and $I_{CQ} = 5$ mA.

(b) Find $1/B$ and LT of the circuit, with $A_v = \infty$.

(c) Draw the Bode diagram of $1/B$.

1.6.5 The instrumentation amplifier

In certain signal processing systems, very low-level signals are superimposed on large interfering CM signals. CMRRs of the order of 100 dB are required in such applications, but cannot be attained using the kind of circuit shown in Fig. 1.27 in which the CMRR is ultimately limited by the properties of the basic amplifier.

Amplifiers answering the most stringent requirements regarding CMRR and gain flexibility are termed *instrumentation amplifiers*. An early version is shown in Fig. 1.40, which consists of a cascade connection between the circuit of Fig. 1.24, whose CM gain is unity, and the circuit of Fig. 1.27. The gain is adjustable by a single resistor R_1. Second-generation instrumentation amplifiers[14,15] exhibit further improved CMRRs

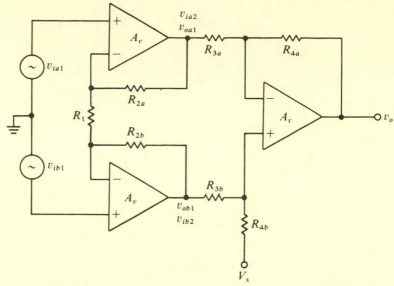

Fig. 1.40. Simple instrumentation amplifier; ideally $R_{jb} = R_{ja}$ for $j = 2, 3, 4$.

and adjustable gains over a wide range. Fig. 1.41 shows symbolically the amplifier described by Van de Plassche,[15] in which the ratio between the two external resistors R_1 and R_2 determines the gain, and the output voltage is referenced to an externally applied sense reference V_s. The various functional blocks and their interconnections are shown in Fig. 1.42, in which the amplifier is seen to consist of two differential transadmittance amplifiers and one transimpedance amplifier. For the circuit-minded reader it may be said that each transadmittance amplifier consists essentially of a differential transistor pair, whose transconductance is

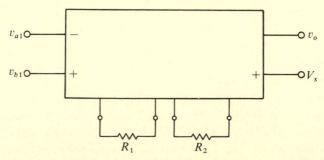

Fig. 1.41. Symbolic diagram of second-generation instrumentation amplifier; $v_o = (v_{b1} - v_{a1})R_2/R_1 + V_s$.

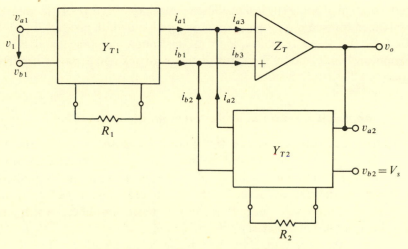

Fig. 1.42. Functional diagram for Fig. 1.41.

determined by resistors R_1 and R_2, respectively, which are connected between the two emitters. Gain accuracy is enhanced by an amplifier suitably connected to each transistor. Inspection of Fig. 1.42 shows that the second transadmittance amplifier is connected in the feedback path of Z_T, stabilizing its transfer function at the inverse value of the feedback transmission – i.e. R_2. Hence, the total transfer function should intuitively be R_2/R_1, which is confirmed by the following derivation:

$$\lim_{Y_{T1}\to\infty} (i_{b1} - i_{a1}) = v_1/R_1;$$

$$\lim_{Y_{T2}\to\infty} (i_{b2} - i_{a2}) = (v_{b2} - v_{a2})/R_2.$$

Hence

$$v_o = Z_T(G_1 v_1 + G_2 V_s)/(1 + G_2 Z_T), \qquad (1.48)$$

and finally

$$\lim_{Z_T\to\infty} v_o = v_1 R_2/R_1 + V_s, \qquad (1.49)$$

with

$$v_1 = v_{b1} - v_{a1}.$$

This kind of amplifier has a gain of 1 to 1000 adjustable by R_1 and R_2, a CMRR of over 100 dB at a gain of 1000, and a bandwidth of 800 kHz independent of gain. The last result is a remarkable feature of a feedback-stabilized transadmittance amplifier, whose bandwidth product is stabilized by the same feedback loop as its transconductance and is not significantly affected by the value of the feedback conductance itself, as long as $LT \gg 1$ (see Section 1.7.3).

1.7 Feedback amplifiers, deviation from ideal behavior

Negative feedback serves the well-known purpose of stabilizing the transfer function and modifying the input and output impedances of an amplifier. The ultimate stability of a feedback amplifier is limited, however, by deficiencies of the feedback network itself, as well as by the effects of undesirable properties of the basic amplifier, which are modified but not eliminated by the feedback.

Unless otherwise stated, we shall deal with the small-signal response for which the amplifier can be assumed to be quasi-linear. Hence, we may apply the principle of superposition and evaluate the effect of each deficiency separately, disregarding the others.

1.7.1 Sensitivity calculations

Two of the main concerns of a design engineer are the insensitivity to the effects of environmental conditions and the reproducibility of the product or system he intends to design. The computation of sensitivity is a convenient technique for calculating deviations from the system's desired performance. These unavoidable deviations are caused by differences between the nominal and actual values of passive components and of parameters of active devices, due to manufacturer's tolerances, ageing and temperature changes. *Sensitivity* can be expressed in *absolute* or *relative* terms, the latter being preferable because changes are more meaningful if expressed as percentages, and also because the mathematical treatment is simpler.

Let $H = f(k)$ be a linear transfer function, whose variation as a function of changes in the value of several parameters k_i is to be investigated. Expansion of H in a Taylor series yields, disregarding terms higher than the first order,

$$H + dH = f(k) + \sum_{i}^{n} \frac{\partial f(k)}{\partial k_i} \, dk_i. \tag{1.50}$$

After cancelling H against $f(k)$ and noting that $\partial f(k)/\partial k = \partial H/\partial k$, (1.50) can be brought into the form

$$\frac{dH}{H} = \sum_1^n \frac{\partial H}{H} \frac{k_i}{\partial k_i} \frac{dk_i}{k_i} = \sum_1^n S_{k_i}^H \frac{dk_i}{k_i}. \tag{1.51}$$

We define S_k^H to be the *relative sensitivity* of H with respect to k. S_k^H can be written in various forms:

$$S_k^H = \frac{\partial H/H}{\partial k/k} = \frac{\text{percentage change in } H}{\text{percentage change in } k}, \tag{1.52}$$

or

$$S_k^H = \frac{\partial H}{\partial k} \frac{k}{H}, \tag{1.53}$$

or

$$S_k^H = \frac{d(\ln H)}{d(\ln k)}. \tag{1.54}$$

Equation (1.52) serves as the definition, whereas (1.53) or preferably (1.54) can be used to calculate sensitivity.

Finally, we consider the case in which $H = H(A)$ where $A = A(k)$. Then

$$\frac{dH}{H} = \frac{\partial H/H}{\partial A/A} \frac{\partial A/A}{\partial k/k} \frac{dk}{k} = S_A^H S_k^A \frac{dk}{k}. \tag{1.55}$$

Relative sensitivity computations provide a powerful tool in the analysis of the effect of parameter changes on system performance. Problem 1.9 introduces the basic concepts.

Problem

1.9 (*a*) For the feedback-stabilized transmittance $H = (1/B)(-LT)/(1-LT)$ show that

$$S_A^H = 1/(1-LT) = 1/\mathscr{F}.$$

The *return difference* $\mathscr{F} = (1-LT)$ is of basic importance in sensitivity calculations.

(*b*) Draw a flow graph for H as defined in (*a*) and show that the *return ratio* LT and the return difference can be obtained by interrupting the loop and feeding it with a signal of unit value.

(*c*) In the general case of a feedback system, the return difference with respect to an arbitrary element (k_0 in our case) can be obtained by bringing the transmittance into the *bilinear form*:[16,17]

$$H = \frac{N(k_0)}{D(k_0)} = \frac{N_1 + k_0 N_2}{D_1 + k_0 D_2}. \tag{1.56}$$

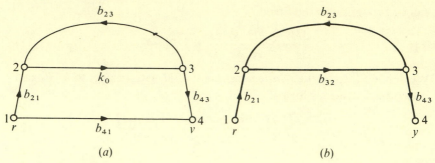

(a)   (b)

Fig. 1.43. (*a*) Bilinear form of feedback system; $H = y/r = f(k_0)$. (*b*) Canonic form of feedback system.

Show that the bilinear form can be further modified as follows:

$$\frac{N_1 + k_0 N_2}{D_1 + k_0 D_2} = b_{41} + \frac{b_{21} k_0 b_{43}}{(1 - k_0 b_{23})}, \qquad (1.57)$$

where $b_{23} = -D_2/D_1$, $b_{21} b_{43} = (N_2 D_1 - N_1 D_2) D_1^2$, and $b_{41} = N_1/D_1 = H_0$, if $k_0 = 0$.

Equation (1.57) is shown in the flow graph of Fig. 1.43(*a*), where the gain factor k_0 appears in a single branch of the feedback loop. This yields the return difference $\mathscr{F}$ by inspection.

(*d*) Show that

$$S_{k_0}^H = 1/\mathscr{F} (1 - H_0/H), \qquad (1.58)$$

where $\mathscr{F} = 1 - k_0 b_{32}$ is the return difference of H with respect to the gain factor k_0. Equation (1.58) is of basic importance in control systems.

The effect of direct transmission

If a control system contains a single gain element or gain factor k_0, then the effect of the direct signal transmission H_0 between r and y on stability is preferably investigated by diverting this branch through nodes 2 and 3 as shown in Fig. 1.43(*b*), in which case

$$b_{32} = k_0[1 + H_0(1 - k_0 b_{23})/k_0 b_{21} b_{43}], \qquad (1.59)$$

and $H_0 = b_{41}$.

In the case of negative feedback, $\mathrm{LT} = k_0 b_{23} < 0$. Furthermore, if $|\mathrm{LT}| \gg 1$, then

$$b_{32} \simeq k_0(1 - H_0 b_{23}/b_{21} b_{43}). \qquad (1.60)$$

Equation (1.60) shows that the direct transmission H_0 introduces a real zero or several complex conjugate ones in LT, whose influence on stability is conveniently visualized from its effect on the root locus.

Depending on whether $b_{23} < 0$ and $k_0 > 0$ (non-inverting amplifier), or $b_{23} > 0$ and $k_0 < 0$ (inverting amplifier), that zero will be located in the LHS or RHS (*left hand side* or *right hand side* of the complex frequency plane), respectively. Its effect on stability will therefore be desirable in the first case (non-inverting), but undesirable in the second (inverting).

Note that, in fact, the branch b_{21} in Figs. B.3 and B.5 of Appendix B also represents direct transmission. It introduces an RHS zero into (B.6), since there we are dealing with an inverting amplifier. The topological differences is merely that, instead of the two node pairs 1, 2 and 3, 4 in Fig. 1.43(a), there appear two single nodes only in Fig. B.3. Hence, there is no need in Fig. B.3. to divert the direct transmission since it already appears in parallel with the branch containing the gain element and may therefore readily be brought into a form similar to (1.59) or (1.60).

To sum up, we shall represent a single-variable control system by the flow graph of Fig. 1.43(b), termed the *canonic form* of the system flow graph. In it, the effect of direct transmission appears as a zero in the LT, which enables its effect on stability to be readily investigated. For such a system, $S_k^H = 1/\mathscr{F}$, since H_0 is incorporated in $\mathscr{F}$. In multivariable systems, in which k is one of two or more gain elements, H_0 contains the remainder and should not be rerouted through nodes 2 and 3.

1.7.2 Common mode rejection ratio

The ideal differential feedback amplifier with infinite CMRR has yet to be built. Deficiencies in the feedback network and in the basic amplifier prevent this ideal from being attained. To illustrate this, we shall evaluate the CMRR due to a non-ideal feedback network connected to an ideal basic amplifier and then, in problem 1.12, the CMRR due to a non-ideal basic amplifier connected to an ideal feedback network. The possibility of employing a feedback which exactly compensates for the deficiencies of the basic amplifier belongs to the realm of science fiction.

Consider the instrumentation amplifier shown in Fig. 1.40, for which we shall assume all amplifiers to exhibit infinite gain.

Identically indexed resistors indicate pairs which should ideally be mutually equal, thereby yielding an infinitely large CMRR. In practice they will slightly differ from each other and therefore cause the CMRR to be finite.

Let δ_j represent the relative deviation of resistor j from its nominal value R_j, and assume that corresponding resistors deviate by equal amounts but in opposite direction. Hence,

$$R_{ja} = R_j(1 + \delta_j) \quad \text{and} \quad R_{jb} = R_j(1 - \delta_j).$$

In order to evaluate the CMRR we shall first write the small-signal transfer function of the two cascaded stages in chain matrix form,[3] and then replace the input driving function by CM and DM signals. The resulting transfer function yields the CMRR directly.

The transfer function for the first stage for $V_s = 0$ becomes

$$\begin{bmatrix} v_{oa1} \\ v_{ob1} \end{bmatrix} = \begin{bmatrix} A_{aa1} & A_{ab1} \\ A_{ba1} & A_{bb1} \end{bmatrix} \begin{bmatrix} v_{ia1} \\ v_{ib1} \end{bmatrix} \tag{1.61}$$

and that of the second

$$v_o = \begin{bmatrix} A_{aa2} & A_{ab2} \end{bmatrix} \begin{bmatrix} v_{ia2} \\ v_{ib2} \end{bmatrix}, \tag{1.62}$$

with the coefficients being defined as

$$A_{aa1} = v_{oa1}/v_{ia1}|_{v_{ib1}=0} = [R_1 + R_2(1+\delta_2)]/R_1 \quad \text{etc.}$$

Noting that $v_{oa1} = v_{ia2}$ and $v_{ob1} = v_{ib2}$, we chain multiply (1.61) by (1.62), which yields the overall transfer function as

$$v_o = \begin{bmatrix} C & D \end{bmatrix} \begin{bmatrix} v_{ia1} \\ v_{ia2} \end{bmatrix}, \tag{1.63}$$

where $C = A_{aa2}A_{aa1} + A_{ab2}A_{ba1}$ and $D = A_{aa2}A_{ab1} + A_{ab2}A_{bb1}$. What remains to be done is to transform the input driving vector into one consisting of the CM and DM input signals v_{ic} and v_{id}, respectively:

$$\begin{bmatrix} v_{ia} \\ v_{ib} \end{bmatrix} = \begin{bmatrix} v_{ic} + v_{id} \\ v_{ic} - v_{id} \end{bmatrix} = \begin{bmatrix} 1 & 1 \\ 1 & -1 \end{bmatrix} \begin{bmatrix} v_{ic} \\ v_{id} \end{bmatrix}, \tag{1.64}$$

where $v_{ic} = (v_{ia} + v_{ib})/2$ and $v_{id} = (v_{ia} - v_{ib})/2$. Substitution of (1.64) in (1.63) yields

$$v_o = [(C+D)(C-D)] \begin{bmatrix} v_{ic} \\ v_{id} \end{bmatrix}, \tag{1.65}$$

giving

$$A_{\mathrm{CM}} = v_o/v_{ic}|_{v_{id}=0} = C + D$$

and

$$A_{\mathrm{DM}} = v_o/v_{id}|_{v_{ic}=0} = C - D.$$

So

$$\mathrm{CMRR} = A_{\mathrm{DM}}/A_{\mathrm{CM}} = (C-D)/(C+D). \tag{1.66}$$

Substitution of the corresponding expressions for C and D yields the CMRR as a function of the various resistor values.

Problems

1.10. A basic differential amplifier has a gain of $v_o = A_b v_b - A_a v_a$, $z_i = \infty$ and $z_o = 0$. It is employed as a non-inverting and inverting feedback amplifier, as shown in Figs. 1.22 and 1.25 respectively.

  (a) Show that the transfer function of the non-inverting amplifier is

$$\frac{v_{o1}}{v_g} \simeq \frac{1}{B_1} \frac{1}{(1 - 2/\text{CMRR} - 1/\text{LT})},\qquad (1.67)$$

and that of the inverting one

$$\frac{v_{o2}}{v_g} = \frac{1}{B_2} \frac{1}{(1 - 1/\text{LT})}.\qquad (1.68)$$

  (b) Explain the physical reason for the difference between (1.67) and (1.68).

1.11. Fig. 1.44 shows the simplified diagram of a differential feedback amplifier, whose CMRR is increased by employing another basic amplifier in a secondary feedback loop.[18]

  With $A_2 = 0$ and $v_{o2} = 0$, the CMRR depends on the equality between R_2 and R_1 only. With A_2 active, the CMRR will be improved if $R_2/R_1 \neq 1$. Assume the amplifiers have ideal input and output impedances.

  (a) Treating the problem as a multivariable feedback circuit (Appendix B.6), find the matrix equation

$$\begin{bmatrix} \mathbf{R} \\ \mathbf{0} \end{bmatrix} = \begin{bmatrix} \mathbf{I}_{11} & \mathbf{B} \\ \mathbf{A} & \mathbf{I}_{22} \end{bmatrix} \begin{bmatrix} \mathbf{E} \\ \mathbf{Y} \end{bmatrix}.$$

  (b) From the matrix elements, show that the CMRR tends to infinity as A_1, A_2 tend to infinity.
  (c) For $A_1 = A_2 = \infty$, find 'by inspection' the expressions for v_{o1}/v_{ga} and v_{o2}/v_{ga} if $v_{gb} = 0$, and those for v_{o1}/v_{gb} and v_{o2}/v_{gb} if $v_{ga} = 0$.
  (d) Find $S_{A_2}^{A_{CM}}$.

1.12. Consider the difference amplifier shown in Fig. 1.27 with an ideal feedback network $Z_1 = Z_3 = R_1$, $Z_2 = Z_4 = R_2$ and a non-ideal basic amplifier for which $v_o = v_b A_b - v_a A_a$, $z_i = \infty$, $z_o = 0$.
  Find the CMRR and draw conclusions.

1.7.3 Frequency response

The closed-loop frequency response is affected by the two factors appearing in (1.29): The factor $1/B(s)$ is solely a function of the feedback network, and its analysis does not present any problems; the second factor will indicate any possible tendency for instability and requires

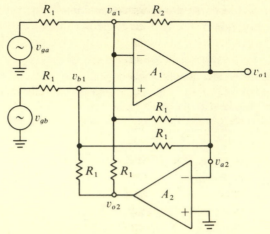

Fig. 1.44. Circuit enhancing CMRR of feedback amplifier by secondary feedback loop.

careful analysis, the most widely used methods being the root locus technique[12,19] and the Nyquist diagram.[20] But, whereas such a rigorous analysis requires detailed information about the gain function, here we shall be content with a simplified model for the amplifier. Such a model yields practical guidelines for feedback amplifier design, and the simplifications which it incorporates will be discussed in what follows.

Most internally compensated amplifiers exhibit a 20 dB/decade roll-off beyond their half-power frequency ω_h (i.e. a 90° phaseshift) over their useful frequency range. Furthermore, the combined additional phaseshift at the unity-gain frequency ω_{c0}, due to the parasitic poles, is sufficiently low to ensure unconditional stability when used as a feedback amplifier provided that the feedback network does not introduce any additional phase lag. The total phaseshift at ω_{c0} is often specified as *phase margin*, which is defined as the difference between the phase lag of the amplifier at ω_{c0} and a phase lag of π radians, which would for $B' = 1$ bring the amplifier to the verge of oscillation.

To emphasize the basic relationships, we shall in most cases simplify the analysis by neglecting secondary effects. To this end we shall disregard the parasitic poles, and apply the gain-bandwidth theorem discussed in Appendix C.3, which replaces the dominant pole by a pole at the origin whilst preserving the value of ω_0. This latter modification retains the asymptotic frequency response and thereby the behavior at high frequencies, but greatly simplifies the computations. In practice, any additional phaseshift introduced by the parasitic poles will have to be allowed for in critical applications. This will invariably require adjustment of ω_0 below the value obtained by the simplified computation, an adjustment easily done on the actual circuit by trial and error or, should this be impracticable, by computer simulation.

The shunt–series feedback amplifier

Let us consider the voltage amplifier discussed in Section 1.6.1 and shown in Fig. 1.22, employing a resistive feedback network. We assume the open-loop gain of the basic amplifier before application of the gain-bandwidth theorem to be

$$A_v(j\omega) = -A(0)/[1 + j\omega A(0)\tau_0], \tag{1.69}$$

$Z_f = R_f$ and $Z_3 = R_3$, with $\tau_0 = 1/\omega_0$, $A(0)\tau_0 = 1/\omega_{hOL}$. Since $B' = R_3/(R_f + R_3)$ designates the feedback attenuation, we evaluate (1.33) as $LT(j\omega) = A(0)B'/(1 + j\omega A(0)\tau_0)$. For $LT(0) = A(0)B' \gg 1$, substitution of (1.33) in (1.29) yields

$$H(j\omega)|_{LT(0)\gg1} = 1/[B(1 + j\omega\tau_0/B')]. \tag{1.70}$$

Fig. 1.45, the Bode plot of (1.70), shows that $\omega_{0LT} = B'\omega_0 = B'/\tau_0 = \omega_{hH}$, i.e. the gain-bandwidth product of the LT becomes the half-power frequency of the closed-loop response, which in turn is proportional to B'. Hence ω_{hH} decreases with increasing gain, since ω_0 remains constant.

For the shunt–series amplifier, $B' = B$. We shall therefore not make any distinction between the two and use B only in the following computations.

Fig. 1.45 is drawn for two closed-loop gain values, $1/B_1$ and $1/B_2$, in order to demonstrate the interplay between $1/B$, $LT(j\omega)$ and $A(j\omega)$ in this kind of amplifier.

Another important observation is the fact that, for $LT(0) \gg 1$, $H(j\omega)$ is neither a function of $A(0)$, nor of ω_{hOL}, the open-loop half-power

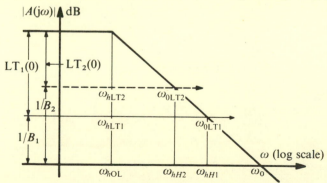

Fig. 1.45. Bode plot of feedback-stabilized voltage amplifier.

frequency. This property is essential for the stabilization by feedback of massproduced basic amplifiers, whose $A(0)$ and ω_{hOL} may exhibit a production spread of an order of magnitude, whereas ω_0 is a reasonably well-controlled and specified parameter. This can be shown by comparing $S_{A(0)}^H$ with $S_{\omega_0}^H$:

$$S_{A(0)}^H = S_{LT}^H S_{A_v}^{LT} S_{A(0)}^{A_v}. \tag{1.71}$$

From (1.29) we obtain $S_{LT}^H = 1/(1-LT)$; from $LT = A_v B$, $S_{A_v}^{LT} = 1$. From (1.69),

$$d \ln A_v = d \ln A(0) - d \ln [1 + j\omega A(0)\tau_0];$$

thus

$$\frac{dA_v}{A_v} = \frac{dA(0)}{A(0)} - \frac{dA(0)j\omega\tau_0}{[1 + j\omega A(0)\tau_0]} = \frac{1}{[1 + j\omega A(0)\tau_0]} \frac{dA(0)}{A(0)},$$

and hence

$$S_{A(0)}^{A_v} = \frac{1}{[1 + j\omega A(0)\tau_0]}.$$

Finally, substituting in (1.71),

$$S_{A(0)}^H = \frac{1}{[1 - LT(j\omega)]} \times \frac{1}{[1 + j\omega A(0)\tau_0]}$$

$$\approx \frac{1}{[1 + A(0)B][1 + j\omega\tau_0/B]}. \tag{1.72}$$

Similarly,

$$S_{\omega_0}^H = S_{LT}^H S_{A_v}^{LT} S_{\tau_0}^{A_v} S_{\omega_0}^{\tau_0}. \tag{1.73}$$

From (1.69),

$$S_{\tau_0}^{A_v} = -j\omega A(0)\tau_0/[1 + j\omega A(0)\tau_0]$$

and

$$S_{\omega_0}^{\tau_0} = d \ln \tau_0/[-d \ln (1/\omega_0)] = -1,$$

yielding

$$S_{\omega_0}^H = j\omega A(0)\tau_0/[1 - LT(j\omega)][1 + j\omega A(0)\tau_0]$$

$$\approx j\omega\tau_0/B(1 + j\omega\tau_0/B). \tag{1.74}$$

Comparison between (1.72) and (1.74) for $\omega \gg B/\tau_0$ shows

$$S_{A(0)}^H \approx 1/j\omega A(0)\tau_0 \quad \text{and} \quad S_{\omega_0}^H \approx 1.$$

This comparison justifies the application of the gain-bandwidth theorem, which replaces the dominant pole in (1.69) by a pole at the origin whilst preserving ω_0. Hence, we could have obtained the same result by a simpler computation, employing

$$A_v(j\omega) \simeq -1/j\omega\tau_0. \qquad (1.75)$$

Fig. 1.45 has been drawn with $B' = B$, which is correct for shunt–series but not for shunt–shunt feedback applied to a basic voltage amplifier in which case $1/B' = 1 + 1/B$. Still, the general nature of Fig. 1.45 is qualitatively preserved also in the latter case, which means that ω_{hH} of shunt–series and shunt–shunt feedback amplifiers employing a basic voltage amplifier decreases as the value of $1/B$ increases.

The series–series feedback amplifier
The frequency response of the series–series feedback-stabilized basic transadmittance amplifier is, however, entirely different. Consider the FET shown in Fig. 1.46, where R_o has been added merely to obtain a dimensionless transfer function for which a Bode plot can be drawn.
We obtain

$$LT(j\omega) = LT(0)/[1 - j\omega LT(0)C_i/\text{gm}]; \qquad (1.76)$$

with $LT(0) = -\text{gm}R_f$. Equation (1.76) shows that in this particular case

$$\omega_{0LT} = \text{gm}/C_i, \qquad (1.77)$$

i.e. the gain-bandwidth product of LT is *not* a function of B. This is a remarkable property, since it follows that the closed-loop half-power frequency ω_{hH}, which equals ω_{0LT}, is also independent of the gain $1/B$ as long as $-LT(0) \gg 1$, and that the closed-loop gain-bandwidth product ω_{0H} actually increases proportionally to $1/B$. This is confirmed by computing

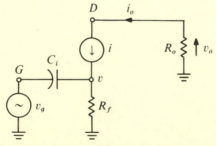

Fig. 1.46 Diagram of feedback-stabilized transadmittance amplifier; $i = \text{gm}(v_g - v)$.

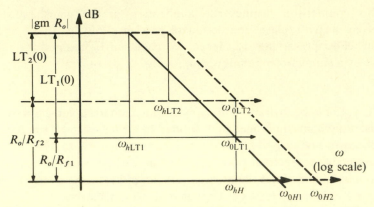

Fig. 1.47. Bode plot for feedback-stabilized transadmittance amplifier.

the transfer function

$$v_o/v_g = H(j\omega) = \frac{-gmR_o}{(1 + gmR_f)[1 + j\omega R_f C_i/(1 + gmR_f)]}$$

which for $-LT(0) = gmR_f \gg 1$ becomes

$$H(j\omega) \simeq [-R_o/R_f]/(1 + j\omega C_i/gm)]. \qquad (1.78)$$

Fig. 1.47 shows the Bode plot for this amplifier for two values of R_f.

The gain independence of ω_{hH} is taken advantage of in the design of the instrumentation amplifier discussed in Section 1.6.5.

In practice, the increase of the closed-loop gain-bandwidth product ω_{0H} by reducing R_f is limited by the condition gm $R_f \gg 1$, where the value of gm is determined by the active device.

The shunt–shunt enhancing feedback amplifier
Consider shunt–shunt feedback applied to a basic transimpedance amplifier, as shown in Fig. 1.48. As usual, we assume mismatch condi-

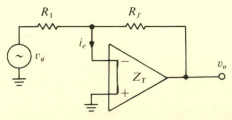

Fig. 1.48. Enhancing shunt-shunt feedback amplifier.

tions to be satisfied and hence the basic amplifier to exhibit idealized input and output impedances, which are both zero in the present case (enhancing combination). With $Z_T(j\omega) = -R_T/[1 + j\omega R_T C_T]$, $\text{LT}(j\omega) = -R_T/R_f(1 + j\omega R_T C_T)$, and $R_T/R_f \gg 1$,

$$H = v_o/v_g G_1 = -R_f \frac{(-\text{LT})}{(1 - \text{LT})} \simeq \frac{-R_f}{(1 + j\omega R_T C_T)}, \qquad (1.79)$$

which differs from shunt–shunt feedback applied to a basic voltage amplifier (non-enhancing) insofar as LT is not a function of the source impedance (see Section 1.6.2, equation (1.38)). Again we note that, although R_1 constitutes the source impedance from the point of view of feedback, functionally it belongs to the feedback network so that in practice we deal with a voltage amplifier whose feedback-stabilized gain equals

$$v_o/v_g \simeq -R_f/R_1.$$

Hence, we may increase the voltage gain by reducing R_1 without affecting LT or ω_{hLT}, which again increases the closed-loop gain-bandwidth product $\omega_{0H} \simeq 1/R_1 C_T$ just as in the case of the trans-admittance amplifier. As a matter of fact, this can be verified by inspection of Fig. 1.48, where R_1 is seen to be completely isolated from the amplifier loop by the short-circuit of its input impedance. The corresponding Bode plot is shown in Fig. 1.49 for two values of R_1. The effect of varying $1/B$ on $H(j\omega)$ in this case is similar to the one shown in Fig. 1.47. Common to both circuits is a resistor injecting current into an active zero admittance node (a CB or common gate connected transistor), whose value does not affect the circuit operation as long as the self-admittance of the node is high compared with that of the resistor (i.e. mismatch conditions are satisfied). This concept will be discussed in greater detail in Section 1.7.4(*a*).

At this point we are ready to appreciate the advantage of feedback-stabilized transimpedance and current amplifiers employing the relevant enhancing combination between basic amplifier and feedback: an accordingly designed transimpedance feedback amplifier (Fig. 1.48) yields for different gains the Bode plot shown in Fig. 1.49 with ω_{hH} independent of gain, whereas use of a voltage amplifier (non-enhancing) produces a response which remains within the boundaries of the Bode plot shown in Fig. 1.45 for any gain, i.e. ω_{hH} inversely proportional to the gain.

A realization of an enhancing transimpedance amplifier by a discrete curcuit has been published by Hatch.[21]

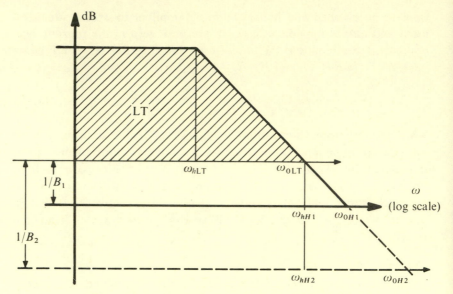

Fig. 1.49. Bode plot for circuit shown in Fig. 1.48.

The series–shunt feedback amplifier

An enhancing current amplifier is shown in Fig. 1.37, fed from a voltage source through a series impedance Z_1. Here we are dealing with two enhancing combinations: one at the input of the transimpedance amplifier, yielding a response as shown in Fig. 1.49 and the other due to the output transistor serving as a transadmittance amplifier, which affects the frequency response of LT as shown in Fig. 1.47. The resulting independence of ω_{0H} on $1/B$ accounts for the simultaneous high-speed and high-gain capabilities of enhancing current amplifiers. One problem encountered in their practical design is the need for level shifting between

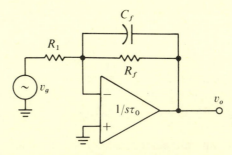

Fig. 1.50. Pole-splitting feedback network.

the transadmittance output stage and the load. A possible solution is indicated in problem 1.7 (Section 1.6.4).

Interplay between frequency-dependent feedback network and frequency response of basic amplifier
Frequently, the feedback network is not purely resistive, but is employed to obtain a feedback-stabilized frequency response. Consider the active *RC* integrator shown in Fig. 1.50 for which

$$LT(s) = -R_1(1 + sR_fC_f)/\{(R_1 + R_f)s\tau_0[1 + s(R_1\|R_f)C_f]\},$$

and

$$H(s) = -R_f/R_1\{1 + s[R_fC_f + \tau_0(1 + R_f/R_1)] + s^2\tau_0R_fC_f\}$$
$$= -R_f/R_1(1 + as + bs^2) = -R_f/R_1(1 - s/s_1)(1 - s/s_2); \tag{1.80}$$

where $a = \tau_1 + \tau_2$, $\tau_1 = R_fC_f$, $\tau_2 = \tau_0(1 + R_f/R_1)$, $b = k\tau_1\tau_2$, and $k = 1/(1 + R_f/R_1)$. Since $k < 1$ is always satisfied, (1.80) always has two real roots. This can be shown from the condition $a^2 \geq 4b$, or

$$\tau_1/\tau_2 + \tau_2/\tau_1 \geq 4k - 2, \tag{1.81}$$

which is true for any ratio τ_1/τ_2 as long as $k \leq 1$.

It is instructive to compute the roots of $H(s)$ for $\tau_1 \gg \tau_2$. For this condition, the resulting poles of (1.80) can be closely approximated by

$$\left.\begin{aligned}
s_1 &\simeq \frac{-1}{a} = \frac{-1}{(\tau_1 + \tau_2)} \\
s_2 &\simeq \frac{-a}{b} = -\frac{1}{k}\left(\frac{1}{\tau_1} + \frac{1}{\tau_2}\right) \simeq \frac{-1}{\tau_0}.
\end{aligned}\right\} \tag{1.82}$$

In this final result, $\tau_1 + \tau_2$ represents the dominant time constant which yields the gain-bandwidth product, whereas s_2 is the value of the parasitic pole which need be known only approximately. Hence, $H(s)$ can be simplified as follows:

$$H(s) \simeq \frac{-(R_f/R_1)}{\{1 + s(R_f/R_1)[R_1C_f + \tau_0(1 + R_1/R_f)]\}(1 + s\tau_0)}. \tag{1.83}$$

Problem
1.13 (*a*) Derive (1.70) and (1.74) employing the gain-bandwidth theorem, i.e. use (1.75) instead of (1.69).

(*b*) Compute $S_{\omega hH}^H$ for the amplifier shown in Fig. 1.22 and compare the result with $S_{\omega 0}^H$.

1.7.4   The effect of feedback on impedances

(*a*) *Active zero immittances*

The effect of feedback on the input immittance (impedance or admittance) of port *j* of a network is discussed in Appendix B.3. In accordance with (B.2), the closed-loop input and output immittances of a negative feedback amplifier approach zero (zero admittance for series and zero impedance for shunt feedback) for a very large LT:

$$I_{\text{in}} \text{ or } I_{\text{out}} = \lim_{|LT| \to \infty} [I_{jj}(1-LT)]^{-1} = 0. \tag{1.84}$$

Active zero immittances do not exist in practice, although they can be created with questionable stability at dc by making LT(0) infinite through local positive feedback. Still, feedback can create immittances which may be considered to equal zero for certain practical purposes.

As an example, consider the case of long interconnecting cables where termination by the characteristic impedance of a transmission line is essential; this termination can be realized by a passive resistance which is connected either in series with an active zero impedance (virtual ground) or in parallel with an active zero admittance.

In particular, the *Miller* and *bootstrapping* effects describe the modification of the input impedance of an amplifier by shunt and series feedback respectively, for the purpose of creating an active zero immittance.

The Miller effect. Consider Fig. 1.51, showing shunt feedback applied to the input of an amplifier. According to (1.84),

$$(Z_{\text{in}})^{-1} = (Y_f + Y_1 + y_i)[1 - A_v Y_f/(Y_f + Y_1 + y_i)] = (Y_1 + y_i) + Y_f(1 - A_v), \tag{1.85}$$

with $A_v = -v_o/v_e < 0$ for negative feedback and assuming mismatch conditions at the output to be satisfied ($r_o = 0$).

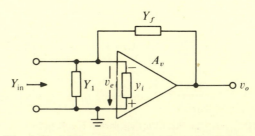

Fig. 1.51. Miller effect due to shunt feedback.

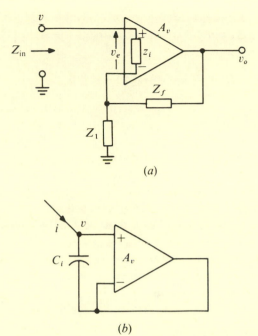

(a)

(b)

Fig. 1.52. (a) Bootstrapping effect due to series feedback. (b) Bootstrapping of capacitor by voltage follower.

The term $Y_f(1 - A_v)$ in (1.85) describes the Miller effect, which evaluates the impedance of the feedback element reflected to the input as a function of the voltage gain of the basic amplifier. The latter may be a voltage, current, transimpedance or transadmittance amplifier, and the resulting voltage gain should be evaluated accordingly.

Bootstrapping. The bootstrapping effect is demonstrated in Fig. 1.52(a), which shows an amplifier with series feedback applied to its input. According to (1.84), with $r_o = 0$,

$$(Y_{in})^{-1} = [z_i + (Z_1\|Z_f)]\{1 - A_v(Z_1\|z_i)/[(Z_1\|z_i) + Z_f]\}$$
$$= (Z_1\|Z_f) + z_i[1 - A_v Z_1/(Z_1 + Z_f)]. \qquad (1.86)$$

A comparison between (1.85) and (1.86) shows that the input impedance is reduced by the Miller effect proportionally to $|A_v|$, for negative feedback $(A_v < 0)$, but increased by bootstrapping proportionally to $|LT|$, again for negative feedback, with $LT = A_v Z_1/(Z_1 + Z_f) < 0$.

Frequency limitations. Feedback techniques appear to be capable of modifying impedances by a factor proportional to A_v; this is however only correct for the frequency range over which the gain of the basic amplifier is not a function of frequency. Consider, for instance, $Y_f = G_f = 1/R_f$ in Fig. 1.51 with $A_v(j\omega) = -A_v(0)/[1+j\omega A_v(0)\tau_0]$. Hence, R_f is reflected to the input as

$$R_f[1+j\omega A_v(0)\tau_0]/[1+A_v(0)](1+s\tau_0), \qquad (1.87)$$

which for $[A_v(0)\tau_0]^{-1} \ll \omega \ll 1/\tau_0$ is an inductance of value $R_f\tau_0$. Similarly, if Y_f is a capacitor C, the corresponding input impedance equals over the same frequency range a resistor valued τ_0/C_f. In both cases, the useful range over which Y_f is increased by the Miller effect by a factor $[1+A_v(0)]$ is restricted to frequencies $\omega \ll 1/A_v(0)\tau_0$. For half-power frequencies $f_h = 1/2\pi A_v(0)\tau_0$ below 100 Hz, a typical value for general purposes IC's, this is indeed a serious restriction.

Similar considerations lead to a modified input impedance by boot-strapping. Consider, as an example, Fig. 1.52(a) with $z_i = 1/sC_i$, $Z_f = R_f$ and $Z_1 = R_1$, which demonstrates the bootstrapping effect of a voltage amplifier on its input capacitance or on a capacitor C_i connected in parallel with its input terminals.

An analysis according to Appendix B.4 yields

$$I_{11}(s) = (R_1\|R_f) + 1/sC_i,$$

$$LT(s) = A_v(s)R_1/(R_1+R_f)[1+s(R_1\|R_f)C_i],$$

and

$$[Y_{\text{in}}(s)]^{-1} = I_{11}(s)[1-LT(s)] = (R_1\|R_f) + [1-A_v(s)R_1/(R_1+R_f)]/sC_i.$$

For $A_v(s) = -A_v(0)/[1+sA_v(0)\tau_0]$ we note that reduction of C_i by a factor $[1+A_v(0)R_1/(R_1+R_f)]$, due to bootstrapping, is only valid for frequencies $\omega \ll 1/A_v(0)\tau_0$. Furthermore, C_i introduces a pole into $LT(s)$, which adversely affects stability (see problem 1.14(a)).

The stability problem may be overcome by connecting a compensating capacitor C_f in parallel with R_f, but this in turn introduces a pole at the frequency $-1/R_fC_f$ into the feedback-stabilized gain $1/B$. In short, a rigorous analysis of the bootstrapping effect reveals rather dubious benefits.

The voltage follower shown in Fig. 1.52(b) is better suited for boot-strapping a capacitance C_i, which in this case merely loads the output but does not introduce a pole in $LT(s)$. Still, reduction of C_i by the factor $[1+A_v(0)]$ is here also limited to frequencies $\omega \ll 1/A_v(0)\tau_0$. This is a rather disappointing result, since the half-power frequency of the voltage follower equals $1/\tau_0$, and we might wrongly expect bootstrapping to be effective over the corresponding frequency range.

Note, however, that over the frequency range $1/A_v(0)\tau_0 \ll \omega \ll 1/\tau_0$ the input admittance of the voltage follower equals

$$Y_{\text{in}} = s^2 C_i \tau_0, \tag{1.88}$$

which is a *second order capacitor* or 'supercapacitor', also dubbed FDNR (*frequency-dependent negative resistance*). This has useful applications in active filters, and its creation over a wide range of frequencies by a bootstrapped capacitor is feasible.

In conclusion, both Miller and bootstrapping effects modify impedances by the factor $[1+A_v(0)]$ over the range of frequencies $\omega \ll 1/A(0)\tau_0$ only. At frequencies $1/A_v(0)\tau_0 \ll \omega \ll 1/\tau_0$ the nature of the reflected impedance is changed and its value is independent of $A_v(0)$. The critical frequency $1/A_v(0)\tau_0$ at which this change-over occurs can be increased by restricting the rise in the gain of the basic amplifier, towards low frequencies, through application of a secondary loop.

Enhancing the Miller effect. In the case of the Miller effect, the choice of the basic amplifier provides another degree of freedom for the reduction of Z_{in}. Interpretation of (1.85) leads to a suitable choice for this purpose. Mismatch conditions for a basic voltage amplifier yield

$$y_i \ll Y_g + Y_f, \tag{1.89}$$

making the term $Y_f(1+A_v)$ dominant, whereas those for a basic trans-impedance amplifier invert the inequality (1.89) and make the term $y_i(1-LT)$ in (1.85) dominant.

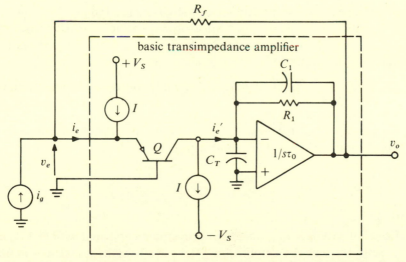

Fig. 1.53. Improved active zero input immittance due to enhancing feedback.

The ensuing advantages of employing a basic transimpedance amplifier will be demonstrated considering the relevant design considerations for the amplifier shown in Fig. 1.53. It consists of a CB connected input transistor serving as current follower, which is employed to convert the low input impedance of the basic transimpedance amplifier into a high-impedance current source feeding the conventional voltage amplifier. The gain of the latter is stabilized by a secondary feedback network R_1C_1. R_1 restricts the low-frequency gain and thereby increases the half-power frequency of the basic amplifier for the purpose of broadening the frequency range, over which the active input impedance of the complete amplifier remains resistive. The capacitor C_1 is required to avoid intro-duction of a dominant pole in the feedback attenuation of the secondary feedback loop, due to the total capacitance C_T between the input terminal of the voltage amplifier and ground.

For the condition $R_1C_1 \gg \tau_0$ we obtain for the feedback-stabilized transimpedance transmission of the voltage amplifier

$$v_o/i'_e|_{\tau_1 \gg \tau'_0} \simeq -R_1/(1+s\tau_1)(1+s\tau'_0), \qquad (1.90)$$

where $\tau_1 = R_1C_1$ and $\tau'_0 \simeq \tau_0(C_1+C_T)/C_1$.

The LT of the complete amplifier is

$$\text{LT}(s) = \frac{-\alpha(0)}{(1+s\tau_\alpha)} \times \frac{v_o}{i'_e} \times \frac{1}{(R_f+r_e)}, \qquad (1.91)$$

with $\alpha(0)$ being defined as the low-frequency current gain, τ_α as $1/\omega_h$ and r_e as the inverse transconductance of the CB connected input transistor Q. In practice, a sufficiently fast transistor may be chosen in most cases to make the time constant τ_α negligibly small; furthermore, mismatch conditions for transimpedance feedback require $R_f \gg r_e$. Hence,

$$\text{LT}(s) \simeq -\text{LT}(0)/(1+sR_1C_1)(1+s\tau'_0), \qquad (1.92)$$

where $\text{LT}(0) \simeq R_1/R_f$.

Finally, applying the mismatch conditions $y_i \gg Y_f + Y_g$ for a trans-impedance amplifier to (1.85), and with $R_fC_1 \gg \tau'_0$,

$$Z_{\text{in}} \simeq \frac{r_e}{(1-LT)} \simeq \frac{r_e(1+sR_1C_1)}{\text{LT}(0)(1+sR_fC_1)}, \qquad (1.93)$$

with $g_e \gg G_g + G_f$ and $\text{LT}(0) \gg 1$.

Equation (1.93) is a significant improvement compared with (1.87); it demonstrates that the amplifier of Fig. 1.53 exhibits a resistive input impedance of r_eR_f/R_1 at frequencies $\omega \ll 1/R_1C_1$. In order to increase

$1/R_1C_1$ to a maximum possible value, the following design procedure should be followed.

(i) *Choice of C_1.* This capacitor prevents excessive phase lag due to C_T. Its effect on stability is recognized from (1.92), whose dominant time constant $R_1C_1 = \mathrm{LT}(0)R_fC_1$ and whose parasitic time constant τ_0' both become infinitely large for $C_1 = 0$. In order to keep the increase in τ_0' due to C_1 within reasonable limits, we shall choose $C_1 \geqslant C_T$ yielding $\tau_0 \leqslant \tau_0' \leqslant 2\tau_0$.

(ii) *Choice of R_1.* Increasing R_1 reduces both the input resistance $r_e R_f/R_1$ and the frequency $1/R_1C_1$. Thus we may sacrifice low input resistance in order to achieve a broader frequency range over which the input impedance remains truly resistive.

(*b*) *Impedance control by multiloop feedback*
Control of closed-loop input impedance. Equation (1.84) indicates the possibility of controlling impedances by using a secondary feedback loop to stabilize the LT. An amplifier exhibiting a negative input impedance controlled by the Miller effect is shown in Fig. 1.54. Assuming $A_v = \infty$, its input impedance is

$$v/i = Z_f/[1 - (R_1 + R_2)/R_1] = -Z_fR_1/R_2. \qquad (1.94)$$

The two feedback loops involved are a negative one due to R_1 and R_2, which controls the voltage gain v_0/v, and a positive one due to Z_f. Practical uses include compensation for source capacitance in biomedical applications (Section 2.2.1). Limitations of (1.94) due to ω_{hH} of the feedback-stabilized voltage amplifier should be taken into account.

Because multiloop feedback can create controlled active impedances, it is capable of replacing impedance matching by a passive resistance,

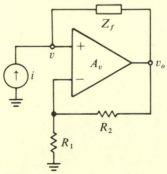

Fig. 1.54. Negative input impedance controlled by positive feedback.

followed by an active zero immittance. The main advantage of active impedance matching is that it eliminates the noise contribution of the passive termination, resulting in an improvement of the noise figure. Furthermore, creation of finite active resistances is possible over a wider frequency range that that of 'infinite' or 'zero' resistance. In short, this technique uses the capabilities of feedback far more efficiently that that of combining active zero immittances with passive elements.

Application of multiloop feedback to the creation of low-noise broadband terminations (*cooled terminations*) for transmission lines has been proposed by Radeka.[22] In this technique, the desired input impedance is obtained by controlling the Miller effect (first loop) by a stabilized gain function (second loop) – hence the term 'multiloop feedback'. Radeka controls the gain function by stabilizing τ_0 of the basic amplifier, which does not necessarily imply the use of a feedback loop. His circuit is shown in Fig. 1.64 and the noise aspects of this technique are discussed in Section 1.7.6.

Control of closed-loop input and output impedances. An elegant extension of multiloop feedback to the control of both input and output impedances of an amplifier has been employed by Aprille.[23] Originally, this technique was used for high-frequency repeaters, whose useful frequency range is extended by the use of reactive components and whose standing wave ratio is optimized over a given passband. It may also be employed in broadband signal processing using terminated transmission lines, in which case the noise figure will be improved because of the elimination of passive terminations.

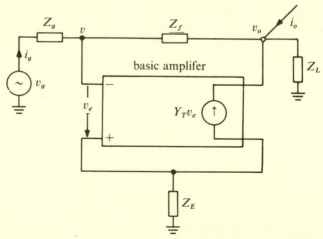

Fig. 1.55. Impedance control by multiloop feedback.

The basic principle is demonstrated in Fig. 1.55. Both input and output impedances are actively controlled. Z_{in} is obtained from a computation of the voltage gain for $Y_T = \infty$ and $i_o = 0$:

$$v_o/v = A_v = -(Y_E - Y_f)/(Y_f + Y_L), \qquad (1.95)$$

which gives

$$v/i_g = Z_{in} = Z_f/(1 - A_v) = (Z_L + Z_f)/(1 + Z_L Y_E). \qquad (1.96)$$

The output impedance with $v_g = 0$ and $Z_L = \infty$ is

$$v_o/i_o = Z_{out} = (Z_g + Z_f)/(1 + Z_g Y_E). \qquad (1.97)$$

An alternative circuit is shown in Fig. 1.56. Assuming again that $Y_T = \infty$ and $i_o = 0$, the voltage gain is

$$v_o/v = A_v = -[Y_4 + Y_L(1 + Z_3/Z_4) - Y_1]/(Y_1 + Y_2), \qquad (1.98)$$

and the input impedance is

$$v/i_g = Z_{in} = Z_1/(1 - A_v). \qquad (1.99)$$

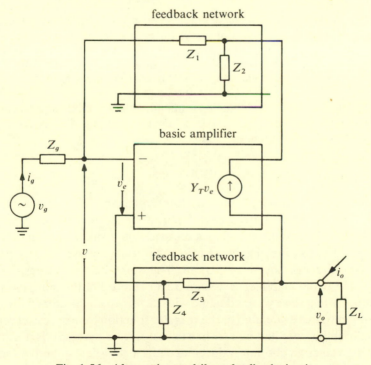

Fig. 1.56. Alternative multiloop feedback circuit.

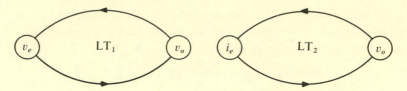

Fig. 1.57. LTs of voltage (LT$_1$) and transimpedance (LT$_2$) amplifiers.

Similarly, if $v_g = 0$ and $Z_L = \infty$, then the output admittance is

$$i_o/v_o = Y_{out} = (Z_3 + Z_4)^{-1} + Z_4(Z_1 + Z_2 + Z_g)/(Z_3 + Z_4)Z_2 Z_g.$$
(1.100)

Computation of the gain v_o/v_g will be left to the reader as a problem.

Problem

1.14. (*a*) Verify (1.86). [Hint: choose the source current i_g as error parameter.]

(*b*) For the amplifier shown in Fig. 1.52(*a*) with $z_i = 1/sC_i$ and $z_o = r_o$, $Z_f = R_f$, $Z_1 = R_1$ and $r_o \ll R_1 + R_f$, show that according to Appendix B.4

$$LT(s) = -A_v(s)R_1/sC_i(R_1 + R_f)[(R_1 \| R_f) + 1/sC_i].$$

(*c*) Compute for the amplifier of Fig. 1.53 the two LTs as defined in Fig. 1.57. Evaluate the exact expression of LT(0) for each case, disregarding mismatch conditions. Note that for the computation of LT$_1$ it is necessary to evaluate the voltage gain v_o/v_e of the transimpedance amplifier. The result shows that LT$_1$ = LT$_2$, i.e. LT is invariant with respect to the chosen parameters. Assume $\omega \ll 1/R_1 C_1$.

(*d*) Compute for Fig. 1.53 the input immittance from (1.85) as a function of LT$_1$ and LT$_2$. Show that in both cases the dominant term is $R_1/R_f r_e$, but that its interpretation is different in each case.

1.7.5 Offset

Consider the feedback-stabilized differential amplifier shown in Fig. 1.58. The dc offset sources have been modelled according to Section 1.2.3, with the two offset voltages lumped into a single one. In order to evaluate ΔV_o as a function of the offset sources, we employ superposition and consider one source at a time. The computation is considerably simplified if the amplifier is assumed to exhibit infinite gain, and little accuracy is lost since we are dealing with dc for which the amplifier gain is maximum. Thus evaluation of the transfer function for each offset source actually amounts to finding $1/B$ in each case. Note that the offset voltage, ΔV, is given as an absolute value, since it is a statistical quantity. On the other hand, I_a and I_b are here base currents of the input transistor pair,

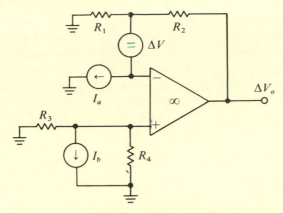

Fig. 1.58. Offset in feedback-stabilized differential amplifier.

whose polarity is defined for npn input transistors as shown.

$$\Delta V_o = |\Delta V|(R_1 + R_2)/R_1 + I_a R_2 - I_b (R_3 \| R_4)(R_1 + R_2)/R_1.$$
$$(1.101)$$

For the recommended condition of equal dc resistances seen by the input terminals of the amplifier – i.e. $(R_1 \| R_2) = (R_3 \| R_4)$ – we obtain in accordance with (1.101)

$$\Delta V_o = |\Delta V|(R_1 + R_2)/R_1 + |\Delta I| R_2, \qquad (1.102)$$

where $\Delta I = I_a - I_b$. We note that in (1.102) both ΔV and ΔI are statistically independent quantities which are frequently given by the manufacturers as 'typical' values. Hence, if we deal with an industrial product and ΔV_o represents the design target for a batch of feedback-stabilized amplifiers, it should be evaluated as

$$\Delta V_o = \{[\Delta V(R_1 + R_2)/R_1]^2 + (\Delta I R_2)^2\}^{1/2}. \qquad (103)$$

Temperature coefficient
An important parameter for dc amplifiers is the temperature coefficient $d\Delta V_{BE}/d\mathscr{T}$ of the differential input pair, which is a physically well-defined parameter:

$$d\Delta V_{BE}/d\mathscr{T} \simeq \Delta V_{BE}/\mathscr{T}, \qquad (1.104)$$

where $\mathscr{T}$ is the absolute temperature in kelvin. Since the maximum value of the input offset ΔV_{BE} is a parameter which is well controlled by the manufacturer, the temperature drift can be predicted with reasonable accuracy for amplifiers in which the dominant offset is associated with the V_{BE}s of the input pair (this is not always true).

On the other hand, the drift of the current offset remains mostly an undefined quantity. As a helpful consideration we note that, in the case of discrete bipolar differential input stages, the emitter current of the input transistor pair is only slightly temperature dependent. Hence, the dominant factor affecting I_B is the dc current gain h_{FE}, whose relative temperature coefficient is roughly $dh_{FE}/h_{FE} \simeq 0.01/°C$, so that dI_B/I_B can be evaluated by a sensitivity computation in accordance with (1.51) and (1.54) as

$$dI_B/I_B = S^{I_B}_{h_{FE}} \, dh_{FE}/h_{FE} = -dh_{FE}/h_{FE}. \qquad (1.105)$$

The temperature coefficient of ΔI can be assumed to be lower in most cases by one order of magnitude.

Some degree of compensation for the positive temperature coefficient of h_{FE} is achieved in IC amplifiers such as the LM 101A. There the emitter current of the input differential stage varies as the absolute temperature, which partly compensates for the positive temperature coefficient of h_{FE} and is in the right direction to reduce the base current variation as a function of temperature.

1.7.6 Noise and the noise figure

Improving an amplifier's noise figure (for the definition of this, see Appendix A.5) by means of feedback is as possible as building a perpetual motion machine. The presence of feedback components invariably affects the noise figure adversely. Minimizing this effect, whilst taking full advantage of the benefits of feedback, is of utmost importance. Feedback can, however, be applied to create active impedances which contribute less noise than passive ones of equal value, which has already been mentioned in Section 1.7.4(*b*).

Basic computations
The practical evaluation of a feedback amplifier's noise performance requires computation of the noise due to the amplifier itself, plus that due to all externally connected components, with the noise preferably referred to the input.

Consider the feedback amplifier shown in Fig. 1.59. For didactic purposes, the amplifier gain is assumed to be independent of frequency, and hence band limitation is due entirely to a filter of midband unity gain and ENB Δf, which is connected in cascade with the amplifier. R_2 represents a resistor used to equalize the resistances seen by both amplifier inputs, which makes the dc drift proportional to ΔI only. G_{pa}

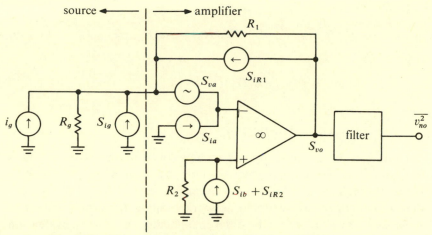

Fig. 1.59. Noise computation for feedback-stabilized amplifier.

and G_{pb} are the parallel noise conductances of the amplifier at each terminal. Hence, $S_{ia} = 4k\mathcal{T}G_{pa}$ and $S_{ib} = 4k\mathcal{T}G_{pb}$ are the mean square noise current spectral densities at the respective input terminals. Similarly, $S_{va} = 4k\mathcal{T}R_s$. Hence, the spectral density S_{vo} at the amplifier output equals

$$S_{vo} = 4k\mathcal{T}[(G_1 + G_{pa})R_1^2 + (R_s + R_2 + R_2^2 G_{pb})(R_g + R_1)^2 / R_g^2], \tag{1.106}$$

and the mean squared noise output voltage is

$$\overline{v_{no}^2} = S_{vo}\,\Delta f. \tag{1.107}$$

In order to refer the amplifier noise output voltage to the input, we must first determine whether the signal generator is a voltage or a current source, which may depend on the kind of transducer we are using. For Fig. 1.59 we assume a current source i_g with an internal resistance R_g. Hence, the total noise must be referred to the input node in parallel with i_g, in order to replace W_{na}/W_{ng} by $\overline{i_{na}^2}/\overline{i_{ng}^2}$, in the noise figure computation.

The feedback-stabilized gain equals $v_o/i_g = R_1$, and the amplifier noise referred to the input node equals

$$\overline{i_{na}^2} = \overline{v_{no}^2}/R_1^2 = 4k\mathcal{T}[G_1 + G_{pa} + (R_s + R_2 + R_2^2 G_{pb})(G_g + G_1)^2]\Delta f. \tag{1.108}$$

The mean squared source noise referred to the same input node is

$$\overline{i_{ng}^2} = 4k\mathcal{T}G_g\,\Delta f.$$

Finally,

$$F = 1 + \overline{i_{na}^2}/\overline{i_{ng}^2} = 1 + (R_g/R_a) + (R_b/R_g) + 2R_bG_1, \qquad (1.109)$$

where

$$R_a = (G_1 + G_{pa} + R_bG_1^2)^{-1},$$

and

$$R_b = (R_s + R_2 + R_2^2 G_{pb}).$$

Evaluation of Δf now appears to have been unnecessary, which is because in this oversimplified case the noise due to all white noise sources passes through the same filter. In the realistic case of frequency-dependent amplifier gain and feedback network, shaping for parallel and series noise sources will in general be different and hence a separate evaluation of Δf for each noise source is essential.

The optimum source resistance for noise matching is obtained from (1.109):

$$R_{g\,\text{opt}} = (R_aR_b)^{1/2}. \qquad (1.110)$$

Hence,

$$F_{\min} = F(R_{g\,\text{opt}}) = 1 + 2(R_b/R_a)^{1/2} + 2R_bG_1. \qquad (1.111)$$

In order to demonstrate the difference in the spectral density at the output due to R_s and R_p if a frequency-dependent feedback network is employed, we shall evaluate S_{vo} for the amplifier shown in Fig. 1.60. In many practical noise computations, the frequency dependence of the amplifier gain can be disregarded, provided that the passband of the complete system falls within the region of high amplifier gain. This is

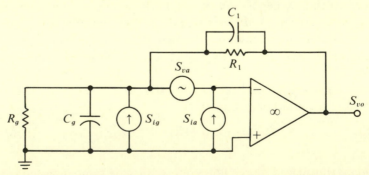

Fig. 1.60. Noise computation for frequency-dependent source and feedback network.

assumed to be the case, and the amplifier gain will again be considered as infinite:

$$S_{vo} = (S_{ig} + S_{ia})R_1^2/(1 + \omega^2 R_1^2 C_1^2) +$$

$$S_{va}(R_1 + R_g)^2[1 + \omega^2(R_1\|R_g)^2(C_1 + C_g)^2]/R_g^2(1 + \omega^2 R_1^2 C_1^2).$$
$$(1.112)$$

We note not only the different frequency dependence of the terms in S_{vo} due to S_{ia} and S_{va}, but also the fact that at high frequencies the output term due to amplifier series noise equals $S_{va}(C_1 + C_g)^2/C_1^2$ and is, therefore, not band-limited (except for the shaping effect due to the amplifier, which has been disregarded).

Noise matching for capacitive signal sources
Several kinds of transducers having practically a purely capacitive source impedance yield comparatively small signals, so that minimization of the noise contribution of the associated amplifiers is of particular importance. We shall discuss here various points to be considered in the design of feedback amplifiers connected to purely capacitive signal sources.

The high impedance levels associated with capacitive sources usually preclude the use of transformers for noise matching. Furthermore, unlike the case with dissipative signal sources, noise is mainly due to secondary effects such as leakage across the source capacitance (parallel noise), due to dc biasing. Capacitive semiconductor detectors create, in addition, Johnson noise due to a resistance in series with the capacitance (series noise).

In the absence of transformer matching, these noise sources may be combined in the noise computations with those due to the amplifier.

Thus we arrive at a noise model, in which a noise current source is connected in parallel and a noise voltage source in series with the detector capacitance. In such a situation two important aspects should be considered.

First, the power spectral density spectrum of the source will be white for series noise, and exhibit a slope of 40 db/decade for mean squared parallel noise. The effect of this on the design of a matched filter will be demonstrated in Section 3.1.6.

Second, a junction FET will be chosen in critical applications as the input gain element, due to its low parallel noise contribution and high transconductance. Technological considerations determine the ratio between transconductance and input capacitance of this device, and an increase in the junction area or parallel connection of several FETs

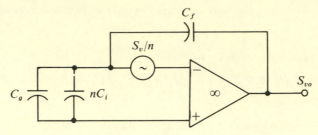

Fig. 1.61. Series noise matching with capacitive source.

makes it possible to minimize the series noise contribution. This effect can be observed considering the charge-sensitive configuration shown in Fig. 1.61, which yields a constant signal irrespective of the number of gain elements connected in parallel. Hence, minimum spectral noise density at the output corresponds to minimum series noise contribution or series noise matching. Parallel noise is disregarded. Only essential components are shown in order to bring out the desired effect without obscuring secondary ones. C_i constitutes the input capacitance of a single FET, and the total input capacitance and series noise resistance of parallel connected FETs equals nC_i and R_s/n, respectively. The noise voltage spectral density at the output, due to series noise, is

$$S_{vo} = (4 \ell \mathcal{T} R_s/n)[1 + (C_g + nC_i)/C_f]^2. \qquad (1.113)$$

Differentiation of (1.113) with respect to n yields a minimum for

$$nC_i = C_f + C_g. \qquad (1.114)$$

This is a corollary of (1.23) for capacitive sources, showing that for minimum series noise contribution, the total input capacitance of all FETs in parallel should equal the self-capacitance of the input node (with the FETs disconnected).

Noise measured at the input of a feedback amplifier
It is instructive to observe the modification of the voltage spectral density at the *input* of an amplifier, due to feedback. This has practical implications for systems employing a transducer such as a position-sensitive radiation detector, which has more than one output terminal and a separate amplifier connected to each of them, in which case the source noise seen by each amplifier includes the noise power transmitted from the inputs of the other amplifiers.

Consider a voltage amplifier connected to a capacitive source as shown in Fig. 1.62. In the equivalent diagram of the amplifier, the terminals v_i

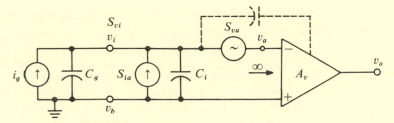

Fig. 1.62. Amplifier noise performance without external feedback.

and v_b represent the actual input, and C_i the input capacitance of the FETs employed.

In the absence of any feedback, the voltage spectral density S_{vi}, at node v_i, is only a function of S_{ia}, the current spectral density of the amplifier at the inverting input:

$$S_{vi} = S_{ia}/\omega^2(C_g + C_i)^2. \tag{1.115}$$

S_{va} has apparently no effect, since it is connected to the infinite input impedance of the noiseless amplifier, in the model.

Physically, however, S_{va} is due to the shot noise created by the dc collector current of the input transistor referred to its own input, which does inject some noise into the input through an internal feedback capacitance. This internal feedback path, shown as a dashed line in Fig. 1.62, has a secondary effect which is noticeable only towards the high-frequency end of the amplifier's useful frequency range, and will be disregarded. Hence, with the above restriction in mind, the feedback capacitorless noise model in Fig. 1.62 is correct and no noise is produced in the input circuit due to S_{va}.

Now consider Fig. 1.63, in which the amplifier is connected as a voltage follower with the immediately apparent effect of increasing its own input impedance seen by the source due to bootstrapping of C_i.

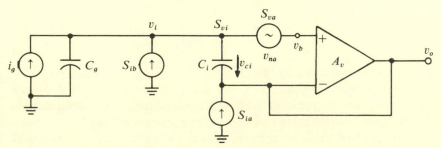

Fig. 1.63. Noise performance for series feedback.

Considering first the voltage spectral density at node v_i due to parallel noise only,

$$S_{vi} = S_{ib}/\omega^2[C_g + C_i/(1 - A_v)]^2, \qquad (1.116)$$

where $A_v < 0$. This is greater than the corresponding value for Fig. 1.62, due to bootstrapping of C_i. S_{ia} has practically no effect, since it is short-circuited by the very low output impedance of the voltage follower.

Turning now to the effect of S_{va}, we shall first assume $\lim |A_v| \to \infty$, for which $v_b - v_o = 0$ and the voltage spectral density across C_i necessarily equals that of S_{va}. This determines the current spectral density S_{inci} through C_i, which also flows through C_g and thereby yields

$$S_{vi} = S_{inci}/\omega^2 C_g^2 = S_{va}\omega^2 C_i^2/\omega^2 C_g^2. \qquad (1.117)$$

This is white noise since we have assumed A_v not to be a function of frequency.

We shall now compute the voltage spectral density at v_i, due to S_{va}, for the more realistic gain function

$$v_o = (v_b - v_o)A_v(0)/(1 + j\omega/\omega_h). \qquad (1.118)$$

This is a classic example of noise modified by a frequency-dependent feedback loop, in which case it is advantageous first to consider the noise as a small-signal instantaneous voltage v_{na}, compute transmission to the desired 'output' $v_{ni} = H(j\omega)v_{na}$ and finally compute $\overline{v_{ni}^2} = S_{va} \int_0^\infty |H(j\omega)|^2 \, df$.

In our case, the voltage across C_i, due to voltage division of v_o between C_i and C_g, is

$$v_{ci} = v_o C_g/(C_i + C_g) = v_o - v_b + v_{na}; \qquad (1.119)$$

substituting $v_b = f(v_0)$ from (1.118) and assuming $A(0)C_g \gg C_i, A_v(0) \gg 1$.

$$v_o \simeq v_{na}(1 + C_i/C_g)/D, \qquad (1.120)$$

where $D = [1 + j\omega\tau_0(1 + C_i/C_g)]$. Finally,

$$\overline{v_{ni}^2} = \overline{v_o^2}C_i^2/(C_i + C_g)^2 = S_v \int_0^\infty [C_i^2/(C_g^2|D|^2)] \, df. \qquad (1.121)$$

The cooled termination technique

The series noise measured at the input of a shunt-shunt feedback amplifier is useful in obtaining 'cooled terminations',[22] a classic example of the useful modification of a noise spectrum by feedback.

The problem is that of terminating both ends of a delay line serving as signal source, by a resistor equal to its characteristic impedance r_0 without

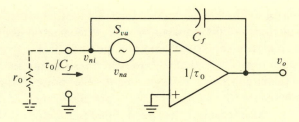

Fig. 1.64. 'Cooled termination' technique.

introducing the full Johnson noise due to the termination. The principle is demonstrated in Fig. 1.64, which shows an amplifier employing capacitive feedback. To simplify the derivation, we first disregard any resistor in parallel with C_f employed for dc stabilization. The amplifier transfer function is assumed to exhibit a single pole at the origin. First, we show that the active input admittance y_i of the circuit is conductive over a wide frequency range:

$$y_i = j\omega C_f(1 - A_v) = (C_f/\tau_0)(1 + j\omega\tau_0), \qquad (1.122)$$

with

$$A_v \simeq -1/j\omega\tau_0.$$

Hence, the input resistance equals τ_0/C_f for $\omega \ll 1/\tau_0$ and will be designed to equal the characteristic impedance r_0 of the delay line:

$$\tau_0/C_f = r_0. \qquad (1.123)$$

The noise voltage $\overline{v_{ni}^2}$ will be evaluated with the practical assumption that the input of the amplifier is connected to a delay line whose opposite end is also terminated by its characteristic impedance. Again evaluating v_{ni} as a function of the small-signal noise voltage v_{na} due to S_{va}, we have

$$v_o = -(v_{ni} + v_{na})/j\omega\tau_0, \qquad (1.124)$$

yielding

$$v_{ni} = v_o j\omega C_f r_0/(1 + j\omega C_f r_0)$$
$$= v_o j\omega\tau_0/(1 + j\omega\tau_0). \qquad (1.125)$$

Finally,

$$v_{ni} = v_{na}/(2 + j\omega\tau_0) = v_{na}/2$$

for $\omega \ll 1/\tau_0$. For the same range of frequencies,

$$S_{vi} = S_{va}/4 = k\mathcal{T}R_s = 4k\mathcal{T}_c r_0. \qquad (1.126)$$

Equation (1.126) is interpreted in such a way that, since the input impedance of the amplifier equals $r_0 = \tau_0/C_f$, the spectral density of the noise $4k\mathcal{T}_c r_0$ measured at the same point appears to be due to r_0, which is at the temperature $\mathcal{T}_c$. Actually, however, there is no such physical resistor and the circuit noise, disregarding parallel noise, is due to R_s in accordance with (1.126). It follows that for practical values of r_0, $\mathcal{T}_c/\mathcal{T} = R_s/4r_0 \ll 1$, for which reason $\mathcal{T}_c$ has been called the *cooled temperature* of r_0.[16]

A practical circuit requires a resistor in parallel with C_f for dc stabilization. Furthermore, a typical gain function of a practical amplifier is

$$A(s) = -A(0)/[1 + j\omega A(0)\tau_0].$$

By making $R_f C_f = A(0)\tau_0$, the active input resistance $Z_f/[1 - A(s)]$ becomes

$$r_i = [R_f \| (\tau_0/C_f)] \tag{1.127}$$

over a frequency range from dc to $\omega \ll 1/\tau_0$.

The noise contribution of dc stabilizing resistors
A problem occurring with high-impedance detectors is that of the Johnson noise introduced by the resistor providing dc stabilization for the amplifier. Consider the amplifier shown in Fig. 1.65, which has two noise current sources connected to its input: one is due to the leakage current I appearing at the amplifier's input, and the other is due to the Johnson noise of R_f. Let it be our design target to find the value of R_f for which the total squared noise current spectral density S_T at the amplifier input is greater by a factor $(1 + b)$ than that due to I alone:

$$S_T = 4k\mathcal{T}(G_p + G_f) = 4k\mathcal{T}G_p(1 + b), \tag{1.128}$$

with

$$b = G_f/G_p = R_p/R_f = 2k\mathcal{T}/qIR_f = 0.052/V_{R_f}, \tag{1.129}$$

where $V_{R_f} = IR_f$, the voltage drop across R_f due to I, and $4k\mathcal{T}/R_p = 2qI$.

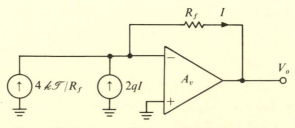

Fig. 1.65. Noise contribution of feedback resistor.

Thus, for a 10% squared noise density contribution or 5% rms noise contribution, the voltage drop across R_f is 0.52 V at room temperature. This is indeed a simple and very practical way of measuring the relative noise contribution of R_f, the only pitfall being that for very high values it may contribute excess noise whose value by far exceeds that predicted by the thermal noise formula. This can be partly avoided by careful choice of the type of resistor employed as R_f.

If the amplifier is connected to a signal source which contributes its own leakage current, and if that current and the leakage current of the amplifier happen to flow in opposite directions, then the total dc current flowing through R_f is due to the difference between the two leakage currents, whereas their noise spectral densities still add. In this case the above method of comparing the noise contribution of R_f with that of the leakage currents breaks down unless one of the two currents is negligibly small.

In charge-sensitive amplifiers, a feedback resistor can be replaced by an electronic switch,[24,25] thereby replacing its noise contribution by that of the switch. In one method employed, the charge accumulating across the feedback capacitor is drained off by shortly illuminating at suitable time intervals the drain gate junction of the input FET. Hence, no additional component is employed since the switch is an integral part of the amplifier. The only price paid are the short 'dead-time' itervals employed for the reset, during which the amplifier is inoperative. A special type of J-FET suitable for '*optoelectronic feedback*' is manufactured by Texas Instruments, England, as Type E8003A.

Problems

1.15. Fig. 1.66 shows a feedback amplifier. The equivalent noise diagram of the basic amplifier is shown in Fig. 1.11(*b*). $R_2C_2 = R_1C_1$. All resistors contribute noise. Find the noise figure F of the circuit.

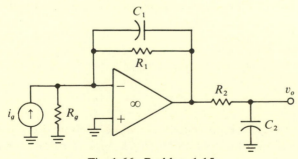

Fig. 1.66. Problem 1.15.

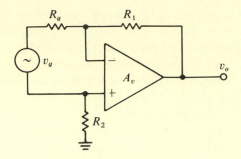

Fig. 1.67. Problem 1.17.

1.16. The noise performance of the instrumentation amplifier shown in Fig. 1.40 is to be evaluated. The equivalent noise diagram of the amplifiers is shown in Fig. 1.11(b). Assume all gain factors $A_v = \infty$, $R_2 = 100R_1$, and $R_3 = R_4 = 10R_1$.

Find S_{vi}, the noise spectral density of the complete amplifier referred to the input (in series with v_{ia} and v_{ib}). Resistor noise should be taken into account. Beware of correlation!

1.17. For the circuit shown in Fig. 1.67, assume for the basic amplifier $A_v = \infty$, $z_i = \infty$, $z_o = 0$ and its noise sources as shown in Fig. 1.11(b) with $S_{ia} = S_{ib}$. All resistors contribute noise.

  (a) Find S_{vo}, the noise spectral density at the output.
  (b) Find the noise figure F of the circuit assuming $R_1 = R_2 = R$.
  (c) Find the condition for which the noise contribution of the resistors, $R_1 = R_2 = R$ is negligible.

1.18. Verify (1.127).

1.19. The purpose of this problem is to compare the noise contributions of the feedback networks employed in the circuits of Fig. 1.68. Make the following assumptions.

  (i) The transimpedance v_o/i_g of the two circuits is the same.
  (ii) The source resistance R_g is the same, in both circuits.
  (iii) The source noise is due to R_g.
  (iv) Only the resistors contribute noise, the amplifier is assumed to be noiseless.
  (v) $R_2 \gg R_4$.

Find F_a, the noise figure of the circuit shown in Fig. 1.68(a), and $F_b = 1 + (F_a - 1)k$, the noise figure for Fig. 1.68(b). Draw conclusions.

1.7.7 Distortion

The effect of feedback on distortion will be evaluated referring to Fig. 1.69, which shows shunt-series feedback applied to an amplifier, with $v_g = V_g \cos \omega t$, and $v_e = V_e \cos \omega t$. Employing the amplifier model for

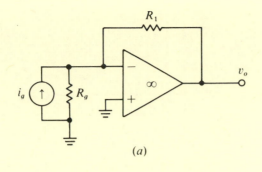

(a)

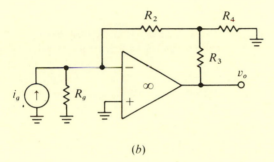

(b)

Fig. 1.68. Problem 1.19.

harmonic distortion discussed in Section 1.2.5, Fig. 1.17, we consider a single distortion term

$$v_{di} = A_1 V_e D_i \cos \omega_i t,$$

with V_e the amplitude of the fundamental frequency term due to v_g, measured at the amplifier's input: $V_e = V_g/(1-\text{LT})$. Hence,

$$v_o(t) = \frac{A_1}{(1-\text{LT})} V_g \cos \omega t + \frac{v_{di}}{(1-\text{LT})}$$

$$= \frac{A_1 V_g}{(1-\text{LT})} \left[\cos \omega t + \frac{D_i}{(1-\text{LT})} \cos \omega_i t \right]. \qquad (1.130)$$

It follows that the distortion coefficient D_i is reduced by feedback by a factor of $1/(1-\text{LT})$.

Problem

1.20. Consider a basic voltage amplifier $A(j\omega) = A(0)/(1+j\omega/\omega_h)$, with $A(0) = 10^5$ and $f_h = \omega_h/2\pi = 25$ Hz. Application of an input signal of 50 Hz yields, at an output amplitude of 5 V, second harmonic distortion (at 100 Hz) of 2%.

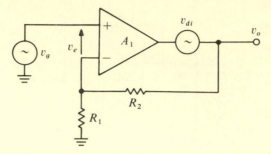

Fig. 1.69. Computation of distortion in feedback amplifier.

(a) Compute the gain at the second harmonic frequency of 100 Hz, and the amplitude of the second harmonic as defined above, referred to the input of the amplifier.

(b) The basic amplifier is connected to a shunt-shunt feedback network, yielding a nominal gain of -9. Find the percentage harmonic distortion at 100 Hz, for a 50 Hz input signal yielding an output amplitude of 5 V.

1.7.8 Settling time,[26] linear and non-linear effects

Settling time τ_s is defined as the time elapsed from application of an ideal instantaneous step input to the time at which the amplifier output has entered and remained within a specified error band. It is determined by a combination of amplifier characteristics, non-linear and linear. The lower limit for the error band is set by noise, which includes, in addition to inherent amplifier circuit noise, interference pick-up. For highest precision signal processing, settling accuracy may be specified to within 0.01% of full scale in which case all possible residual effects must be taken into account. These effects can be divided into those occuring within the linear range of amplifier operation, and others occuring within its non-linear range.

Linear effects

The linear response of an amplifier may be affected by the following transient effects:

(a) The closed-loop transient response in the case of a feedback amplifier, to be discussed in detail in this section.

(b) Dielectric absorption (hysteresis) in capacitors employed in the signal path (compensating capacitors) and external bypass capacitors. When a capacitor is suddenly charged or discharged, it requires a period of 'soaking' to return to internal charge equilibrium. This transient may cause long 'tails' of at least 10 to 10^3 µs, often greatly exceeding the settling time of otherwise fast amplifiers.

(*c*) Thermal after-effects due to non-linear slewing or saturation. During slewing or overload, the normally equal distribution of dissipations, particularly in the input stage, may become grossly unbalanced. In addition, a large change in load current may significantly change dissipations in the output stage (and hence temperatures inside the amplifier package), and if these temperature changes are unequally conducted to the two sides of the input stage, transient, as well as steady-state, unbalances can result. CM input voltage swing is usually negligible in inverting amplifier configurations, but is equal to FS output voltage swing in unity-gain voltage followers, where it may cause substantial changes in the input stage dissipation. Such thermal transients may cause tails exceeding milliseconds in duration, affecting the settling time long after the amplifier has recovered from overload and the linear transient has died out. In practice, thermal transients will be noticeable for settling time accuracies better than 0.1%.

Computation of settling time

The foregoing definition of τ_s is illustrated in Fig. 1.70, which shows a normalized monotonic step response $h_{-1}(t)$ approaching unity asymptotically. The error band is defined by $1 \pm \varepsilon(\tau_s)$, τ_s being the settling time. For τ_s to exist, $h_{-1}(t)$ must be asymptotically stable. Another restriction is the requirement for $h_{-1}(\infty)$ not to equal zero (differentiation in the signal path causes a zero value).

The normalized error $\varepsilon(t)$ is obtained from Fig. 1.70 as

$$\varepsilon(t) = [h_{-1}(\infty) - h_{-1}(t)]/h_{-1}(\infty), \tag{1.131}$$

where $h_{-1}(t)$ is the step response of the amplifier. Since $\varepsilon(\tau_s)$ will in a

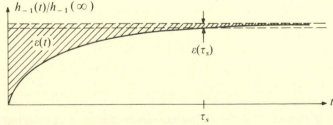

Fig. 1.70. Monotonic system response and associated error band.

practical case be defined by the specified maximum permissible error, τ_s can be obtained by solving (1.131) for t.

As an example, assume that the step response of an amplifier is

$$h_{-1}(t) = V[1 - \exp(-t/\tau)], \tag{1.132}$$

and let τ_s be defined as the time it takes the amplifier response to settle within 0.001 of the final value; then substituting (1.131) in (1.132),

$$\varepsilon = 0.001 = \exp(-\tau_s/\tau) \tag{1.133}$$

or $\tau_s = \tau \ln(1/\varepsilon) \simeq 6.9\tau$.

In spite of the above restrictions, τ_s may still be defined for bandpass systems exhibiting a flat frequency response over a reasonably broad region of frequencies and a low-frequency doublet with the zero at the origin, due to differentiation, if the doublet is excluded in the computation of τ_s. Fig. 1.71 shows the normalized response $h_{-1}(t)$ of such a system including a zero at the origin, and $\overline{h_{-1}(t)}$, its response with the doublet excluded. In this case, $\varepsilon(t)$ will be meaningful only for $\tau_s < \tau_{s\,max}$, and with that restriction τ_s can be computed from $\overline{h_{-1}(t)}$.

A general expression for ε can be derived from the response $H(s)$ of a feedback amplifier as defined in (1.29), employing the final value theorem $h_{-1}(\infty) = \lim_{s \to 0} H(s)$ and the identity $[-LT(s)]/[1 - LT(s)] = 1 - 1/[1 - LT(s)]$, and noting that, for negative feedback, $\lim_{s \to 0} LT(s) = -LT(0)$:

$$\varepsilon(t) = [h_{-1}(\infty) - h_{-1}(t)]/h_{-1}(\infty)$$

$$= \mathscr{L}^{-1}\left\{\frac{1}{s} \frac{\{1 - 1/[1 + LT(0)]\}/B(0) - \{1 - 1/[1 - LT(s)]\}/B(s)}{\{1 - 1/[1 + LT(0)]\}/B(0)}\right\},$$

in which for $LT(0) \gg 1$ the denominator becomes approximately $1/B(0)$.

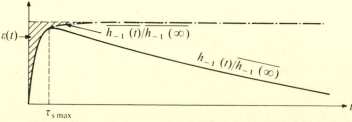

Fig. 1.71. Response of system including differentiation, and meaningful error band.

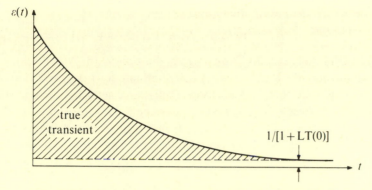

Fig. 1.72. Resulting error if term $[1+\text{LT}(0)]^{-1}$ is omitted.

Hence,

$$\varepsilon(t)|_{\text{LT}(0)\gg 1} \simeq \mathscr{L}^{-1}\left\{\frac{1}{s}\left[1-\frac{B(0)}{B(s)}+\frac{B(0)}{B(s)[1-\text{LT}(s)]}-\frac{1}{[1+\text{LT}(0)]}\right]\right\}.$$

(1.134)

Here we might be tempted to omit the term $1/[1+\text{LT}(0)]$, but its importance in (1.134) is demonstrated in Fig. 1.72: its omission results in an asymptotic error of $1/[1+\text{LT}(0)]$ for ε. However, by applying the gain-bandwidth theorem (i.e. replacing $\text{LT}(s)$ by $\text{LT}'(s)$ in accordance with Appendix C.3), $\text{LT}'(0)$ becomes infinite and the term $1/[1+\text{LT}'(0)]$ becomes zero, which greatly simplifies computation of the remaining terms. Hence, finally,

$$\varepsilon'(t) \simeq \mathscr{L}^{-1}\left\{\frac{1}{s}\left[1-\frac{B(0)}{B(s)}+\frac{B(0)}{B(s)[1-\text{LT}'(s)]}\right]\right\}.$$

(1.135)

Since the root locus for $\text{LT}'(s)$ is practically the same as that for $\text{LT}(s)$, the difference between (1.134) and (1.135) is negligible.

An interpretation of (1.135) shows that the term following 1 gives the effect of the feedback network, and the last term that of $\text{LT}(s)$ on the settling time.

With the exclusion of possibly present low-frequency doublets (i.e. no differentiation), (1.135) becomes in the time domain

$$\varepsilon(t) = \sum^{n} a_i \exp\left(-t/\tau_i\right).$$

(1.136)

The dominant terms in (1.136) are easily recognizable; in case of doubt, the relative contribution of two terms as a function of time can be compared by computing their ratio as

$$\varepsilon_i(t)/\varepsilon_j(t) = (a_i/a_j) \exp\left[-t(1/\tau_i - 1/\tau_j)\right]. \tag{1.137}$$

Setting (1.137) equal to unity yields the time t_1 for which both terms contribute equally. If $a_i > a_j$, $\varepsilon_i(t)$ will dominate for $t < t_1$, and vice versa.

In the analysis of the closed-loop small-signal settling time one should distinguish between 'fast' and 'slow' transients. Fast transients are due to singularities of the LT beyond ω_0, resulting in a closed-loop response which may be composed of overdamped, critically damped or oscillatory

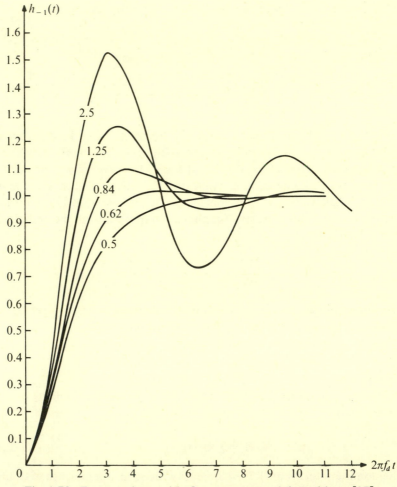

Fig. 1.73. Fast transients with Q as parameter. Adapted from [27].

terms. Typical fast transients are shown in Fig. 1.73. Presence of a reasonably well-controlled fast transient may even be desirable, since a slightly underdamped second order response will improve the settling time.

The effect of doublets

A doublet occurring at frequencies below ω_0 is anathema to short settling times, since it causes slow transients. This effect has been treated in the literature.[28,29]

As an example, consider an amplifier employing frequency-independent feedback, whose LT exhibits a doublet at frequencies $1/\tau_p$ and $1/\tau_z$, with $\tau_p = m\tau_z$, and τ_p, $\tau_z > 1/\omega_0$. Figs. 1.74(a), (b) and (c) show typical step responses for three values of m. For $m = 1$, Fig. 1.74(a), the pole cancels the zero and no slow transient is present. For $m = 1/2$, Fig. 1.74(b), the zero occurs at a lower frequency than the pole $(1/\tau_z < 1/\tau_p)$; response is monotonic, but exhibits a slowly decaying portion. For $m = 2$, the pole occurs at a lower frequency than the zero and response is non-monotonic, exhibiting a slow overshoot as shown in Fig. 1.74(c). The deviation of m from unity in Figs. 1.74(b) and (c) is much greater than normally encountered; this is to demonstrate the effect clearly. In practice, the value of m will be very close to unity.

Doublets occur in amplifiers employing two independent gain stages, such as fast-slow amplifiers discussed in Section 3.1.4, and in amplifiers employing the feedforward technique, where the slow input stage is bypassed by a capacitor in order to improve their speed of response. While this latter technique greatly improves the gain-bandwidth product, the zero produced by the feedforward path cannot be made to cancel exactly with the corresponding pole of the slow amplifier, because of the production spread in monolithic capacitors, resistors and transistor parameters. The resulting poor settling time may completely offset any speed advantage gained by the feedforward technique.

A circuit using feedback to reduce the open-loop doublet separation has been designed.[30] This *doublet compression technique* allows the full benefit from the increased gain-bandwidth product to be derived without introducing slow transients. The principle of doublet compression is based on the fact that a pole at $-1/\tau_p$ whose root locus terminates at the zero $-1/\tau_z$ for an infinite gain factor k_0 will approach the zero very closely for high value of k_0 (see problem 1.23).

The effect of a doublet on the closed-loop frequency response can be evaluated in terms of Q of its denominator, as demonstrated in problem 1.21(e).

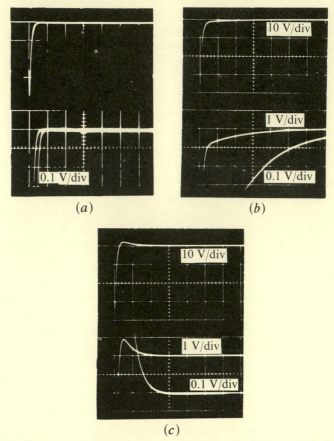

Fig. 1.74. Slow transients for $\tau_p = m\tau_z$: (a) $m = 1$, (b) $m = 1/2$, (c) $m = 2$. Adapted from [26].

The rise time as criterion for the settling time

If applied to a monotonic response, rise time and settling time are intimately connected. Although their optima do not necessarily coincide, the basic design criteria are the same for both. This is fortunate, since the rise time τ_R of a monotonic system is conveniently obtained from the Laplace transform of its impulse response in terms of the second time moment T_2, as shown in Appendix D:

$$\tau_R = (2\pi)^{1/2} T_2.$$

In what follows we shall use the rise time τ_R to define a figure of merit for a monotonic system. Consider first the simplified response of a typical

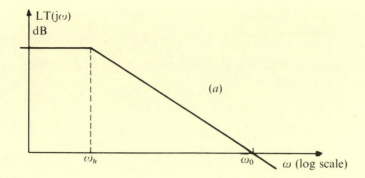

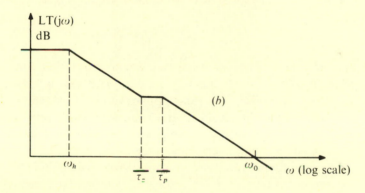

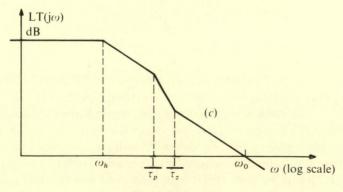

Fig. 1.74 (cont.)

compensated IC amplifier:

$$A(s) = A(0)/[1 + sA(0)\tau_0].\tag{1.138}$$

The response of (1.138) is characterized in the frequency domain by the gain–bandwidth product

$$\omega_0 = A(0)\omega_h = 1/\tau_0\tag{1.139}$$

and in the time domain by the gain/rise time quotient

$$A(0)/\tau_R = 1/(2\pi)^{1/2}\tau_0.\tag{1.140}$$

Both expressions are linearly related to the amplifier's settling time and may serve as a figure of merit.

It is instructive to consider the effect on $H(0)/\tau_R$ of choosing individual values of $A_j(0)$ in a cascade of n non-interacting gain stages.[13] If

$$H(s) = \prod_{}^{n} \{A_j(0)/[1 + sA_j(0)\tau_0]\},$$

i.e. τ_0 is identical for all cascaded stages, it can be shown that, for a given value of $H(0)$, $H(0)/\tau_R$ will be maximum if the $A_j(0)$ are all equal. The proof is left to the reader. The same applies to ω_h of cascaded stages, which is quite obvious since both τ_R and $1/\omega_h$ are linearly related to T_2.

Having found the optimal relationship between the individual values of $A_j(0)$ in a cascade of n amplifiers, what is the number of amplifiers which will yield the minimum overall rise time for a given total gain $H(0)$? This has been evaluated by Elmore[31,32] for the case in which each stage's gain-bandwidth product remains constant as gain and bandwidth are exchanged. If each stage is of equal gain,

$$H(0) = [A_j(0)]^n,$$

and

$$\tau_R = (2\pi n)^{1/2}[H(0)]^{1/n}\tau_0.\tag{1.141}$$

Equation (1.141) exhibits a minimum for $n = 2\ln H(0)$. Hence, the gain for each amplifier for the optimal rise time is $A_j(0)' = e^{1/2} = 1.65$ for $j = 1, 2, \ldots, n$, which is not an encouraging result, since it indicates that a relatively large number of cascaded stages is required to obtain a minimum τ_R for a given gain. In order to investigate the loss in rise time if fewer than the optimum number of stages is employed, (1.141) is plotted in Fig. 1.75 with τ_R normalized with respect to that of unity gain, as a function of n, with the total gain $H(0)$ chosen as parameter. It appears

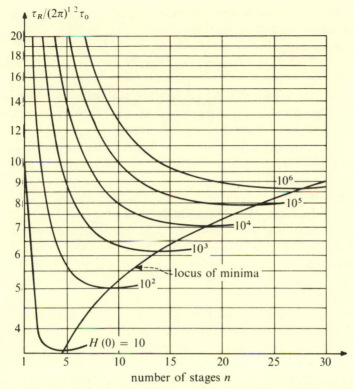

Fig. 1.75. Plot of normalized rise time versus number of stages, with total gain as parameter. Adapted from [31].

that (1.141) has a very broad minimum, so that a sacrifice of, say, 10% in the rise time will significantly reduce the number of stages required.

A related problem arises if the τ_0 of the individual stages is shorter than the minimum required for the specified rise time and total gain of the cascaded stages. If an individual stage gain of 1.65 is used, the resulting total gain is larger than is required, and so fewer stages need be used to achieve the actual gain and rise time required. This problem has been dealt with by Comer and Griffith.[33]

The effect of delay on the rise time
The effect of delay on the root locus is demonstrated in Appendix C.2, where its presence in the LT is shown to reduce the value of the breakaway factor $k_0(\sigma_b)$. Here we shall investigate the effect of delay on the rise time, and show that introduction of a suitably placed zero leads to a significant improvement in the performance.

A simplified LT exhibiting the basic properties of a feedback system and including a stabilizing zero is

$$\mathrm{LT}'(s) = (1 + s\tau_2)\exp(-s\tau_D)/s\tau_0(1 + s\tau_1). \qquad (1.142)$$

The exponential term in (1.142) represents the total signal delay, and $-1/\tau_1$ a single parasitic pole. In practice there is, of course, more than one parasitic pole, but a single one is sufficient to model the typical behavior of a feedback system in the presence of delay. After this simplified model has been investigated, interpretation of a more accurate computer simulation will be greatly facilitated.

The speed of the amplifier employed is in fact limited by the parasitic pole at the frequency $1/\tau_1$, and the delay will be specified by the ratio τ_D/τ_1. The gain-bandwidth product $1/\tau_0$ is a designable parameter, since it is part of the multiplicative gain factor in the root locus, which should be adjusted to ensure the shortest possible settling time.

Table 1.3 compares the rise times of various closed-loop responses, including that due to (1.142).

Case 1 deals with a simple LT without delay, also tabulated in Table C.1. Case 2 deals with the same case, but includes delay with the ratio τ_D/τ_1 chosen as 2. Both cases are adjusted for a closed-loop double pole on the real axis. Note the significant reduction in the value of $|\sigma_b|$ and ω_0, and the resulting increase in τ_R, due to the delay in case 2.

Table 1.3

	Case 1	Case 2	Case 3
$-\mathrm{LT}(s)$	$\dfrac{1}{s\tau_0(1 + s\tau_1)}$	$\dfrac{\exp(-s\tau_D)}{s\tau_0(1 + s\tau_1)}$	$\dfrac{(1 + s\tau_2)\exp(-s\tau_D)}{s\tau_0(1 + s\tau_1)}$
$-\sigma_b$	$\dfrac{1}{2\tau_1}$	$\dfrac{2 - 2^{1/2}}{2\tau_1}$	—
$\dfrac{1}{\tau_0}$	$\dfrac{1}{4\tau_1}$	$\dfrac{1}{8.7\tau_1}$	$\dfrac{1}{3.6\tau_1}$ (optimized)
τ_R	$4\pi^{1/2}\tau_1 = 7.1\tau_1$[a]	$\dfrac{4\pi\tau_1}{2 - 2^{1/2}} = 12.1\tau_1$[a]	$3.1\tau_1$[b]
Condition	$\tau_D = 0$	$\tau_D = 2\tau_1$	$\tau_D = 2\tau_1$ (measured) $\tau_2 = 1.4\tau_1$ (optimized)

[a] According to Elmore's definition.
[b] Computed between 10% and 90%.

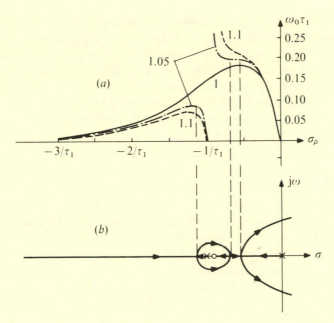

Fig. 1.76. Effect of zero on closed-loop response of (1.142). Adapted from [34]. (a) Normalized gain-bandwidth product versus σ_p with τ_2/τ_1 as parameter. (b) Root locus for $\tau_2/\tau_1 = 1.05$.

In case 3, a zero at the frequency $1/\tau_2$ is introduced to alleviate the effect of the delay. For this case, the nature of the root locus is strongly affected by the ratio τ_2/τ_1. A plot of k_0 versus σ is helpful in demonstrating the nature of the root locus (see Appendix C.1). Fig. 1.76(a) shows a plot of

$$\omega_0 = -\sigma(1+\sigma\tau_1)\exp{(\sigma\tau_D)}/(1+\sigma\tau_2). \qquad (1.143)$$

The real roots of (1.143) represent the location of the migrating poles along the real axis. Values corresponding to three trajectories are shown, with τ_2, normalized with respect to τ_1, serving as parameter. The gain-bandwidth product ω_0 is also normalized with respect to τ_1, so that preserving relationships employed in drawing Fig. 1.76(a) leads to identical behavior of systems exhibiting different speeds.

A maximum in the plotted curves indicates a point σ_b from which the migrating poles branch out into the complex frequency plane with increasing ω_0, whereas a minimum corresponds to a point at which a complex conjugate pole pair returns to the negative real axis.

If τ_2/τ_1 is chosen as unity, the zero cancels the pole and the root locus degenerates into the one shown in Fig. C.3. If the parameter τ_2/τ_1 is chosen as 1.05, (1.143) yields two maxima and one minimum. Part of the resulting root locus is plotted in Fig. 1.76(b). The trajectory of a single pole only, due to the delay, is shown. An increase in τ_2/τ_1 to 1.1 prevents the complex conjugate pole pair from returning to the real axis, in which case the pole at the origin migrates towards the zero without encountering another pole, and (1.143) exhibits only a single maximum.

Here we are dealing with a system exhibiting a non-monotonic response which can be designed for minimum settling time by optimizing the values of ω_0 and a corrective zero with respect to the characteristic system parameters. The closed-loop response of (1.142) has been realized by a high-speed amplifier, and simulated by computer program.[34,35] The factor τ_D/τ_1 was measured as 2, and the response optimized both in reality and by simulation, adjusting the ratios τ_2/τ_1 and τ_1/τ_0. Agreement between calculated and measured performances was good. The rise time computed for 2% overshoot was 3.1 τ_1, and settling time to within 0.5% was 11 τ_1. Inspection of Fig. 1.76(a) reveals qualitatively that the root locus of (1.142) yields for the optimized ratios $\tau_2/\tau_1 = 1.4$ and $\tau_1/\tau_0 = 0.28$ a complex conjugate pole pair and one negative real pole close to the zero at $-1/\tau_2$ (see problem 1.23).

The closed-loop step function response of (1.142) is shown in Fig. 1.77 for three values of the ratio τ_2/τ_1, all other parameters being kept constant. Traces (A) and (B) show the responses with the zero located to the left of the pole at $-1/\tau_1$; trace (B) shows the optimal response with $\tau_2/\tau_1 = 1.4$. For trace (C), the zero has been shifted to the right of $-1/\tau_1$. The low amplitude of the overshoot of trace (B) indicates the closeness of

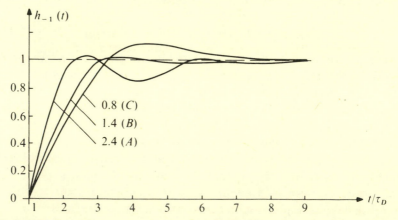

Fig. 1.7.7. Closed-loop step response of (1.142) with τ_2/τ_1 as parameter; $\omega_0\tau_1 = 0.28$. Adapted from [35].

the complex conjugate pole pair to the negative real axis, if $\tau_2/\tau_1 = 1.4$. The ratio between the rise time and the delay is 1.55 in this case. This compares favorably with the trivial condition of $\tau_2 = \tau_1$, in which case this ratio is 3.3. This example demonstrates the beneficial effect of introducing a zero in the LT of a control system which includes significant delay.

Improving control-loop stability in the presence of delay can also be important in industrial control. For example, a chemical reaction could be part of the loop. Another example of a control system incorporating delay is the temperature controlled oven, in which a dominant pole due to thermal capacity and a delay due to heat flow are introduced. In all these cases, the transient response may be significantly improved by introducing a judiciously placed zero, as has been done in case 3 of Table 1.3.

Slewing and saturation effects[26]

Non-linear transient effects are due to slewing rate limitations and overload. At output amplitudes approaching FS, most amplifier responses to an ideal input step enter the slewing region where the rate of output voltage change is limited by an internal capacitor and/or a capacitive load, and the feedback loop is broken. Typical slewing responses of two amplifiers are compared in Fig. 1.78, which shows that amplifier A, which slews much faster than amplifier B, has a longer small-signal settling time. Hence, fast slewing does not necessarily go together with a fast settling time.

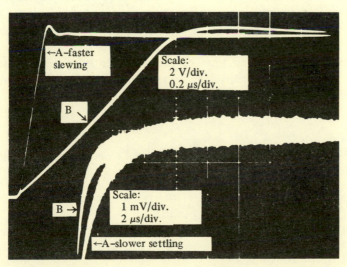

Fig. 1.78. Comparison of two amplifiers having similar settling times, but differing slewing rates. Adapted from [26].

For steady-state output amplitudes not exceeding FS, a slewing amplifier will finally enter the linear region. For steady-state output amplitudes exceeding the linear region, the amplifier will saturate. After the overloading signal has been removed, the amplifier reenters the linear region only after large charges away from normal operating values on the circuit capacities (including minority carrier storage in semiconductors) have been discharged back to equilibrium values. The same effect occurs after slewing. The total time it takes an amplifier to reenter the region of linear response, after an overloading signal has been removed, or after applying a non-overloading signal which causes slewing, is called its *recovery time*. The charges stored during overload are proportional to its duration, for short overload signals; hence the recovery time increase with the duration of the overload but reaches a maximum for overload durations beyond the time necessary for the stored charges to reach their saturation value.

Measurement of settling time

Sixteen-bit D/A converters are commercially available. Full utilization of fourteen-bit accuracy 'only' makes it necessary to specify an error band of 1 mV for the output amplifier, at 10 V FS.

Settling time is affected by non-linear and linear characteristics of the amplifier, by input and output loading, and by the signal itself. As an illustration, an output step of 10 V amplitude could produce a drift several milliseconds in duration, due to thermal feedback from the output to the input transistors, whereas this kind of drift would be practically absent for a step output of, say, 0.1 V.

Measuring the settling time of an amplifier is an art in itself. A satisfactory method of measuring it to within an error band of 1 mV has been published by Teledyne Philbrick.[36]

Problems

1.21. We investigate the effects of various LTs on the settling time of an amplifier, with $1/B \neq f(\omega)$.

Compute τ_s for an error band $\varepsilon = 10^{-3}$ with:

(a) $LT_1(s) = -1/s\tau_0$.

(b) $LT_2(s) = -A(0)/[1 + sA(0)\tau_0]$, $A(0) \gg 1$.

(c) $LT_3(s) = -A(0)(1 + s\tau_1)/[1 + sA(0)\tau_0\tau_1/\tau_2](1 + s\tau_2)$, $A(0) \gg 1$ and $\tau_1 = 10\tau_2 = 100\tau_0$. [Hint: Assuming that the characteristic equation $[1 - LT(s)] = 1 + as + bs^2$ exhibits two widely spaced poles (one dominant), it can be factorized into $s_1 \approx -1/a$, $s_2 = -a/b$.]

(d) $LT_4(s) = -A_1(0)A_2(0)(1 + s\tau_2)/[1 + sA_1(0)\tau_{01}][1 + sA_2(0)\tau_{02}]$.

(e) For $LT = -(1 + s\tau_z)/s\tau_0(1 + s\tau_p)$, and a frequency-independent feedback network, show that the quality factor of the denominator of the closed-loop response is

with $m = \tau_p/\tau_z$. $Q = m^{1/2}/[(\tau_z/\tau_0)^{1/2} + (\tau_0/\tau_z)^{1/2}]$ (1.144)

1.22. In this problem it will be demonstrated that, for a monotonic feedback-stabilized basic amplifier whose gain-bandwidth product is adjustable by external compensation, the minimum rise time τ_R is a function of the parasitic poles only and independent of the feedback-stabilized gain over the range for which the gain-bandwidth product can be suitably adjusted. We shall make the simplified assumption that the basic amplifier exhibits only a single parasitic pole at τ_1:

$$A(s) = -1/s\tau_0(1 + s\tau_1).$$

The gain-bandwidth product $\omega_0 = 1/\tau_0$ can be adjusted by external compensation to be less than or equal to a maximum value $\omega_{0\max}$. The basic amplifier is connected to a feedback network a shown in Fig. 1.22, the value of τ_{0LT} is adjusted by choosing the feedback-stabilized gain $1/B$ for the limiting condition of a double pole on the real frequency axis, in the closed-loop response.
   (a) For the above condition (double pole), find the maximum gain as a function of $\omega_{0\max}$.
   (b) For the same condition, find the rise time τ_R of the closed-loop response.
   (c) Show that the maximum possible gain/rise time quotient equals $\omega_{0\max}/\pi^{1/2}$.

1.23. In this problem we demonstrate the effect of doublet compression. Assume that

$$LT(s) = -LT(0)(1 + s\tau_z)/(1 + s\tau_p),$$

$m = \tau_p/\tau_z$. Show that, with m denoting the ratio between the open-loop pole and the zero, the corresponding closed-loop ratio equals $1 + m/LT(0)$, if $LT(0) \gg 1$.

1.24. Draw the two root locus plots of (1.142) for $\tau_2/\tau_1 = 0.9$ and $\tau_2/\tau_1 = 1.4$, respectively. For a graphical method of drawing the root locus in the presence of delay, see reference [37].

1.8 Gain stabilization by feedforward error correction

*Feedforward error correction**** was invented before feedback.[38] Its main virtue is that it reduces distortion without the stability problems inherent in the latter. Feedback, because it compares input with output, works only for small transit times in comparison with the information rate; this means that it requires a much larger bandwidth than that of the information processed by the amplifier. Feedforward, in contrast, makes all comparisons in a forward sequence. Hence, the bandwidth need only be

---

* This is not to be confused with the 'feedforward technique', in which part of a slow amplifier is bypassed in order to improve its speed of response.

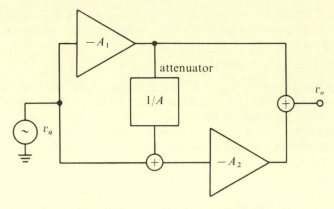

Fig. 1.79. Feedforward error correcting system.

that required for the actual information contained in the signal. Neglected for many years, feedforward has recently found an increasing number
of applications.[39,40,41]

The basic principle of feedforward error correction is shown in Fig.
1.79. The transmission of this circuit is

$$v_o/v_g = H = -(A_1 + A_2 - A_1A_2/A), \qquad (1.145)$$

which shows that if either A_1 or A_2 equals A, then $H = A$ irrespective of
the gain value of the other, inaccurate amplifier. How does this come
about? Inspection of Fig. 1.79 shows that if A_1 deviates from A, and the
gain of A_2 equals A, then A_2 amplifies and subtracts the error of A_1 from
the output, thereby cancelling it exactly. If A_2 deviates from A, and A_1 is
accurate, then the input to A_2 is simply zero and the gain is due to the
perfect amplifier A_1 only.

In practice, both amplifiers will deviate from their nominal value, and
the total error of the transmission is evaluated as follows.

If A is the nominal gain of the amplifier and $A_1 = A(1 + \delta_1)$, $A_2 = A(1 + \delta_2)$ with δ_1 and δ_2 designating the relative deviation of A_1 and A_2
from their nominal value, then

$$S_{A_1}^H = \frac{A_1 - A_1A_2/A}{A_1 + A_2 - A_1A_2/A} = \frac{-\delta_2 - \delta_1\delta_2}{1 - \delta_1\delta_2}, \qquad (1.146)$$

and

$$S_{A_2}^H = \frac{A_2 - A_1A_2/A}{A_1 + A_2 - A_1A_2/A} = \frac{-\delta_1 - \delta_1\delta_2}{1 - \delta_1\delta_2}. \qquad (1.147)$$

For small deviations, $\delta_1\delta_2 \ll \delta_1$, $\delta_1\delta_2 \ll \delta_2$; hence

$$S_{A1}^H \simeq -\delta_2 \quad \text{and} \quad S_{A2}^H \simeq -\delta_1,$$

and finally

$$\frac{\mathrm{d}H}{H} = S_{A1}^H \frac{\mathrm{d}A_1}{A_1} + S_{A2}^H \frac{\mathrm{d}A_2}{A_2} \simeq 2\delta_1\delta_2, \tag{1.148}$$

with $\mathrm{d}A_1 \simeq -\delta_1 A_1$, $\mathrm{d}A_2 \simeq -\delta_2 A_2$. Thus, a maximum deviation from its nominal value of 5% by each amplifier will result in a total deviation by no more than 0.5%.

Feedforward connected amplifier pairs can be iterated as shown in Fig. 1.80.

Feedforward error correction has been applied to the stabilization of amplifiers at very high frequencies, at which the use of feedback is no longer feasible.[40] A more recent application[41] is that of combining a highly non-linear class C power amplifier with a linear low-level amplifier, which is merely required to supply the signal restoring the linearity of output.

Problem
1.25. Verify (1.146), (1.147) and (1.148). Note that the Taylor series (1.148) should be expanded as $[H(A_1+\mathrm{d}A_1,\ A_2+\mathrm{d}A_2)-H(A_1, A_2)]/H(A_1, A_2)$. Hence, $\mathrm{d}A_1 \simeq -\delta_1 A$ and $\mathrm{d}A_2 \simeq -\delta_2 A$.

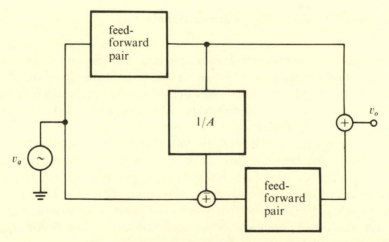

Fig. 1.80. Iterated feedforward amplifiers.

2

Transducers

2.1 Classification of transducers

A *transducer* is a device that converts one form of energy to another. Hence, all possible kinds of transducers can be tabulated in a matrix[1] as shown in Table 2.1. The diagonal elements of this matrix list modifiers only, i.e. devices which act upon an energy form without changing its nature. All other elements of the matrix are true transducers.

Transducers may be classified with respect to their input (temperature, pressure, etc.), or to the type of signal they produce. Several treatments of transducers have been published.[2,3,4] Here we deal mainly with analog signal processing, hence we shall focus on transducers providing an electric output (see Table 2.1, row 4), and in particular on the signal-carrying parameter which they provide, which may be voltage, current, charge, resistance or carrier modulation. The techniques employed in amplifying these signals will be discussed.

2.2 Amplification of transducer signals

Most transducer signals are too weak and not optimally shaped for extracting the desired information directly. The choice of amplifier configuration depends upon the nature and level of the signal which the transducers provide. Very low level signals require careful design of the amplifier for low-noise contribution. With this aim in mind, various decisions have to be made by the designer.

First, there is the choice between designing an amplifier for maximum power gain, and choosing one of the four basic amplifiers defined in Section 1.1. In the first case the real part of the product of the signal parameters – voltage and current – is maximized, whereas in the second case a single parameter is chosen to carry the signal at the input and at the output, the other parameter being undesirable and hence minimized through mismatch.

Amplifier gain has no bearing on the noise figure of a single amplifier but reduces the noise contribution of the second in a cascade of two.

Table 2.1[1]

out \ in	1 Mechanic	2 Thermal	3 Magnetic	4 Electric	5 Electromagnetic	6 Molecular
1 Mechanic	Gravity balance, Ballistic pendulum, Bellows	Thermometer, Bimetal strip	Magnetometer, Magnetostriction	Electrometers, Electrostrictive systems	Crooke's light mill, Radiation pressure systems	Hydrometer, Electrodeposition cell
2 Thermal	Adiabatic systems, Friction calorimeter		Eddy current systems	Thermal converter	Bolometer	Calorimeter, Thermal conductivity cell
3 Magnetic	Rowland disk, Magnetoelastic transducers		Magnetic storage systems	Electromagnetic meters, recorders, relays	Curie effect radiation meter	Magnetic balance, Nuclear magnetic resonance systems
4 Electric	Potentiometric, Inductive, capacitive, piezoelectric systems, Strain gauge	Thermistor, Thermocouple	Magnetoresistance, Hall effect systems	Langmuir probe, Charge collectors, Skin electrodes	Photoelectric, photoresistive, photogalvanic cells	Potentiometry, Conductimetry, Polarography
5 Electromagnetic	Interferometer, Photoelastic systems	Thermal radiation systems	Faraday cell	Kerr cell, Luminous gas discharge systems	Scintillator	Emission and absorption spectroscopy, Scintillator
6 Molecular		Thermal dye indicator			Photographic emulsion	

Furthermore, the maximum power gain of a particular amplifier can always be shown to be greater than its voltage or current gain. Hence, wherever reduction of the second gain stage noise contribution is of primary importance, the first amplifier stage should be designed for maximum power gain. This technique has important implications for passband amplifiers operating at high frequencies, at which the noise contribution of the second stage can be significant because of the comparatively low gain available. Furthermore, complex impedance matching is practicable at these frequencies.

In view of the available signal levels, such a degree of sophistication is neither essential, nor is it feasible with most kinds of transducers, whose signals may cover a broad range of frequencies. Moreover, an information-carrying signal parameter is defined for most transducers, which will therefore be operated under mismatch conditions in which the desirable signal parameter is enhanced, whereas the other one is being made negligibly small.

The second question to be settled regards the choice between a feedback amplifier and a basic one, i.e. one not employing overall feedback (it may still use internal feedback to stabilize the gain-bandwidth product). The noise figure of the former will always be worse than that of the latter because of the presence of the feedback network, but the difference can be made negligible in most cases through careful design.

2.2.1 Voltage transducers

Several kinds of transducers present a voltage source output to the signal processing channel. The electromagnetic transducer produces an output by movement of a permanent magnet with relation to a coil system (hence the signal is self-generating). Tachogenerators provide both voltage and frequency proportional to their angular velocity. Rotary transformers (selsyns) convert angular position into carrier modulation by means of a varying magnetic field inducing voltage in a coil. All the foregoing types of transducers require straightforward voltage measuring techniques.

In the variable reluctance transducer, the force or displacement to be measured causes a change in the magnetic flux linkage of a coil. The resulting variation in the inductance can be measured either as an amplitude change in a bridge, or as a change in the resonant frequency of an oscillator.

The bridge circuit frequently employs a differential arrangement of two coils. Source impedances are low, and no particular design problems are encountered in the signal processing channel.

Amplifier design for voltage sources
We now turn our attention to the practical design of amplifiers for the voltage source type of transducer. Clearly, a transducer whose output voltage is proportional to the measured parameter should be connected to a voltage amplifier whose input impedance is high compared with the transducer's output impedance. Fig. 2.1 shows a suitable circuit. The voltage source requires $z_i(1-LT) \gg Z_g$; for negligible noise contribution of the feedback network compared with that of the amplifier, $(R_1 \| R_2) \ll R_s$ and $S_{ia}(R_1 \| R_2)^2 \ll S_{ib}R_g^2$ should be satisfied, with $S_{va} = 4k\mathcal{T}R_s$ and $R_g = \mathrm{Re}\,(Z_g)$.

Measurement of bioelectric signals, and correct interpretation of the waveforms observed, is an art in itself.[5,6,7] In measuring a bioelectric signal, potential is frequently the quantity of interest. Furthermore, the equivalent circuit of the source impedances encountered may consist of a combination between capacitances and resistors – all frequency dependent. Thus, failure to employ an amplifier exhibiting adequately high input impedance may result in serious waveform distortion. Fig. 1.54 in Section 1.7.4 shows a voltage amplifier which not only exhibits high input impedance, but also can provide some compensation for the source impedance, according to (1.94).

This kind of circuit is used to compensate for the combined capacitance of the source, electrode and interconnecting cable. A better arrangement is to connect the cable screen to a guarding voltage (see Section 2.3), and to compensate only for the source and electrode capacitances.

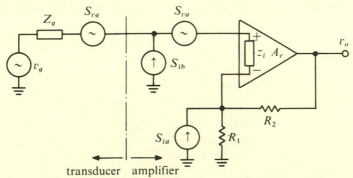

Fig. 2.1. Connection between voltage source type transducer and amplifier.

2.2.2 Current and charge transducers

Both current and charge transducers exhibit high output impedance; the difference between them lies in the signal-carrying parameter. Their output impedance is in most cases capacitive. Piezoelectric and pyroelectric transducers, semiconductor radiation detectors and ionization chambers belong to this category. As an example, consider a piezoelectric transducer frequently employed as accelerometer for vibration measurement, converting pressure F into charge Q, with Q proportional to F. Here we have two options. We may measure either the voltage across the transducer or its current into a virtual ground approaching a short-circuit. In the latter case we obtain the first derivative of the pressure, unless the current is integrated in order to recover a signal which is proportional to the charge, or to the pressure. We shall investigate these two alternatives in greater detail.

Connection to an ideal voltage-sensitive amplifier yields an output voltage $v_o = Q_s A_v / C_T$, under the unrealistic assumption that the input resistance of the amplifier is infinite. A more realistic situation is shown in Fig. 2.2, where the input impedance of the signal processor is shown to consist of r_i in parallel with the total capacitance C_T. We obtain

$$v_o(j\omega) = i_g(j\omega) r_i A_v / [1 + j\omega r_i (C_g + C_T)], \tag{2.1}$$

which, for $\omega \gg 1/r_i(C_g + C_T)$, becomes in the time domain

$$v_o(t) = A_v \left[(C_g + C_T)^{-1} \int_0^t i_g(t)\,\mathrm{d}t + V(0) \right]$$

$$= A_v \left[\frac{Q(t)}{C_g} \times \frac{C_g}{(C_g + C_T)} + V(0) \right]. \tag{2.2}$$

Hence, for frequencies limited at the low end by the input time constant and at the high end by the response of the amplifier, the output voltage

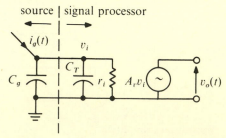

Fig. 2.2. Capacitive transducer connected to signal processor.

will be proportional to the signal charge across the transducer capacitance, attenuated by a factor $C_g/(C_g+C_T)$. Considering $Q(t)$ as the signal-bearing and hence desirable parameter, we recognize r_i and C_T as undesirable elements, with r_i making the frequency range and C_T the amplitude of the signal dependent upon the signal processing circuit. Series or shunt feedback may be applied to the input of the signal processor to reduce the effects of r_i and C_T.

Reduction of transducer loading by bootstrapping (series feedback)
The technique of increasing an undesirable impedance by bootstrapping was discussed in Section 1.7.4, and the limited range of frequencies over which it is effective was pointed out. Applied to Fig. 2.2 it is important to recognize that bootstrapping r_iC_T will reduce the signal attenuation to $C_g/\{C_g+C_T/[1+\mathrm{LT}(0)]\}$ over the appropriate range of frequencies, but if $C_T \gg C_G[1+\mathrm{LT}(0)]$, the time constant r_iC_T of the transfer function remains unaffected since the bootstrapping factor $[1+\mathrm{LT}(0)]$ cancels out.

Transducer loading by Miller effect (shunt feedback)
The effect of shunt feedback on the output signal of a capacitive transducer will be shown considering Fig. 2.3, initially disregarding the frequency dependence of A_v. The feedback network can be replaced by the equivalent diagram of Fig. 2.4, which shows the feedback network, modified by the Miller effect, connected in parallel with the transducer. For large values of A_v, shunt feedback is seen to place a low impedance in parallel with the detector, thereby reducing the effect of the detector impedance:

$$v_o(j\omega) = \frac{i_g(j\omega)R_f/(1+A_v)}{\{1+j\omega[C_g+C_T+C_f(1+A_v)]R_f/(1+A_v)\}} \times (-A_v)$$

$$\approx \frac{-i_gR_f}{\{1+j\omega[(C_g+C_T)/A_v+C_f]R_f\}}. \qquad (2.3)$$

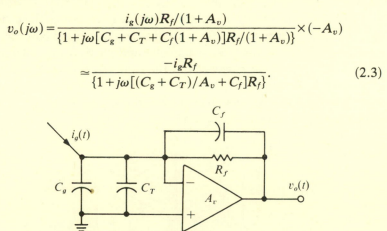

Fig. 2.3. Loading of capacitive transducer by shunt feedback.

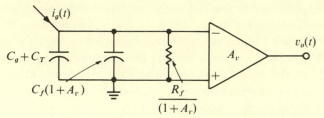

Fig. 2.4. Equivalent diagram of Fig. 2.3.

For $A_v C_f \gg (C_g + C_T)$, v_o is practically independent of $C_g + C_T$ and the transfer function is primarily a function of the feedback network $R_f C_f$. Hence, the difference between shunt and series feedback applied to a detector is such that with shunt feedback the source impedance is shunted by the low input impedance of the feedback amplifier, thereby reducing its effect on the transfer function, whereas with series feedback the effect of stray and amplifier input capacitance is reduced by the feedback, but the transfer function is primarily a function of the source impedance.

Just as in the case of bootstrapping, restrictions on the effectiveness of the Miller effect due to the frequency dependence of A_v are serious. This aspect has also been discussed in Section 1.7.4 and should be carefully taken into account in a practical design.

Measurement of displacement by capacitive transducer
Electrostatically charged capacitance transducers are employed in vibration measurements, in biomedical engineering,[7,8] and as condenser microphones. Shown schematically in Fig. 2.5, the effect to be measured is a change in the transducer's capacitance as a function of the distance x between the capacitor plates. Two modes of operation are possible: one is to charge the capacitor electrostatically and measure the effect of the resulting displacement current, and the other is to modulate a carrier. Both amplitude and frequency modulation are feasible, the latter being particularly attractive since it can be directly converted into a digital representation by using a counter.

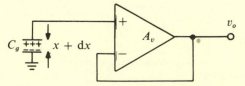

Fig. 2.5. Charged capacitive transducer.

Here we shall concern ourselves with the first mode of operation and discuss the effect of the change in the distance x between the plates of the electrostatically polarized transducer capacitance on the resulting signal. In this computation we shall make the simplified assumptions that the transducer capacitance C_g is equal to $A\varepsilon/x$ (ε is the dielectric constant), that the capacitor plates of area A remain parallel to each other during displacement, and that the end-effect in computing the capacitance may be neglected.

Although the foregoing simplifications do not apply to the general case, they serve well to emphasize the basic relationships of differential changes of x, and thereby aid the designer to choose a mode of operation best adapted to his requirements.

All following derivations will be based upon (2.4), obtained from $v_c = Q/C_g$ as

$$\frac{dv_c}{dt} = \frac{1}{C_g}\frac{dQ}{dt} + Q\frac{d(1/C_g)}{dt}. \tag{2.4}$$

From (2.4) we shall derive the responses of shunt-series and shunt–shunt feedback amplifiers to a change dx in the distance between the capacitor plates, corresponding to C_g connected to a high or a low input impedance feedback amplifier, respectively.

Operation of a capacitance transducer in the voltage mode
Consider Fig. 2.5 again, showing the simplified case of a voltage follower with infinite input impedance being connected to the capacitive transducer. Hence, the charge Q does not vary as a function of x, and the factor dQ/dt in (2.4) vanishes:

$$\frac{dv_c}{dt} = Q\frac{d(1/C_g)}{dx}\frac{dx}{dt} = \frac{Q}{A\varepsilon}\frac{dx}{dt}. \tag{2.5}$$

In practice, however, the input impedance of the voltage follower is not infinite, and a resistor will be used to keep the average voltage across the capacitor C_g at the suitable voltage V_c for polarization by dc. The corresponding resistance R in parallel with C_g, in the small-signal equivalent circuit, introduces a time constant RC_g which restricts the validity of (2.5) to time intervals $\Delta t \ll RC_g$.

The dc polarization causes the steady-state value of Q to be a function of V_c: $Q = V_c A\varepsilon/x$, which can be substituted in (2.5).

$$dv_c|_{\Delta t \ll RC_g} = V_c\, d(\ln x). \tag{2.6}$$

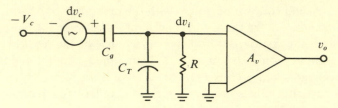

Fig. 2.6. Voltage division due to parasitic capacitance.

Equation (2.6) shows that dv_c depends logarithmically on x, but is independent of A! This is remarkable, since it shows that reduction in the size of the tranducer will not affect sensitivity. The only factor limiting the ultimate size of the transducer is now the time constant RC_g which is proportional to A and restricts the transducer reponse to frequencies $\omega \gg 1/RC_g$. We also note that sensitivity is proportional to V_c.

No parasitic capacitance is shown in Fig. 2.5 in parallel with C_g. A more realistic case is shown in Fig. 2.6, in which the transducer signal reaching the amplifier is shown to be reduced by the voltage divider consisting of C_g and C_T, the total parasitic capacitance of the circuit connected to the transducer. For a steady-state charge across C_g of value Q and a displacement dx of the plates, the resulting equivalent voltage source in series with C_g equals $dv_c = Q\, dx/A\varepsilon$. The resulting incremental voltage at the input of the amplifier equals for time intervals short compared with $R(C_g + C_T)$

$$\Delta v_i = \Delta v_c C_g/(C_g + C_T) = V_c\, \Delta x/x(1 + xC_T/A\varepsilon). \qquad (2.7)$$

Equation (2.7) shows that the capacitance C_T in parallel with C_g not only reduces the available signal, but also affects the logarithmic relationship.

Low input impedance operation (current mode)
Shunt feedback applied to a capacitive transducer is shown in Fig. 2.7. The low input impedance of the amplifier keeps the voltage across C_g practically constant, and hence the term dv_c/dt in (2.4) vanishes:

$$\frac{1}{C_g}\frac{dQ}{dt} = -Q\frac{d(1/C_g)}{dt} = \frac{Q}{C_g^2}\frac{dC_g}{dt},$$

or

$$\frac{dQ}{dt} = -i_c = \frac{Q}{C_g}\frac{dC_g}{dx}\frac{dx}{dt} = -V_c\frac{A\varepsilon}{x^2}\frac{dx}{dt}. \qquad (2.8)$$

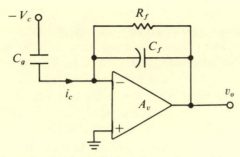

Fig. 2.7. Shunt feedback applied to charged capacitive transducer.

Note that, although both the charge Q and the value C_g of the capacitor are functions of x, their ratio is constant:

$$Q(x)/C_g(x) = V_c \neq f(x).$$

Finally, the incremental output voltage is obtained for $\Delta t \ll R_f C_f$:

$$\Delta v_o = -i_c \, \Delta t/C_f = -(V_c A \varepsilon / C_f) \, \Delta x/x^2. \tag{2.9}$$

As with series feedback, the output signal is again proportional to the voltage V_c across the transducer, but the input–output relationship is inversely proportional to x^2.

Again it should be emphasized that the preceding treatment ignores practical restrictions on the nature of active impedances, due to the frequency-dependent gain of the basic amplifier employed.

2.2.3 Resistive transducers

An important class of transducers comprises those exhibiting variable resistance: wire and semiconductor strain gauges measure strain, pressure, force, acceleration or displacement.

Resistance thermometers or photoconductors detect the incidence of electromagnetic radiation by measuring the temperature rise of a substance caused by absorption of that radiation. Termed *bolometers*, they depend on the positive temperature coefficient of the resistance of metals or on the negative one of semiconductors (thermistors), whose conductivity increases with the number of charge carriers generated in the material by the internal photoelectric effect. Finally we mention potentiometric type transducers whose wiper is linked mechanically to the displacement.

Bridges

A change in resistance may be measured by applying a constant voltage to it and measuring the current, or a constant current and measuring the voltage, or by the balanced bridge technique. The first method is highly non-linear, as everybody will have noticed who is familiar with the ancient ohmmeter employing a moving-coil instrument; the second method yields a linear measurement of resistance change, but is insensitive; finally, the Wheatstone bridge method and its variations are capable of measuring directly the change ΔR in the value R of a resistance and hence are extremely sensitive. Furthermore, in the bridge method the unknown resistance can be referred to another resistance, which makes it possible to compensate a secondary effect common to both.

The number of bridge circuits is legion.[4] Here we shall only discuss a few salient features. A bridge may be fed from an ac or dc voltage source. Excitation by ac has the advantage that zero offset and drift are 'completely' absent. However, in order to utilize this dubious advantage fully, the reactive components of the bridge must be carefully balanced. Availability of highly stable dc operational amplifiers has rendered this method obsolete.

A linear bridge is shown in Fig. 2.8. This is in fact a circuit in which the operational amplifier keeps the voltage drop across R_3 equal to that across R_2, thereby feeding the transducer R_4 by a constant current $i_4 = VR_2/(R_1 + R_2)R_3$. If we choose $R_2/R_1 = R_3/R_4$, v_o can easily be seen by inspection to equal zero, and, since i_4 is constant, any change in the value of R_4 results in a proportional change of v_o. More specifically, if $R_1 = R_2$,

$$v_o = V(R_3 - R_4)/2R_3 \qquad (2.10)$$

To demonstrate the versatility of the bridge method, we shall assume

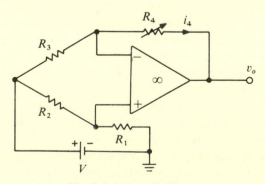

Fig. 2.8. Linear bridge.

$R_1 = R_2$, $R_{03} = R_{04}$, $R_3 = R_{03}(1 + \delta_{32})$ and $R_4 = R_{04}(1 + \delta_{41} + \delta_{42})$; δ_{41} denotes the resistance variation to be measured and $\delta_{32} = \delta_{42}$ describes a secondary effect common to both resistors R_3 and R_4.

In practice, the unknown resistance R_4 could be that of a strain gauge, δ_{41} its relative variation as a function of the strain to be measured, and δ_{42} its relative variation due to temperature changes, which is compensated by an identical resistance element R_3 to which no strain is applied but which is mounted near the loaded element so that both elements are at the same temperature. R_1 and R_2 are identical high-precision resistors. The difference amplifier should be of the high input impedance type and exhibit a high CMRR in order to reject possible interference pick-up by the interconnection between amplifier and bridge. The lead connecting R_4 to the output of the amplifier is not susceptible to pick-up because of the low output impedance of the amplifier.

Anemometer design

The design of a hot wire anemometer[9] presents an interesting implementation of an air flow meter, employing a self-balancing bridge.

The purpose of the instrument is to measure air speed by its cooling effect upon an electrically heated platinum filament that exhibits a high positive temperature coefficient of resistivity.

The filament characteristic is shown in Fig. 2.9, for two values of air speed. Two classical ways of operating it would be at constant voltage or constant current.

If a constant current is applied (say, 0.6 A), the sensitivity is reasonably good – about 0.4 V change for $\Delta S = S_2 - S_1$. However, there is every possibility that in still air the filament could burn itself up: resistance increases as temperature increases, due to lack of air flow, which causes the voltage to increase (constant current), increasing the dissipation, the temperature, and again the resistance, etc.

Applying a constant voltage, the increase of resistance with temperature causes operation to be quite safe, but also relatively insensitive, especially at low air speeds.

Another factor that makes both modes of operation unsatisfactory is the necessity for the temperature of the filament to change to detect a change in air speed; this necessarily causes delay. If the air speed indicator is part of a control loop, the measurement delay could cause slow response or instability of the control loop.

An approach that answers all these objections satisfactorily is to operate the filament as though it had constant resistance (i.e. constant temperature).

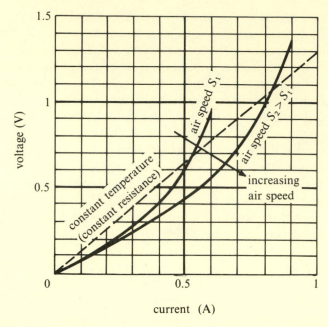

current (A)

Fig. 2.9. Volt-ampere characteristic of $\frac{7}{16}$'L × 0.002' D straight filament of pure platinum in the presence of moving air, for two values of air speed. Adapted from [9].

If, for example, the resistance were maintained at 1.3 Ω, as indicated by the dashed line in Fig. 2.9, one could obtain a current change of 0.3 A and a voltage change of 0.4 V, with no danger of overheating in normal operation. Response would be quite speedy, since temperature changes are momentary and small.

The basic circuit for achieving constant temperature operation is the feedback circuit of Fig. 2.10, consisting of a bridge, an operational amplifier, and a power amplifier. The operational amplifier continuously adjusts the flow of current (through the power transistor) to maintain the equality of its two inputs. This can be done only by keeping the voltage across the filament R_f equal to that across R_2 and the filament current equal to the current through R_1. However, since the current through R_1 is proportional to the current through R_0 (which has the same voltage drop as R_1), and the current through R_0 is determined by the voltage drop across R_2, it can be seen that the resistance of the filament, R_F, must be equal to that of R_2, multiplied by the ratio of R_1 to R_0.

Suppose now that, starting from a given operating equilibrium point, the air flow increases. This will take heat away from R_F, causing its

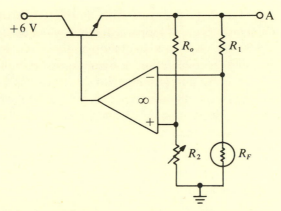

Fig. 2.10. Constant temperature control circuit for anemometer; R_f(avg) = $R_2 R_1 / R_0$. Adapted from [9].

voltage to drop. The amplifier's output voltage increases, which increases the current through the power transistor, and thus makes more power available for the filament to dissipate to maintain its temperature (and hence its resistance) constant.

The output voltage is measured at terminal A, which provides an amplified version of the filament voltage, at low impedance level.

The voltage at zero air speed is backed off by means of an auxiliary constant voltage. The scale is a non-linear function of air speed, expanding toward the lowest values; low air speeds can be read with high sensitivity.

Circuit notes
(1) A voltage offset must be deliberately introduced into the operational amplifier (or elsewhere) to ensure that the output is positive with zero differential input; otherwise, the circuit might remain dead when switched on.
(2) The power transistor must have ample current-handling capacity; the filament requires several hundred milliamperes.
(3) High-frequency oscillations are possible in some physical layouts, especially where the filament is some distance away from the electronic circuit and the connecting leads are twisted. Though these oscillations are not visible with low-frequency readout devices (however, rectification can cause voltage offsets), they may be observed on an oscilloscope screen. A 0.1 μF capacitor between the base and collector of the power transistor and a small resistor in series with the base can often prevent these oscillations.

(4) The filament is a physical device with thermal lag. Although the circuit is fast enough to prevent loop oscillations when used in a larger control loop, it is itself a process control loop and may require the usual compensation techniques to maintain its own internal stability.

(5) R_0 and R_2 form a trim potentiometer to set the operating temperature (e.g. resistance) of the filament. If R_2 is a variable resistance, one should start with $R_2 = 0$ at zero air speed, and increase it until the filament just starts to glow, then back down a little. This will give optimal sensitivity.

2.3 The guarding technique

An *electrostatic guard* is defined to be a screen connected to an electrostatic potential to prevent leakage and interference currents from flowing between signal and ground. Hence, its main area of application is in certain kinds of transducers with a high output impedance, for which leakage between the output electrode and the cable screen or ground is undesirable. Capacitive leakage reduces the signal and may, in addition, introduce interferences. Bootstrapping, which effectively reduces capacitive leakage, belongs to the category of guarding techniques and has been discussed in Section 1.7.4(*a*). More will be said on this subject in the chapter dealing with interferences and ground-loops.

Resistive leakage can be particularly damaging, if it occurs between high-voltage electrodes and ground. The resulting leakage current produces shot noise and may seriously impair the SNR. In this case, a *guard ring* is commonly employed, which surrounds the high-voltage electrode and is held at its dc potential. No small-signal bootstrapping is usually required with high-voltage electrodes.

Problem

In Fig. 2.7, replace the dc voltage V_c by a carrier voltage $V_c \cos \omega_c t$. Assume that C_g is a fixed capacitor and C_f the displacement-varying capacitor. Find the amplitude V_o of the carrier at the output as a function of x.

This circuit is discussed in reference [7], pp. 63–5.

3

Analog signal conditioning

When designing a system employing both analog and digital signal processing, it must be decided which is the best point to convert the analog signal by an A/D converter. From this point any further processing will be entirely digital, although a later reconversion to analog is, of course, another alternative.

Digital processing is often somewhat loosely stated to yield any desired accuracy, provided a sufficiently large number of bits is employed. However, when applied to electronic signal processing, it should not be forgotten that most signals must first be converted by a transducer into an electrical signal, whose SNR is inherently limited. Furthermore, unless the transducer output is digital, the signal must in most practical cases first be amplified and then converted into digital form, each step contributing its own error. At this point, the greater accuracy available by digital processing may well lose its significance, and an error budget will have to be made in order to decide which alternative to choose.

Compared with digital processing, analog processing is cheaper and superior in speed, although inferior in *potential* accuracy. It is capable of processing frequencies up to several gigahertz, and the accuracy of feedback-desensitized circuits depends mainly on the passive components employed. Nuclear spectroscopy systems bear witness to the fact that judicious use of matched components makes it possible to keep errors well below 10^{-3}.

Wherever accuracy is sufficient, analog processing is usually less complex than digital. The ever increasing variety of available ICs performing, in addition to amplification, functions such as multiplication, division, A/D and D/A conversion, linear gating, analog multiplexing, etc., and the possibility of combining analog and digital techniques in hybrid designs provide continual challenges to the creative designer.

3.1 Linear amplification, noise considerations, shaping and routing

There are several up-to-date treatments of linear analog signal processing.[1-9] The range of applications for available IC modules can be significantly extended beyond the uses suggested by their manufacturers by connecting additional components externally or to the compensating terminals which are external in certain amplifiers, and/or combining them with other modules in a variety of interconnections. Such an approach requires a thorough understanding of the modules' internal structures. A detailed description of IC circuit techniques is given in references [3], [10] and [11]. Here we restrict ourselves to the 'black box' approach.

3.1.1 Precision voltage amplifiers

Response and stability considerations for feedback-stabilized voltage amplifiers have been discussed in Chapter 1. In accordance with their transfer function

$$H(j\omega) = 1/B(1 + j\omega\tau_0/B') \tag{3.1}$$

for an amplifier gain $A(j\omega) = 1/j\omega\tau_0$ and employing a frequency-independent feedback network, their deviation from $1/B$ in the frequency domain is for both the non-inverting and inverting cases

$$-\omega\tau_0/B' \tag{3.2}$$

for the excess phaseshift, and

$$-(\omega\tau_0/B')^2/2 \tag{3.3}$$

for the amplitude, over the frequency range $\omega \ll (B'/\tau_0)$, where (B'/τ_0) is the frequency of the closed-loop dominant pole. In certain applications such as highly selective filters, it may then be necessary to trim the passive elements in order to compensate for deviations from the ideal phase response at the operating frequency. The gain errors are usually of minor importance.

Wilson[12] has proposed a transfer function of the form

$$H(s) = A_f(0)(1 + s\tau_1)/(1 + s\tau_1 + s^2\tau_1^2) \tag{3.4}$$

which exhibits for the frequency range $\omega \ll (1/\tau_1)$ an excess phaseshift of approximately $-\omega^3\tau_1^3$ and a gain error of about $-\omega^2\tau_1^2$, with $\tau_1 = \tau_0/B'$.

Compared with (3.2) and (3.3), the excess phaseshift is reduced for the given frequency range by the third power, at the cost of doubling the gain error which was, however, small to begin with.

Fig. 3.1 shows Wilson's modification of the feedback network for a non-inverting voltage amplifier, yielding (3.4) if we choose $\tau_1 = A_f(0)\tau_0 = RC/A_f(0)$. $A_f(0)$ denotes the feedback-stabilized dc gain.

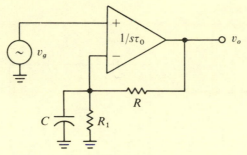

Fig. 3.1. Wilson's non-inverting VCVS; $R_1 = R/[A_F(0)-1]$.

Fig. 3.2 shows his realization for an inverting voltage amplifier. The condition yielding the closed-loop response defined by (3.4) is now

$$\tau_1 = [1 + A_f(0)]\tau_0 = RC/A_f(0).$$

A disadvantage of Wilson's modified feedback networks is that, in effect, a capacitor is connected between the two input terminals of the amplifier, giving rise to excessive series noise at high frequencies. An alternative circuit with improved noise performance employs an externally compensated amplifier and yields the same closed-loop response as given by (3.4) for the inverting and non-inverting cases. The circuit is shown in Fig. 3.3 for the inverting case.

It employs an IC amplifier designed for external compensation, and the compensating network consists of one resistance and two identical capacitors satisfying $R = (R_1 + R_2)/4$ gm R_1, $C \gg \tau_0/R$, with $\tau_1 = 2RC$. Series noise is roughly the same as in the uncompensated case, since the feedback network is purely resistive and the closed-loop frequency response is controlled by the active gain element alone. The parameter gm of an IC is sufficiently stable to satisfy the above relationship with reasonable accuracy.

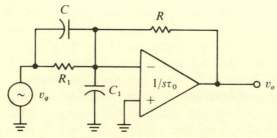

Fig. 3.2. Wilson's inverting VCVS; $R_1 = R/A_f(0)$, $C_1 = C/A_f(0)$.

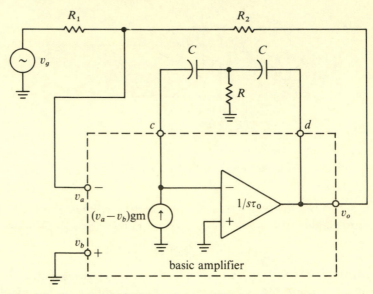

Fig. 3.3. Inverting VCVS based on modified basic amplifier response.

3.1.2 Active integration

Fig. 3.4 shows an ideal integrator, whose transfer function yields a pole at the origin:

$$\lim_{A(s)\to\infty} H(s) = -1/sRC. \qquad (3.5)$$

Disregarding practical limitations for $A(s)$ and the finite bandwidth of the physical power spectrum at the amplifier input, noise power at the output is infinite. But, even taking into account the practical limitations of amplifier dc gain, noise at the output will still be excessive. This fact restricts the use of such an ideal integrator to applications such as:

a part of a major feedback loop in an active filter employing several feedback-stabilized operational amplifiers such as shown in Fig. 3.5. In this case, noise is limited by the closed-loop dc gain.

a low noise preamplifier for capacitive transducers, as discussed in section 3.1.6. Noise is limited by the following differentiation.

a gated integrator, employed in linear ramp and dual slope A/D conversion, or signal shaping in nuclear electronics, in which case noise is limited due to the relatively short time interval during which the integrator is active.

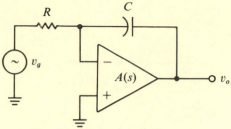

Fig. 3.4. Operational integrator.

Noise performance

The noise performance of a gated integrator will be evaluated with reference to Fig. 3.6. The switch S opens during the interval of integration τ_i. The frequency dependence of the signal processing channel following the integrator is modelled by R_2C_2. In the case of a ramp or dual slope A/D converter, the time constant R_2C_2 is related to the speed of response of the following discriminator.

We compute the noise voltage at the output, due to R_s and R_p of the amplifier, as a function of τ_i, with $R_1C_1 = \tau_1$, $R_2C_2 = \tau_2$. If S is closed, noise at the output is negligible and will be disregarded. In accordance with Appendix A, we first compute the weighting function $w(t)$ for the series and parallel noise of the amplifier, which is conveniently obtained for the small-signal noise sources v_{ns} and i_{np} as

$$v_{no2} = [v_{ns}(\tau_1/\tau_2 - 1) - i_{np}R_1]w_1(t) + (v_{ns} + i_{np}R_1)w_2(t), \qquad (3.6)$$

yielding

$$\overline{v_{no2}^2} = 2kT\left[R_s\left(\frac{\tau_1}{\tau_2} - 1\right)^2 + \frac{R_1^2}{R_p}\right]\int_0^{\tau_i} w_1^2(t)\,dt$$

$$+ \left(R_s + \frac{R_1^2}{R_p}\right)\int_0^{\tau_i} w_2^2(t)\,dt, \qquad (3.7)$$

where $w_1(t) = (1/\tau_1)\exp(-t/\tau_2)$ and $w_2(t) = 1/\tau_1$.

For $\tau_i = \infty$, i.e. an ungated integrator, rms noise at the output becomes infinite since we have assumed the voltage amplifier to have infinite gain.

Deviation from ideal performance

The performance of practical integration falls short of that indicated by (3.5). We shall consider the amplifier gain for the integrator of Fig. 3.4 to be $A(s) = -A(0)/[1 + sA(0)\tau_0]$, yielding, for $A(0) \gg 1$, $RC \gg \tau_0$, $RC' = RC(1 + \tau_0/RC)$ and $\tau_0' = \tau_0/(1 + \tau_0/RC)$,

$$H(s) = v_o/v_g = -A(0)/[1 + sA(0)RC'](1 + s\tau_0'). \qquad (3.8)$$

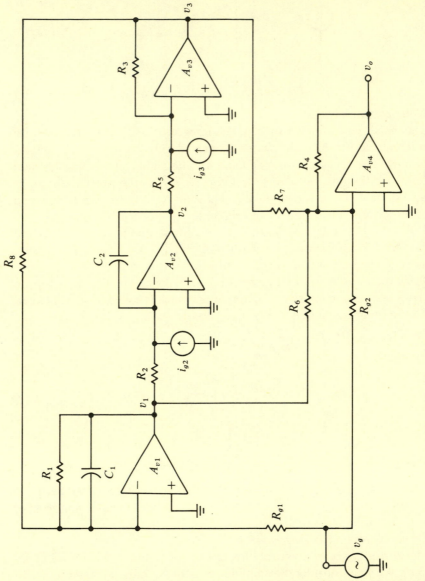

Fig. 3.5. Use of operational integrator in active filter.

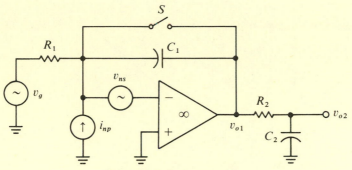

Fig. 3.6. Gated integrator.

Deviation of (3.8) from (3.5) in the frequency domain occurs at low frequencies (no increase in gain for frequencies below $1/A(0)RC$ and excess phaseshift due to the pole at the frequency $1/\tau_0'$). In addition, the difference between RC and RC' results in a gain error, although in practice $\tau_0/RC \ll 1$ and the difference between RC and RC' is very small.

With regard to stability, the feedback network attenuation introduces phase lead in the LT, which does not, however, improve stability since this occurs at frequencies far below $1/\tau_0$.

In the time domain, with a step of amplitude V applied at the input, (3.8) becomes

$$v_o(t) = -A(0)V\left\{1 - \frac{\exp\left[-t/A(0)RC'\right]}{[1-\tau_0'/A(0)RC']}\right.$$
$$\left. + \frac{\exp\left(-t/t_0'\right)}{[1-\tau_0'/A(0)RC']A(0)RC'/\tau_0'}\right\}. \quad (3.9)$$

Expansion of (3.9) in the vicinity of $t = 0$, for $t \ll A(0)RC'$, yields for $A(0) \gg 1$ and $RC' \gg \tau_0'$

$$v_o(t) \approx -V\{t - \tau_0'[1 - \exp\left(-t/\tau_0'\right)]\}/RC'. \quad (3.10)$$

Equation (3.10) is plotted in Fig. 3.7.

Similarly, we may investigate the behavior of (3.9) for times for which $\exp\left(-t/\tau_0'\right) \ll 1$, yielding the response

$$v_o(t) \approx -A(0)V\{1 - \exp\left[-t/A(0)RC'\right]\}. \quad (3.11)$$

Equation (3.11) is plotted in Fig. 3.8.

The dashed line in Fig. 3.7 intersecting the origin shows the response of an operational integrator employing an amplifier whose response is not a

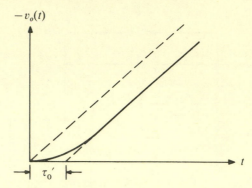

Fig. 3.7. Delay of active integrator; $t \ll A(0)RC'$.

function of frequency, i.e. $\tau_0 = 0$. The effect of τ_0 on the integration, shown by the solid line, is such that the linearly rising response (or ramp) is approached with a time constant of τ_0' and delayed by τ_0'. On the compressed time scale of Fig. 3.8 the delay τ_0' is too small to be recognized, but the exponential decay to $A(0)V$ is clearly visible. In practice, the voltage $A(0)V$ will be far beyond the linear range of the amplifier employed, but the non-linearity of the ramp may still be recognizable over the linear range of the amplifier and can be approximated as

$$v_o(t) \approx \frac{Vt}{RC'}\left[1 - \frac{t}{2A(0)RC'}\right] \qquad (3.12)$$

by expanding the exponential term in (3.11). Equation (3.12) is a good approximation for $\tau_0' \ll t \ll 3A(0)RC'$

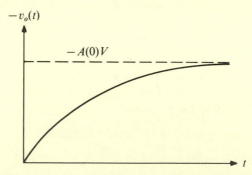

Fig. 3.8. Non-linearity of active integrator, disregarding saturation limit of basic amplifier.

Improving the performance

What are the practical limitations of analog integration as regards the delay τ_0' and the non-linearity due to a finite $A(0)$?

Discussing the delay first, this can be improved by using a basic amplifier whose unity-gain time constant τ_0 can be controlled by external compensation as shown, for example, in Fig. 1.5, for which $\tau_0 = C/gm$. The transfer function for such an amplifier is

$$A(s) = A(0)/[1 + sA(0)\tau_0] \prod^{n} (1 + s\tau_j). \qquad (3.13)$$

The parasitic time constants τ_j in (3.13) are always present but have been ignored in (3.8) and (3.10), since there τ_0 dominated the delay.

Comparing (3.13) with (3.8), we note that here the time constant of integration is determined by τ_0, and that the delay of the integrator is a function of the parasitic poles only, or of the ultimate speed of the amplifier, and not of τ_0 as in (3.8).

One could object to the use of a basic amplifier as an integrator on the grounds that absence of feedback would adversely affect stability. This argument, however, is only partly valid since control of τ_0 by a compensating capacitor implies feedback; furthermore, the parameter gm in Fig. 1.5 is a function of the internal dc biasing of the IC employed and exhibits a reasonable degree of stability. Hence, this method may successfully be applied where reduction of delay is more important than ultimate stability. The same circuit also eliminates the error in the integrating time constant, i.e. the difference between RC and RC', in (3.8).

As to linearity, this may be improved in accordance with (3.12) by increasing $A(0)$, using positive feedback. Fig. 3.9 shows an amplifier for

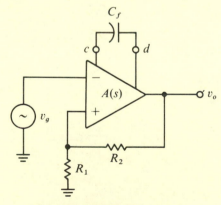

Fig. 3.9. High-gain integrator.

which $A(s) = -A(0)/[1 + sA(0)\tau_0]$, $\tau_0 = C_f/\text{gm}$, and

$$v_o = -v_g/[1/A(0) - R_1/(R_1 + R_2) + s\tau_0]. \tag{3.14}$$

Equation (3.14) shows that, by adjusting $(1 + R_2/R_1) = A(0)$, gain at dc becomes infinite. However, in practice this technique requires individual adjustment of the ratio R_1/R_2, since $A(0)$ is not a well-defined parameter. Moreover, sensitivity to ambient temperature changes will be high (typical for positive feedback) mainly due to the temperature sensitivity of $A(0)$. The gain of some amplifiers even increases beyond infinity (i.e. it reverses polarity at infinity and then becomes smaller) at low frequencies, because positive thermal feedback affects the offset voltage of the differential input stage.

If applied with caution, the circuit shown in Fig. 3.9 is capable of increasing $A(0)$ by at most an order of magnitude. Its main application is with high-speed integrators employing a fast amplifier. Such amplifiers exhibit comparatively low dc gain, in which case a significant improvement in linearity may be achieved by using positive feedback.

Increased dc gain through positive feedback can be combined with stabilization of the integrating time constant by negative feedback (Fig. 3.10 with $z_2 = R_2$). The transfer function is

$$v_o = -v_g \bigg/ \left\{ \frac{1}{A(0)} - \frac{R_1}{(R_1 + R_2)} + sRC' \right\} \{1 + s\tau_0'\}, \tag{3.15}$$

which for $1 + R_2/R_1 = A(0)$ becomes identical to (3.8), with $A(0) = \infty$.

In applications such as active filters, it may be desirable to compensate for the difference between RC and RC' in (3.8). To this end, Wilson[12] proposed the circuit shown in Fig. 3.10 with $Z_2 = 1/sC_2$. For $R_1C_2 = \tau_0$

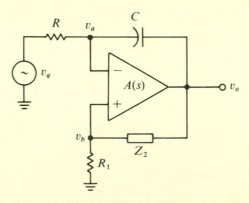

Fig. 3.10. Modified feedback-stabilized integrator.

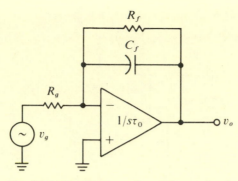

Fig. 3.11. Realization of negative real pole.

and for frequencies $\omega \ll 1/A(0)\tau_0$,

$$H(s) = \frac{-1}{sRC}\frac{(1+s\tau_0)}{[1+s\tau_0(1+\tau_0/RC)+s^2\tau_0^2]}. \tag{3.16}$$

Unlike that in (3.8), $1/B$ in (3.16) is exactly equal to $1/sRC$; its undesirable term, with $\tau_0 \ll RC$, is nearly the same as in (3.4), whose excess phaseshift compares favorably with that of (3.8).

Finally, we shall consider two possible realizations of a negative real pole. First consider Fig. 3.11, for which we obtain with $(R_f \| R_g)C_f \gg \tau_0$

$$H(s) = -R_f/[Rg(1+s\tau_f')(1+s\tau_0')], \tag{3.17}$$

with $\tau_f' = R_fC_f[1+\tau_0/(R_g\|R_f)C_f]$, and $\tau_0' = \tau_0/[1+\tau_0/(R_g\|R_f)C_f]$.

As in (3.8), we have a 'gain error' due to the difference between R_fC_f and τ_f'. The similarity of (3.17) and (3.8) becomes obvious if we consider (3.8) as a special case of (3.17), with $R_f = \infty$. We therefore expect that having a purely resistive feedback network and introducing the dominant time constant through controlling τ_0 as in the amplifier described by (3.13) will lead to a similar improvement.

The corresponding circuit yielding a negative real pole is shown in Fig. 3.12. With the open-loop gain of the amplifier given again by (3.13), the closed-loop transfer function becomes for $A_f(0)\tau_0 \gg \sum_1^n\tau_i$

$$\frac{v_o}{v_g} \simeq \frac{A_f(0)}{[1+sA_f(0)\tau_0]\prod_1^n(1+s\tau_i)}, \tag{3.18}$$

with $A_f(0) = (R_1+R_2)/R_1$ and τ_0 controlled by external compensation. Equation (3.18) shows (without proof), that, for a large ratio between the closed-loop dominant pole and the parasitic poles, migration of the parasitic poles in the root locus is insignificant. As in the case of realizing a

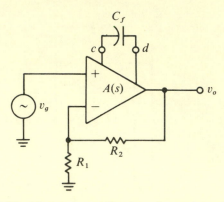

Fig. 3.12. Alternative realization of negative real pole.

pole at the origin, there is no 'gain error' in this realization of a negative real pole by a basic amplifier, i.e. the dominant pole is exactly $-1/A_f(0)\tau_0$.

Problems

3.1. Verify the transfer function (3.4) for the amplifiers shown in Figs. 3.1, 3.2, and 3.3. Take note of the various relationships between the gain-bandwidth of the amplifier and the feedback components, as indicated in the text.

3.2. For (3.18), find the physical significance of $A_f(0)\tau_0$ related to the Bode plot of $LT(j\omega)$.

3.3. Verify the statement relating to (3.18) that, for $A_f(0)\tau_0 \gg \sum^n \tau_j$, migration of the parasitic poles in the root locus is insignificant. Assume $n = 2$, $\tau_1 = \tau_2$.

3.1.3 Active differentiation

Active differentiators play an important role in signal processing, and a thorough understanding of their limitations is essential for their design. Fig. 3.13 shows the Bode plots for three kinds of differentiators. An ideal differentiator (plot (a)) would yield a response $j\omega\tau$ which increases by 20 dB/decade beyond limits. The response of a practical differentiator is $j\omega\tau/(1 + j\omega\tau)$ (plot (b)), which flattens out beyond the frequency $1/\tau$ up to the next corner frequency, which could be arbitrarily given by the limited high-frequency response of the system. But the flat (frequency-independent) portion of such a practical differentiator serves no useful purpose, since differentiation is only performed over the frequency range for which the response rises by 20 dB/decade; it merely amplifies any noise present and thereby impairs the noise performance. It is therefore good design practice to add at least another pole at the frequency $1/\tau$ (plot (c)), which

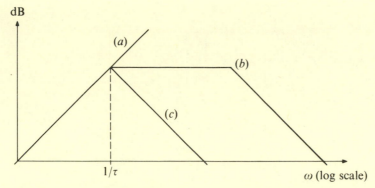

Fig. 3.13. Bode plots of differentiators: (a) ideal, (b) practical, (c) band-limited.

eliminates the flat portion with minimum interference over the frequency range of differentiation.

Fig. 3.14(a) shows a conventional operational differentiator which can be designed to be band-limited as defined in the preceding paragraph. The resistor R is essential to eliminate integration due to R_fC at high frequencies, which would otherwise occur in the feedback path and bring the circuit to the verge of oscillation. Proper design conditions will create a double pole on the negative real axis. These conditions and the resulting response are

$$H(j\omega) = v_o/v_g = -2Aj\omega\tau/(1+j\omega\tau)^2,\qquad(3.19)$$

where

$$A \simeq R_f/R \gg 1 \quad\text{and}\quad \tau = RC/2 = 2A\tau_0.\qquad(3.20)$$

Verification of (3.19) will be left to the reader as a problem.

Noise performance
The noise v_{ni}^2 referred to the input of the differentiator shown in Fig. 3.14(a) will be evaluated in this section. The flow graph of Fig. 3.14(c), derived from the small-signal equivalent noise diagram shown in Fig. 3.14(b), will be used to compute the various transfer functions for the noise signals to the output.

$$v_{no} = \left[\frac{Y_g v_{nR}}{(Y_g+G_f)} + \frac{(i_{nf}+i_{np})}{(Y_g+G_f)} + v_{ns}\right]\frac{(-1/j\omega\tau)}{(1-LT)}$$

$$= \frac{v_{nR}j\omega R_fC + (i_{nf}+i_{np})R_f(1+j\omega RC) + (1+j\omega R_fC)v_{ns}}{(1+j\omega RC - \omega^2R^2C^2/4)},$$

with $Y_g = 1/(R+1/j\omega C)$ and $G_f = 1/R_f$.

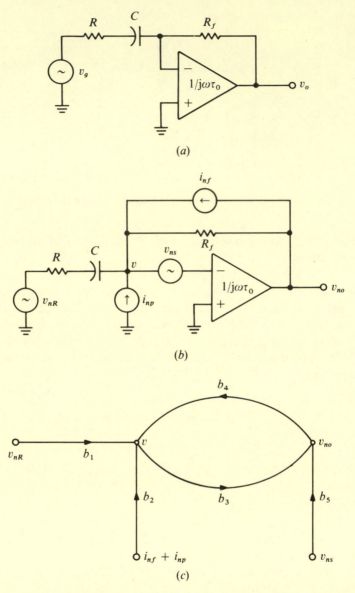

Fig. 3.14. Active differentiator: (a) circuit diagram, (b) noise sources, (c) flow graph for noise computations;

$$b_1 = Y_g/(Y_g + G_f), \qquad b_2 = 1/(Y_g + G_f),$$
$$b_3 = -1/j\omega\tau_0, \qquad b_4 = G_f/(Y_g + G_f),$$
$$b_5 = 1/j\omega\tau_0.$$

Substituting the various spectral densities of the noise sources,

$$d(\overline{v_{nR}^2})/df = 4k\mathcal{T}R, \qquad d(\overline{i_{nf}^2})/df = 4k\mathcal{T}/R_f,$$

$$d(\overline{i_{np}^2})/df = 4k\mathcal{T}/R_p, \qquad d(\overline{v_{ns}^2})/df = 4k\mathcal{T}R_s,$$

and multiplying by the corresponding squared absolute values of the transfer functions, we obtain

$$\frac{d(\overline{v_{no}^2})}{df} = 4k\mathcal{T}\frac{[R\omega^2 A^2 4\tau^2 + R_f^2(G_f + G_p)(1 + \omega^2 4\tau^2) + R_s(1 + \omega^2 A^2 4\tau^2)]}{(1 + \omega^2\tau^2)^2}.$$

Since the 'midband gain' equals A, $\overline{v_{ni}^2}$, the squared noise voltage referred to the input, becomes

$$\overline{v_{ni}^2} = \overline{v_{no}^2}/A^2 = \frac{4k\mathcal{T}}{RC}\left\{R_s + R\left[1 + \frac{5}{4}\left(\frac{R}{R_p} + \frac{1}{A}\right)\right]\right\} \qquad (3.21)$$

with $1/RC$ having the physical significance of the ENB.

Equation (3.21) is typical for the noise performance of active differentiation. It shows that series noise (due to R_s and R) will contribute dominantly, with parallel noise (due to R_p and R_f) becoming negligible when $R \ll R_p$, $A \gg 1$. Optimal noise performance will be obtained by making $R = R_f/A \ll R_s$, which also implies $A \gg 1$ since the practical minimum value for R_f is limited to about 1 kΩ. This gives

$$\overline{v_{ni}^2}(\text{opt}) = 4k\mathcal{T}R_s/RC. \qquad (3.22)$$

Note that reduction of R does *not* increase $\overline{v_{ni}^2}$, since RC is kept constant.

Equation (3.21) has important implications regarding the design of active differentiators. Most gain stages can be 'noise matched' to the source – i.e. their noise figure becomes minimum, with parallel and series noise contributing equally, for a certain source resistance. Active differentiators behave differently: the series noise contribution of the basic amplifier tends asymptotically to a minimum with parallel noise becoming negligible if the differentiator gain is sufficiently increased. From this consideration it can be concluded that a bipolar transistor is much better suited as input stage for an active differentiator than an FET, since the series noise contribution of the former can be made smaller than that of the latter. Furthermore, the relatively high parallel noise of the bipolar transistor does not significantly contribute if A in (3.21) is chosen large enough (see also the discussion on the circuit shown in Fig. 3.18).

Normalization
Another important aspect in the design of active differentiators is the comparison between various configurations with respect to their gain,

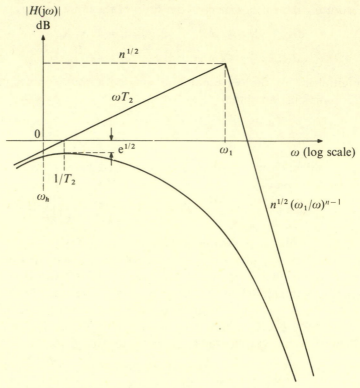

Fig. 3.15. Bode plot of active differentiator normalized with respect to $1/T_2$; $n \gg 1$, $\omega_h \approx (\ln 2)^{1/2}/T_2$, $\omega_1 = n^{1/2}/T_2$.

speed and noise performance. Monotonic response – a reasonable assumption in signal processing – enables us to characterize their response by the second time moment T_2 (see Appendix D), which is directly related to the rise time in the time domain and to the half-power frequency in the frequency domain. In the calculations that follow we shall use a transfer function with n identical negative real poles and express results in terms of T_2. Accordingly, conclusions will also be applicable to different monotonic differentiators whose transfer function exhibits non-identical poles and includes finite zeros but which are characterized by the same T_2.

For n cascaded integrators with identical time constants τ, $T_2 = n^{1/2}\tau$. Hence, the normalized transfer function of an ideal differentiator in cascade with n integrators is

$$H(j\omega) = A_D j\omega T_2/(1 + j\omega T_2/n^{1/2})^n, \tag{3.23}$$

with A_D being the nominal gain of the differentiator. The Bode plot of (3.23) is drawn in Fig. 3.15. It shows that the resulting response can be approximated by two asymptotes, one rising by 20 dB/decade and the other falling by 20 $(n-1)$ dB/decade. For $n \gg 1$, maximum response occurs at $\omega \simeq 1/T_2$, at an amplitude $1/e^{1/2}$ below 0 dB. Note that the excess phase for $\omega \ll n^{1/2}/T_2$ is $\phi \simeq -\omega n^{1/2} T_2$, i.e. it increases proportionally to $n^{1/2}$; the amplitude error for $\omega \ll n^{1/2}/T_2$ is $-\omega^2 T_2^2/2$, which is not a function of n.

Equation (3.19) can be expressed in terms of (3.23) with $n = 2$, if we identify $A_D = 2^{1/2} A$, $T_2 = 2^{1/2} \tau$. The squared noise voltage referred to the input for $R \ll R_s$ and negligible parallel noise contribution is then

$$\overline{v_{ni}^2} (\text{opt}) = 2^{3/2} \mathscr{k} \mathscr{T} R_s / T_2. \tag{3.24}$$

Alternative designs
An alternative differentiator is provided by the circuit shown in Fig. 3.16(a), in which the creation of a double pole on the real axis does not depend upon a proportional relationship between R, R_f, C, and τ_0 as for

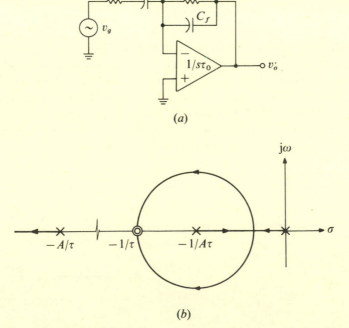

(a)

(b)

Fig. 3.16. Active differentiator: (a) circuit diagram, (b) root locus.

Fig. 3.14(*a*), but upon the inequality $RC \gg A\tau_0$. This is an advantage, since τ_0 itself is not a narrowly controlled parameter. A further condition is $R_f C_f = RC$.

The desirable response is now entirely due to the passive components of the feedback network, but the price paid for this advantage is the slower response of the circuit due to the large span required between RC and τ_0.

The response for the above condition is

$$H(j\omega) = v_o/v_g \approx -j\omega A\tau/(1+j\omega\tau)^2(1+j\omega\tau_0)$$

$$= -A_D j\omega T_2/(1+j\omega T_2/2^{1/2})^2(1+j\omega\tau_0) \qquad (3.25)$$

with $T_2 = 2^{1/2}RC$, $A_D = A/2^{1/2}$, $\tau = RC$, $A = R_f/R$, and an LT

$$LT(j\omega) \approx -(1+j\omega\tau)^2/(1+j\omega A\tau)(1+j\omega\tau/A)j\omega\tau_0. \qquad (3.26)$$

Substituting s for $j\omega$ in (3.26), the root locus becomes as shown in Fig. 3.16(*b*). The contribution of the root locus for $RC \gg A\tau_0$ will be merely a pole at $-1/\tau_0$: the two migrating poles which originate at $-1/A\tau$ and at the origin, respectively, approach the double zero at $-1/RC$ sufficiently to make the effect of the resulting two doublets on the overall response negligible. Furthermore, the open-loop pole at $-A/\tau$ approaches the value of $-1/\tau_0$ in the closed-loop response. Hence, the condition $RC \gg A\tau_0$ results in twin doublet compression.

Another solution of the stability problem of active differentiators is shown in Fig. 3.17. It employs a basic transimpedance amplifier (enhancing combination), whose low input impedance effectively shunts the connection between R_f and C to ground and thereby separates the feedback loop from the input network R in series with C.

The low open-loop input impedance $h_{ib}(s)$ of the basic transimpedance amplifier is provided by a CB connected transistor as shown in Fig. 3.18,

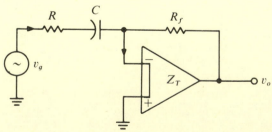

Fig. 3.17. Differentiator employing basic transimpedance amplifier.

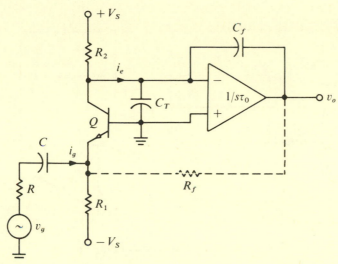

Fig. 3.18. Circuit diagram of differentiator employing basic transimpedance amplifier.

whose collector feeds a conventional voltage amplifier connected as an operational integrator. The dc stability is ensured by R_f, which provides the difference between emitter and collector currents of the input transistor supply by dc biasing.

The transfer function of this differentiator is

$$H(\mathrm{j}\omega) = v_o/v_g = -\mathrm{j}\omega ARC/(1 - s/s_1)(1 - s/s_1^*), \qquad (3.27)$$

with s_1, $s_1^* = -(1 \pm \mathrm{j})/RC$, for the following conditions: $h_{ib}(0)C_f \gg \tau_0'$, $A \gg 1$, $R_f C_f = RC$ and $\tau_\alpha \leqslant \tau_0'$, where $\tau_0' = \tau_0(C_T + C_f)/C_f$, $A = R_f/R$, $R = h_{ib}(0)$ and $1/\tau_\alpha$ is the CB half-power frequency of the input transistor Q.

The input transistor of this circuit is operating at a dc emitter current I_E of comparatively high value (of the order of 1 mA). This has a beneficial effect on the noise performance defined by (3.21), since R_s is inversely proportional to I_E (at currents below 1 mA). Furthermore, the parallel noise due to the base current and passive circuit components can be rendered negligible by proper design.

Fig. 3.19 shows an interesting circuit which gets around the stability problem by subtracting the inverted and integrated signal from itself.[2] The result is obviously negative phase lag or phase lead, which is differentiation. However, the noise performance of such a circuit has the same basic limitation as the other circuits discussed. Nature cannot be

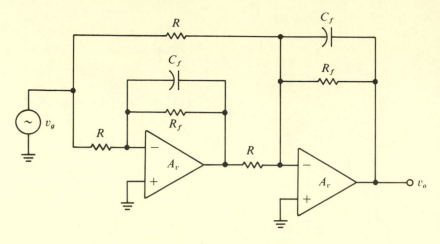

Fig. 3.19. Alternative realization of differentiator.

fooled with respect to noise, although we can get around the stability problem.

Problems

3.4. (*a*) For Fig. 3.18, show that

$$v_o/i_e = -1/j\omega C_f(1 + j\omega\tau'_0),\qquad(3.28)$$

with $\tau'_0 = \tau_0(C_T + C_f)/C_f$, and that

$$LT = -G_f/Y_{11}sh_{ib}(0)C_f,\qquad(3.29)$$

with $Y_{11} = G_f + 1/h_{if}(0) + (R + 1/sC)^{-1}$, $\tau_\alpha = 0$.
(*b*) Verify (3.27).

3.5. In this problem we compute the noise performance of the differentiator shown in Fig. 3.18. Design considerations are as follows:

We aim at achieving a noise performance similar to that given by (3.24), with $R \ll R_s$ being a desirable condition. This mismatch condition can be naturally applied to the circuit of Fig. 3.14(*a*), which employs a voltage amplifier. However, for the circuit of Fig. 3.18, mismatch conditions do not coincide with those for optimal noise performance. Hence, we choose as a practical design condition $R = r_e$, in accordance with (3.27).

Finally, in order to obtain a band-limited noise performance (compare with Fig. 3.13), a passive integrator of timeconstant $\tau = R_f C_f$ must be cascaded with the differentiator. This completes the practical design.

In the computation, assume that the input transistor Q and the input stage of the voltage amplifier are operating under identical conditions, i.e. the values of their noise parameters R_s and R_p are equal. Express the result in a form similar to (3.21) and show that the noise contribution of R_1, R_2 and of the voltage amplifier can be made negligible compared with that of $R + R_{s1}$, where R_{s1} is the series noise resistance of the input transistor Q.

3.6. Compare the circuits shown in Figs. 3.14(a), 3.16(a), 3.18 and 3.19 with regard to speed and noise referred to input. Normalize with respect to A_D and T_2.

3.7. Show that for Fig. 3.19, if $R_f \gg R$, $RC_f \gg \tau_0$ and $A_v = 1/j\omega\tau_0$, then

$$v_o/v_g \simeq -j\omega R_f C_f/(1+j\omega R_f C_f)^2(1+j\omega\tau_0). \qquad (3.30)$$

3.1.4 Fast-slow amplifiers

Amplifiers may be optimized either for dc performance or for speed. In most amplifiers' design, a compromise is struck between these two. Several improved designs employ the *feedforward* technique in which the signal path for high frequencies is shortened by bypassing one or more slow stages. More recently, techniques combining bipolar devices with FETs as input stages yield an increase in the speed/bias ratio of several orders of magnitude, compared with 'bipolar only' techniques. However, for ultimate performance a separate amplifier should be chosen for each frequency range. The resulting combination can be treated as a single amplifier, termed the *fast-slow amplifier*. We will now investigate the response and dc offset of fast-slow amplifiers in greater detail.

Design for minimum settling time
The open-loop frequency response of fast-slow amplifiers must be carefully designed to avoid lengthening the settling time τ_s by undesirable closed-loop transients. In Section 1.7.8, the dependences of τ_s on LT(s) was demonstrated in Figs. 1.74(a), (b) and (c). Fig. 1.74(a) shows the step response and LT of an amplifier exhibiting an ideally shaped LT, and Figs. 1.74(b) and (c) show the step response and LT with two possible imperfections of LT.

Let $A_1(s)$ be the gain of the slow amplifier and $A_2(s)$ that of the fast one with

$$A_1(s) = -A_1(0)/[1+sA_1(0)\tau_{01}], \qquad (3.31)$$

and

$$A_2(s) = -A_2(0)/[1+sA_2(0)\tau_{02}]; \qquad (3.32)$$

then the LT of the combined amplifier is typically

$$LT(s) = A_2(s)[1 - A_1(s)]B'(0),   \tag{3.33}$$

with $A_2(0)\tau_{02} = \tau_p$, $\tau_{01} = \tau_z$.

Conditions are

$$\tau_{01} = A_2(0)\tau_{02} \quad \text{for Fig. 1.74}(a),$$

$$\tau_{01} > A_2(0)\tau_{02} \quad \text{for Fig. 1.74}(b),$$

$$\tau_{01} < A_2(0)\tau_{02} \quad \text{for Fig. 1.74}(c).$$

The response shown in Fig. 1.74(a) is optimal for short settling time, but the condition for accurate matching between τ_{01} and $A_2(0)\tau_{02}$ may be difficult to achieve, unless $A_2(0)$ is controlled by a local feedback loop. For example, a resistor may be connected between the corresponding terminals of an IC designed for external compensation such as shown in Fig. 1.5.

Discussion of various circuits

The classical fast-slow amplifier was proposed by Goldberg.[13] The circuit, an inverting amplifier, is shown in Fig. 3.20. From the corresponding flow graph in Fig. 3.21 we obtain

$$LT = (b_2 b_3 + b_5 b_6)b_4 \simeq -A_1(0)A_2(0)R_2/(R_1 + R_2)$$

$$\times [1 + sA_1(0)A_2(0)\tau_{02}],   \tag{3.34}$$

if $R_3 \gg (R_1 \| R_2)$ and $\tau_{01} = A_2(0)\tau_{02}$.

Introducing the correspondingly indexed offset voltages ΔV and bias currents I of A_1 and A_2, respectively, the combined total offset V_t of

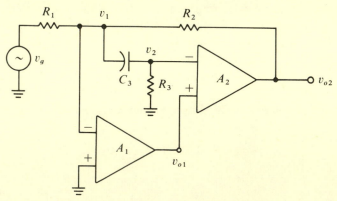

Fig. 3.20. Goldberg's fast-slow amplifier.

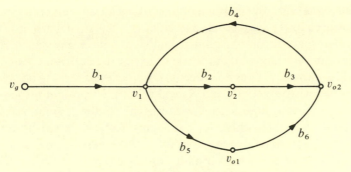

Fig. 3.21. Flow graph for Goldberg's amplifier.

both amplifiers can be referred to the input of A_1 as

$$\Delta V_t = \Delta V_1 + I_1(R_1 \| R_2) + (\Delta V_2 + I_2 R_3)/A_1(0), \quad (3.35)$$

yielding an offset referred to the signal input (in series with v_g) of

$$\Delta V_{in} = \Delta V_t (1 + R_1/R_2). \quad (3.36)$$

Capacitors are detrimental to fast recovery under heavy overload conditions. A capacitorless inverting fast-slow circuit is shown in Fig. 3.22.[14] An intuitive consideration shows that for high frequencies the gain of the slow amplifier A_1 is negligible and the virtual ground v is held

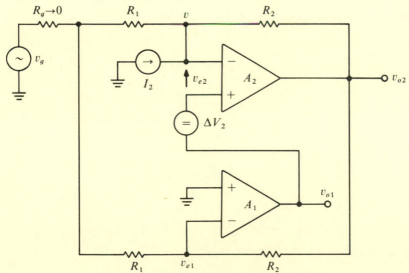

Fig. 3.22. Young's fast-slow amplifier.

at small-signal zero by the fast amplifier A_2. The slow amplifier picks up gain towards lower frequencies and stabilizes the virtual ground voltage v_{e1} with respect to dc. Assuming negligible source resistance R_g and negligible closed-loop output resistance of A_2, the dc offset at node v due to the bias current I_2 does not affect v_{e1}. Imperfect matching between the respective resistor pairs R_1 and R_2 will affect the transient response and thereby the settling time, since this causes the feedback-stabilized gain at low frequencies, over which A_1 dominates, to be different from that at higher frequencies, where A_2 dominates. Non-zero source and output impedances limit the dc performance of this circuit.

The LT of this amplifier is

$$LT(s) = A_2(s)[1 - A_1(s)]R_1/(R_1 + R_2). \qquad (3.37)$$

Its design for a smooth 20 dB/decade roll-off and evaluation of the offset referred to the input will be left to the reader.

Common to most fast-slow amplifiers is the requirement of matching the dominant pole of the fast amplifier with the gain-bandwidth product of the slow one, in order to achieve a smooth slope of 20 dB/decade over the dominant frequency range of LT. But whereas the gain-bandwidth product of an amplifier is a reasonably stable parameter, the frequency of its dominant pole is inversely proportional to the dc gain which may vary significantly for individual amplifiers of a certain type. If the effect due to the resulting doublets cannot be accommodated, it may be necessary to limit the dc gain of the fast amplifier by applying local feedback, as mentioned before, to stabilize the dominant pole.

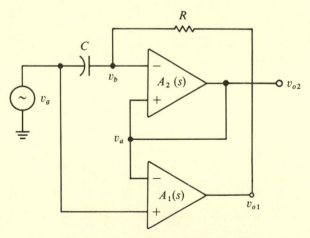

Fig. 3.23. Fast-slow voltage follower.

The fast-slow voltage follower[15] shown in Fig. 3.23 behaves basically differently insofar as the desirable closed-loop frequency response is obtained if the ratio between two time constants is much greater than unity.

In order to simplify the basic relationships, we shall divide the frequency range into a high-frequency region governed by A_2, and a low-frequency one governed by A_1. Considering first the low-frequency region $|s| \ll 1/\tau_{o2}$, A_2 can be considered as an ideal voltage follower whose transfer function v_{o2}/v_b equals unity, in which case

$$v_{o2}/v_g = [-A_1(s) + sRC]/[1 - A_1(s) + sRC], \qquad (3.38)$$

with $A_1(s) = -A_1(0)/[1 + sA_1(0)\tau_{01}]$. Equation (3.38) can be written as

$$\frac{v_{o2}}{v_g} = \frac{A_1(0)(1 + s\tau_{z1})(1 + s\tau_{z2})}{[1 + A_1(0)](1 + s\tau_{p1})(1 + s\tau_{p2})} \qquad (3.39)$$

It exhibits two doublets τ_{z1}, τ_{p1} and τ_{z2}, τ_{p2}, which for

$$RC \gg [A_1(0)]^2 \tau_{01} \qquad (3.40)$$

become widely separated, with $m_1 = \tau_{z1}/\tau_{p1} = [1 + A_1(0)]/A_1(0)m_2$, and $m_2 = \tau_{z2}/\tau_{p2} = 1 + A_1(0)\tau_{01}/RC = 1 + \varepsilon$, $\varepsilon \ll 1$.

Hence, with (3.40) being satisfied, we have an effective twin doublet compression scheme, which ensures that the effect of the doublets on the settling time is negligible if the LT is sufficiently high.

At high frequencies, neglecting parasitic poles, the voltage follower A_2 contributes a pole at the frequency $1/\tau_{02}$.

In practice, condition (3.40) may be impractical to guarantee due to the large production spread in the dc gain of IC amplifiers, which for high-precision amplifiers is very high to begin with. Hence, it is advantageous to limit the gain of the slow amplifier, stabilizing it with a secondary feedback loop as shown in Fig. 3.24. We define

$$v_{o1}/v_{e1} = A_1'(s) = -A_1'(0)/[1 + sA_1'(0)\tau_{01}'], \qquad (3.41)$$

where $A_1'(0) = R_2/R_1$ and $\tau_{01}' = R_1 C_2$. Hence, (3.40) becomes $RC \gg R_2^2 C_2/R_1$.

Equation (3.41) is valid provided that the local LT of $A_1'(s)$ is high over the frequency range extending beyond $1/\tau_{01}'$, which is ensured if the gain-bandwidth product of $A_1(s)$ is much higher than $1/\tau_{01}'$ - i.e. $\tau_{01} \ll \tau_{01}'$.

The circuit shown in Fig. 3.24 also provides gain. Note that in this case, however, the bias current associated with the inverting input terminal of A_2 flows through R_4. If the resulting offset voltage is prohibitively high,

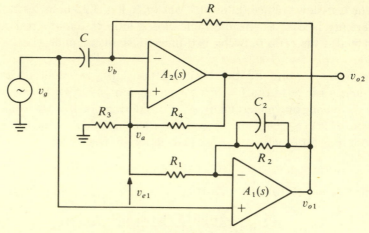

Fig. 3.24. Fast-slow amplifier.

one may connect a high-transconductance source follower between the junction R_3–R_4 and the input of A_2. In this circuit, the total gain is due to the fast amplifier A_2, with the slow amplifier merely providing unity gain between v_g and v_a at low frequencies and dc.

3.1.5 Active filters

Filters are an essential part of most signal processing systems. In the early stages of sampled data systems they reduce aliasing; later they shape the signal in such a way as to optimize the SNR. Both analog and digital filters are widely employed, the latter being more versatile and more capable of preserving accuracy, but also slower and usually more expensive.

General design aspects of analog filters are well covered in the literature.[1,3,16,17,18] Here we shall concern ourselves merely with a very limited range of topics.

(a) Classification
The transfer function of an active filter can be described in terms of a scale factor K, zeros s_i and poles s_j as

$$H(s) = K \prod_{}^{a} (s - s_i) \bigg/ \prod_{}^{b} (s - s_j). \qquad (3.42)$$

The factorized singularities may each consist of first order (real singularities) or second order terms, with the latter being of the form

$$(s - s_i)(s - s_i^*) = s^2 + (\omega_n/Q)s + \omega_n^2, \qquad (3.43)$$

with $s_i, s_i^* = -(\omega_n/2Q)[1 \pm (1-4Q^2)^{1/2}]$ being real singularities if $2Q \leq 1$, and complex conjugate if $2Q > 1$. ω_n is termed the *undamped natural frequency of oscillation*, and Q the *magnification* or *quality factor*.

The *damped natural frequency of oscillation* is

$$\omega_d = \mathrm{Im}\,\{s_i\} = \omega_n(1-1/4Q^2)^{1/2}. \tag{3.44}$$

For $4Q^2 \gg 1$, the minimum absolute value of (3.43) occurs approximately at ω_d, as can be readily seen from its vectorial representation in the complex frequency plane.

For a second order term in (3.42) appearing in the denominator or numerator, $H(\omega_n)$ is respectively directly or inversely proportional to Q. Even the most complex filter may be considered to consist of several sections in cascade, each with a term of not higher than second order in the numerator and denominator. In what follows we shall discuss the five possible kinds of sections, assuming in all cases that the singularities due to the second order term are complex conjugate, i.e. damping is less than critical.

A *low-pass section* is described by

$$H(s) = d/(s^2 + as + b) \tag{3.45}$$

whose typical response is shown in Fig. 3.25 on a linear scale.

A *high-pass section* is described by

$$H(s) = ms^2/(s^2 + as + b) \tag{3.46}$$

whose typical response is shown in Fig. 3.26.

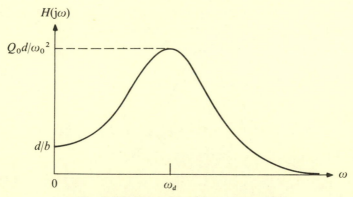

Fig. 3.25. Response of low-pass section.

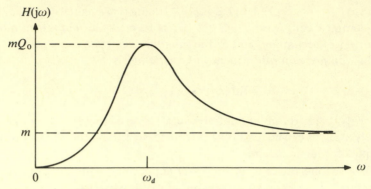

Fig. 3.26. Response of high-pass section.

A *band-pass section* is described by

$$H(s) = cs/(s^2 + as + b) \tag{3.47}$$

whose typical response is shown in Fig. 3.27.

An *all-pass section* is of the form

$$H(s) = K(s^2 - as + b)/(s^2 + as + b) = N(s)/D(s) \tag{3.48}$$

where $N(s) = D(-s)$ and hence $|N(j\omega)| = |D(j\omega)|$ or $|H(j\omega)| = K$.

Only the phase of an all-pass section changes as a function of ω, but the amplitude remains constant.

Finally we consider a *notch filter* or *band elimination filter*. For infinite Q of the numerator,

$$H(s) = (ms^2 + d)/(s^2 + as + b). \tag{3.49}$$

This response is shown in Fig. 3.28.

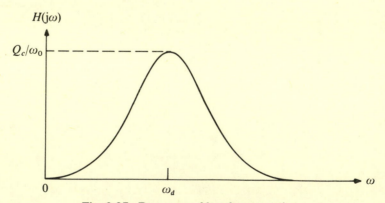

Fig. 3.27. Response of band-pass section.

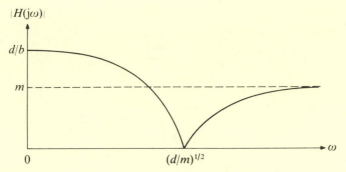

Fig. 3.28. Response of notch filter; $Q_0 = \infty$, $\omega_n^2 = d/m$.

(b) Preprocessing filters for sampled data systems

Preprocessing filters attenuate the signal's high-frequency components prior to sampling in order to reduce aliasing.

Since a sampled data system operates in the time domain, such filters should be designed for minimum distortion of the time response, which implies monotonicity. Hence, due attention to both their amplitude and phase response is essential. This consideration eliminates designs which are optimized solely with respect to amplitude, such as the maximally flat Butterworth or equiripple Tchebycheff filters. Low-pass filters whose phase versus frequency characteristic is linear over the passband have a constant group delay and therefore satisfy the requirement for low distortion.

After the basic properties of a filter have been defined, one frequently arrives at the prototype which satisfies the design target but has an infinite number of singularities. The Gaussian response is an example. Realization by a filter of order n (where n depends on the number of singularities employed) approaches the ideal response as n increases, at the expense of some other property. In the case of linear phase filters, the signal delay becomes greater with increasing order n. Delay in the signal path of a sampled data system is, however, of no importance provided that the system is not part of a closed loop.

Two possible realizations of linear phase networks have been proposed. A network approaching linear phase response has been described by Storch,[19] who employs the Bessel polynomials to synthesize constant time delay ladder networks. Bessel filters of various orders are available as standard building blocks from several manufacturers. A possible realization is given in reference [20].

Another approach has been taken by Guillemin,[21] who proposes a linear array of equidistant poles as shown in Fig. 3.29. The phase slope

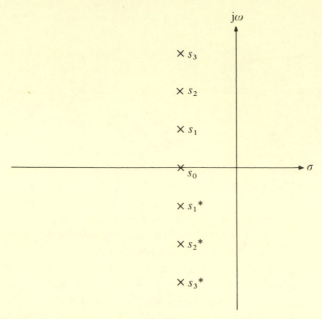

Fig. 3.29. Pole pattern with nearly linear phase response over passband.

of such a filter, for a large number of poles, has an equiripple character. Babić[22] and Mosher[23] have investigated this kind of filter for use in nuclear pulse spectrometry. A practical realization has been proposed by Babić, consisting of a cascade of a single RC integrator with several second order sections, each introducing a complex pole pair. The circuit of such a section providing the second order transfer function is shown in Fig. 3.30.

Fig. 3.31 shows the Bode plot of various order Bessel filters, normalized with respect to ω_h. Such a plot is useful in the comparison of

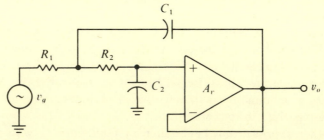

Fig. 3.30. Realization of second order low-pass filter section.

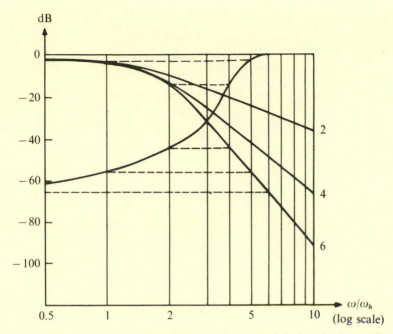

Fig. 3.31. Normalized Bode plot of second, fourth and sixth order Bessel filters. Also shown is the upper first order alias of the sixth order filter for a sampling rate of $6f_h$.

various filters with regard to their capability to suppress aliases, although the attenuation factor obtained is correct only for impulse signals or white noise. Actual attenuation factors may be obtained if the spectrum of the input is known. Evaluation of the alias suppression factor is shown here for a six pole Bessel filter. Assuming a white signal spectrum, the filter response $|H(f)|$ also represents the spectrum of the data to be sampled. With the sampling rate being a multiple (n) of $2f_h$, i.e. $T = 1/2nf_h$, and i the order of the aliases, the amplitude spectrum of the sampled data is for ideal sampling impulses $|H_i(\Delta f)|$ with $\Delta f = 2nif_h - f$, $i = 0, \pm 1, \pm 2, \ldots$, $-\infty \leqslant f \leqslant \infty$.

The frequency attenuation of the first alias ($i = 1$) over the passband of $H_0(f)$ can be indicated on the log–log scale of Fig. 3.31 by noting that the absolute values of $|H_0(f)|$ and $|H_1(\Delta f)|$ are equal at the frequency pairs f and $(2nf_h - f)$, respectively. The factor n has been chosen in Fig. 3.31 as 3. Accordingly, attenuation of the upper first order alias, whose center frequency is at $6f_h$, drops from -56 dB at f_h to -65 dB at zero.

Owing to the symmetry of the sampled spectrum around zero, attenuation of the lower first order alias, whose center frequency is at $-6f_h$,

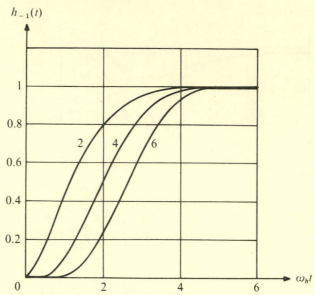

Fig. 3.32. Normalized step response of second, fourth and sixth order Bessel filters.

drops from −65 dB at zero to −73 dB at f_h. Since the responses of the two first order aliases add linearly, total attenuation is halved, or reduced to −59 dB, at $f = 0$, but is not much affected at $f = f_h$ by the lower first order alias.

The time response of various filters is preferably compared by using a time scale normalized with respect to rise time, which is proportional to $(1/\omega_h)$. Fig. 3.32 shows the step response of various order Bessel filters.

(c) *Transversal filters*[24]

Transversal filters[25] are versatile building blocks, ideally suited to the design of an impulse response in the time domain. The basic principle is shown in Fig. 3.33(a), in which a transversal filter is shown to consist of a tapped delay line, whose weighted outputs are connected to a summing amplifier.

Their great versatility is apparent if the impulse response of a delay line with infinite bandwidth is examined. Such a case is illustrated in Fig. 3.33(b), in which the attenuator settings have been picked to generate a specific time function. The attenuators can assume both positive and negative values. As can be seen, the sample values of an arbitrary time function can be specified at the tap-spacing intervals over a time period

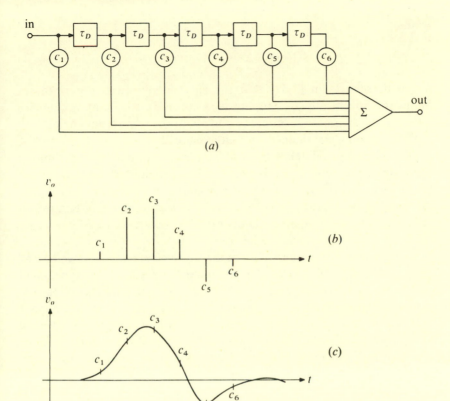

Fig. 3.33. Ideal transversal filter: (a) basic configuration, (b) impulse response, (c) smoothed impulse response. Adapted from [24].

equal to the length of the delay line. According to the sampling theorem, a waveform band-limited to a frequency of f_0 hertz can be completely specified by samples in the time domain at intervals of $\tau < 1/2f_0$ seconds. For a band-limited signal and with the preceding relation satisfied, the smoothed waveform shown in Fig. 3.33(c) might be obtained.

The time domain impulse response of the equalizer is related through the Fourier transform to the frequency response. The impulse response may be written

$$c(t) = \sum_{n=-N}^{N} c_n \delta(n\tau_D), \qquad (3.50)$$

where the number of taps is $2N + 1$, c_n is the attenuator setting of the nth tap, and $\delta(n\tau_D)$ is the Dirac delta function. The Fourier transform can be written

$$C(\omega) = \sum_{n=-N}^{N} c_n \exp{(jn\omega\tau_D)}. \tag{3.51}$$

It can be seen from (3.51) that the spectral response is periodic in frequency. Specifically, the amplitude response is even about frequencies $2n\pi/\tau_D$ and the phase response is odd about the same frequencies.

The foregoing statements are mathematically true, but they must be modified for physical delay lines, which themselves have a band-limited response. If the input signal is properly band-limited, the bandwidth of the delay line need not be a design consideration. A good rule of thumb for lumped parameter delay lines is that the total delay-bandwidth product is limited to approximately 40. Cost rises very rapidly if larger delay-bandwidth products are desired.

Some insight into the flexibility of the transversal equalizer can be had from the following discussion. The attenuators associated with the tapped delay line can be manipulated in pairs to achieve useful effects (as viewed) in the frequency domain. If the center tap attenuator on the delay line is set at unity and symmetrically located pairs of attenuators are given equal values (but magnitudes much smaller than unity), it is possible to adjust the amplitude response, leaving the phase unchanged. For this case, $c_0 = 1$ and

$$c_{-n} = c_n. \tag{3.52}$$

Conversely, if symmetrically located pairs of attenuators are given equal absolute values (again much smaller than unity) but opposite signs, it is possible to adjust the phase response, leaving the amplitude response flat. For this case, $c_0 = 1$ and

$$-c_{-n} = c_n. \tag{3.53}$$

This property has been recognized by several investigators, including Wiener and Lee in an early patent.[26]

In view of the ease with which a transversal filter can be adjusted, it can be employed as an adaptive filter, in which the attenuator settings or weights are adjusted by an adaptive process that automatically seeks an optimal response.[27] Feedback can be applied to transversal filters, leading to the design of *recursive* filters.

As alternatives to the electromagnetic delay line we shall mention the mechanism of acoustic delay provided by elastic waves in solids,[28] leading to the design of SAW (*surface acoustic wave*) filters. Charge transfer devices provide another possible realization and will be discussed briefly in the next section.

(d) Filters employing charge coupled devices

Analog charge coupled devices[29] are essentially analog shift registers, capable of storing and transferring analog signals under the control of externally applied clock pulses. These devices perform discrete-time processing while keeping analog amplitude information. The flexibility and capability of discrete-time processing, combined with the extreme simplicity and efficiency of analog multiplication and addition, makes for a simple implementation of correlation, convolution, filters, electronically variable delay lines and parallel-to-series or series-to-parallel conversion of analog values.

This component, although afflicted with the basic limitations of analog signal processing plus those of sampled data systems, promises to become a building block equalling the monolithic operational amplifier in its low cost and performance, but exceeding it as far as versatility is concerned in applications such as transversal filters with all the peripheral circuitry on-chip, which include clock drivers, anti-aliasing filters[30,31] and output amplifiers.

(e) Matched filter approximations for causal functions

The matched filter for a causal function $f(t)$ is not accurately realizable but can be very closely approximated if the same shape, reversed in time, is delayed by a sufficiently long time to make $f(-t-\tau_D)|_{t<0} = 0$ a reasonable assumption. For this condition, $f(-t-\tau_D)$ is a causal function and hence realizable. Miller[32] has described the realization of a matched filter for pulse spectroscopy, employing a transversal filter. An alternative realization employing a lumped network transfer function has been described by Watson.[33]

Problems

3.8. The active filter shown in Fig. 3.34 employs a VCVS exhibiting $z_i = \infty$, $z_o = 0$ and $v_2 = -A_v v_3$.

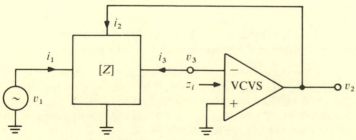

Fig. 3.34. Problem 3.8; $z_i = \infty$, $v_2 = -A_v v_3$.

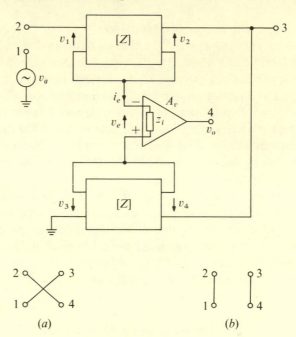

Fig. 3.35. Problem 3.9; $v_o = -A_v i_e z_i$, $A_v = \infty$.

The feedback network is described by the y parameter matrix

$$
\begin{bmatrix} i_1 \\ i_2 \\ i_3 \end{bmatrix} = \begin{bmatrix} y_{11} & -y_{12} & -y_{13} \\ -y_{21} & y_{22} & -y_{23} \\ -y_{31} & -y_{32} & y_{33} \end{bmatrix} \begin{bmatrix} v_1 \\ v_2 \\ v_3 \end{bmatrix}
$$

with $y_{ij} = y_{ji}$.

Show that $1/B = -y_{31}/y_{32}$ and $LT = -A_v y_{32}/y_{33}$.

[Hint: The result may be obtained directly from the matrix equation above, taking into account the impedances and the gain of the VCVS.]

3.9. An active filter employing two two-port networks is shown in Fig. 3.35. The networks are described by their z parameters. Find $1/B$ with the interconnections between the amplifier and the filter as shown in (a) and in (b).

3.10. Realize the second order sections for the pole pattern shown in Fig. 3.29. Use the circuit shown in Fig. 3.30 with $R_1 = R_2 = R$ and $C_2 = f(C_1, Q)$.
   (a) Find $1/B$ for the circuit shown in Fig. 3.30 in terms of (3.43).
   (b) We define $D = \text{Re} \, (s_1)/\text{Im} \, (s_1)$, and Q_i as the quality factor of section i, whose poles are $s_i = \sigma_i + j\omega_{di}$, $s_i^* = \sigma_i - j\omega_{di}$, $i = 1, 2, \ldots, n$, where n is the order of the filter. Show that $Q_i = (1 + i^2/D^2)^{1/2}/2$.

(c) Find $S_{A_v}^Q = S_H^Q S_{a_v}^H$ for a single section as a function of R, C and A_v for $|LT| \gg 1$, $Q \gg 1$, $\omega = \omega_n$.

[Hint: In the computation of S_H^Q, assume $H \simeq 1/B$. Furthermore, $S_{A_v}^H = 1/\mathcal{F}$, where $\mathcal{F} \simeq -(AB')$.]

3.11. For Fig. 3.5, we define with the principal feedback loop open $v_1/v_g = b_1$, $v_2/v_1 = b_2$, $v_3/v_2 = b_3$, $v_o/v_1 = b_4$, $v_o/v_g = b_5$, $v_o/v_3 = b_6$, $v_1/v_3 = b_7$, $v_2/i_{g2} = b_8$, and $v_3/i_{g3} = b_9$.

(a) Draw the flow graph for Fig. 3.5.
(b) From the flow graph, obtain $\mathbf{B}^{-1}$ as defined in Appendix B. Consider v_g, i_{g2} and i_{g3} as inputs, and v_1, v_2, v_3 and v_o as outputs.
(c) Compute $S_{a_1}^H$, where H equals v_o/v_g and a_1 is the gain of A_{v1}. Assume that the voltage amplifier A_{v1} exhibits infinite input and zero output impedances.

[Hint: Use $S_{a_1}^H = S_{b_1}^H S_{a_1}^{b_1}$ and $S_{b_1}^H \simeq S_{1/B_1}^H$, for large $LT_{ii}(s)$; $1/B_1$ can be obtained from the flow graph by inspection, and $S_{a_1}^{b_1}$ from the bilinear form of the flow graph.]

3.1.6 Signal processing in nuclear spectroscopy

The system described in this section is a typical example of broadband signal processing in which a transformer cannot be used to optimize the noise figure of the input amplifier. It furthermore provides a simple example of the SNR computation of a matched filter.

Nuclear spectroscopy measures the energy of particles or photons encountered in research and in a variety of practical applications. This energy is measured by observing the interaction between radiation and a suitable detector. In most detectors this interaction releases a charge proportional to the energy absorbed by the detector.

In what follows, we shall make several simplifying assumptions. Although this impairs the generality of the treatment, the results thus obtained bring out typical properties and basic limitations of performance for the majority of systems. The assumptions are as follows.

(a) The charge is released by the detector during a time interval which is short compared with the second time moment T_2 of the signal filter employed. In other words, compared with the time required for processing, the signal can be considered to be an impulse.

(b) Amplifier noise sources will be considered as white, i.e. low-frequency noise is assumed to be negligible (in practice, it can be reduced to negligible proportions by using a baseline restorer).

(c) The detector is assumed to be purely capacitive.

Fig. 3.36 shows the basic elements of a typical nucleonic signal processing system.

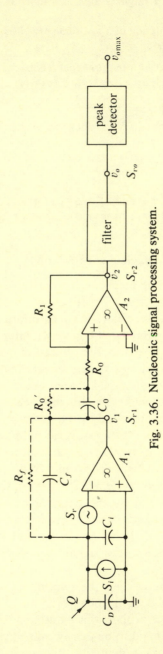

Fig. 3.36. Nucleonic signal processing system.

Amplifiers A_1 and A_2 provide the gain needed to raise the signal to a level at which the additional noise contributed by further processing and shaping is insignificant. They are followed by a filter designed to improve the SNR, and the following peak detector extracts the desired amplitude information from the signal. C_D is the detector capacitance. Detector and amplifier parallel noise are lumped into a single noise generator of spectral density $S_i = 4k\mathcal{T}/R_p$, since the absence of a noise matching transformer makes it irrelevant to distinguish between noise sources of the amplifier and those of the signal source. Series noise of the detector is assumed to be negligible, and that of the amplifier is modelled by $S_v = 4k\mathcal{T}R_s$. Assuming an FET is used as input stage, C_i comprises its gate source capacitance plus any stray capacitance present.

Since the use of feedback for stabilizing the gain of A_1 is essential, and since the use of a feedback resistor for gain stabilization is excluded because its noise contribution would be prohibitive (a typical value for R_p is of the order of $10^9 \ \Omega$), a feedback capacitor C_f serves as gain-stabilizing element. Hence, the signal current due to Q deposited across C_D is integrated by the first amplifier, which necessitates differentiation by the following stage. In practice, a feedback resistor R_f provides the necessary dc stabilization for the integrator, and R_0' compensates for the pole introduced by $R_f C_f$ (pole zero compensation). If properly dimensioned, these resistors introduce merely second order effects in the computation of the SNR and may therefore be safely disregarded.

In what follows, we shall first derive the matched filter response in the frequency domain, disregarding limitations on the filter's realizability, and compute the best theoretically attainable SNR. We shall then consider a practical filter and compute its actual performance.

Design of matched filter
Consider the block diagram shown in Fig. 3.37, in which H_1 represents the transfer function of the preamplifier, H_2 that of the following differentiator, and $H_3 H_4$ the matched filter. In accordance with matched filter design praxis,[34,35,36] H_2 is designed as a noise whitening filter (see Appendix A.6), namely

$$|H_2(j\omega)|^2 = 1/H_{n1}^2(\omega),\qquad(3.54)$$

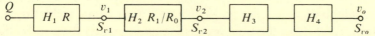

Fig. 3.37. Block diagram of nucleonic signal processing system.

where $H_{n1}^2(\omega)$ is the dimensionless noise power density spectrum at the output of the preamplifier.

Furthermore, since the input signal is assumed to be an impulse, the signal spectrum

$$v_2(j\omega) = QRH_1(j\omega)H_2(j\omega), \tag{3.55}$$

where R is merely a normalizing resistance making $H_1(j\omega) = 1/j\omega C_f R$ dimensionless.

Finally, the frequency response of the matched filter is

$$H_3(j\omega)H_4(j\omega) = H_1^*(j\omega)H_2^*(j\omega). \tag{3.56}$$

This completes the theoretical design of the matched filter.

We may also derive from Fig. 3.37 the general case of designing a matched filter, when the input noise spectrum to the matched filter is not white. We now define $H_2(j\omega)H_3(j\omega)H_4(j\omega)$ as the matched filter, which is to be designed from knowledge of the signal spectrum $H_1(j\omega)$. In order to satisfy (3.56), and in accordance with (3.54), we choose $H_3(j\omega) = H_2^*(j\omega)$ and $H_4(j\omega) = H_1^*(j\omega)$; hence

$$H_2(j\omega)H_3(j\omega)H_4(j\omega) = |H_2(j\omega)|^2 H_1^*(j\omega)$$

$$= H_1^*(j\omega)/H_{n1}^2(\omega). \tag{3.57}$$

Computation of optimal SNR
From Fig. 3.36 we compute

$$S_{v1}(f) = S_i R^2 H_{n1}^2(\omega) = \frac{4k\mathcal{T}R^2}{R_p} \frac{(1+\omega^2\tau_c^2)}{\omega^2 C_f^2 R^2}, \tag{3.58}$$

where $C_T = C_D + C_i + C_f$ and $\tau_c = (R_s R_p C_T^2)^{1/2}$; τ_c is commonly designated the *noise corner time constant*. Hence, the noise spectrum at the output of the preamplifier is

$$S_{v1}(f) = \frac{4k\mathcal{T}R^2}{R_p} \frac{(1+\omega^2\tau_c^2)}{\omega^2\tau_c^2}, \tag{3.59}$$

where we have chosen $C_f R = \tau_c$. Equation (3.59) defines the noise whitening filter $H_2(j\omega)$:

$$H_2(j\omega) = j\omega\tau_c/(1+j\omega\tau_c), \tag{3.60}$$

making $S_{v2} \neq f(\omega)$, if we choose $R_0 C_0 = \tau_c$. The following derivations will show that the noise corner time constant τ_c is associated with the second time moment T_2 of the sinal processing filter, and hence is indicative of the signal processing time required.

Finally, the signal spectrum at the preamplifier output is found to be $v_1(j\omega) = H_1(j\omega) = 1/j\omega C_f R = 1/j\omega\tau_c$.

We may now formally synthesize the optimal filter:

$$|H_2(j\omega)|^2 H_1^*(j\omega) = \frac{\omega^2\tau_c^2}{(1+\omega^2\tau_c^2)} \frac{1}{(-j\omega\tau_c)} = \frac{j\omega\tau_c}{(1+\omega^2\tau_c^2)}. \tag{3.61}$$

Hence, the spectral noise density at the output is

$$S_{vo}(f) = 4 k \mathcal{T}\left(\frac{1}{R_p\omega^2 C_f^2} + R_s\frac{C_T^2}{C_f^2}\right)|H_2(j\omega)|^4|H_1(j\omega)|^2$$

$$= \frac{4 k \mathcal{T}}{R_p}\frac{C_0^2 R_1^2}{C_f^2}\frac{(1+\omega^2\tau_c^2)}{(1+\omega^2\tau_c^2)^2}, \tag{3.62}$$

giving

$$\overline{v_{no}^2} = \frac{k \mathcal{T}}{R_p}\frac{C_0^2 R_1^2}{C_f^2\tau_c}. \tag{3.63}$$

In Laplace notation, the signal is

$$v_o(s) = \frac{QC_0R_1}{C_f\tau_c^2}\frac{1}{(s+1/\tau_c)(s-1/\tau_c)}, \tag{3.64}$$

$$v_o(t) = \begin{cases} V\exp(t/\tau_c), & t<0, \\ V\exp(-t/\tau_c), & t>0, \end{cases}$$

$$v_o(0) = V = v_{o\,max} = QC_0R_1/2C_f\tau_c.$$

Hence, the optimal $(SNR)^2$

$$\frac{v_{o\,max}^2}{\overline{v_{no}^2}} = \frac{Q^2}{4 k \mathcal{T}C_T}\left(\frac{R_p}{R_s}\right)^{1/2}. \tag{3.65}$$

The inverse Laplace transform of (3.64) is commonly termed the *cusp*. It was derived for the first time by Den Hartog.[37] Since for a realizable filter the impulse response must be zero for $t<0$, the cusp is not realizable.

Hence, the matched filter merely serves in this case as a measure for the quality of practically realizable filters. In fact, a single *RC* integration following the noise whitening filter yields a SNR which is smaller than the optimal one merely by a factor of $2/e$ – a remarkable result, considering the simplicity of the filter.

Computation of SNR for simple filter
It is instructive to compute the performance for a single RC integration following H_2 for the general case in which the shaping time constant τ_s is not equal to τ_c. Let

$$H_2(j\omega)H_3(j\omega)H_4(j\omega) = j\omega\tau_s/(1+j\omega\tau_s)^2, \qquad (3.66)$$

where $\tau_s = R_0C_0$. Then

$$\overline{v_{no}^2} = \frac{k\mathcal{T}C_0^2R_1^2}{2C_f^2\tau_s^2}\left(\frac{R_sC_T^2}{\tau_s}+\frac{\tau_s}{R_p}\right) \qquad (3.67)$$

and

$$v_o(t) = \frac{Q}{C_f}\frac{R_1}{R_0}\frac{t}{\tau_s}\exp\left(\frac{-t}{\tau_s}\right) \qquad (3.68)$$

yields

$$(\text{SNR})^{-2} = \frac{\overline{v_{no}^2}}{v_{o\,\text{max}}^2} = \frac{k\mathcal{T}e^2}{2Q^2}\left(\frac{R_sC_T^2}{\tau_s}+\frac{\tau_s}{R_p}\right). \qquad (3.69)$$

Equation (3.69) shows that for $\tau_s > \tau_c$ parallel noise is dominant and for $\tau_s < \tau_c$ series noise. For equal series and parallel noise contributions, $\tau_{s\,\text{opt}} = \tau_c$, and

$$(\text{SNR}_{\text{opt}})^2 = \frac{Q^2}{k\mathcal{T}C_T\,e^2}\left(\frac{R_p}{R_s}\right)^{1/2}. \qquad (3.70)$$

Table 3.1 compares the SNR obtained with filters performing n integrations of time constant $RC = \tau_c/n^{1/2}$, with the normalized SNR obtained with a matched filter. The noise whitening filter is given by (3.60).

3.1.7 Signal transmission and routing

Linear analog signal conditioning involves not only amplification and shaping, but also transmission over various distances and electronic

Table 3.1

$H_3(j\omega)H_4(j\omega)$	n	SNR
$1/(1+j\omega\tau_c/n^{1/2})^n$	1	0.73
	2	0.82
	4	0.86
	∞	0.89
$1/(1-j\omega\tau_c)$	—	1

switching or gating, for the purpose of signal routing. The unavoidable spurious signals thereby introduced are error components inseparable from the output data. At best, these error components can be made insignificantly small by reducing them at the source,[38,39,40,41] and/or by filtering; at worst, they can completely obliterate the measurement data.

This section deals with interference problems encountered in the transmission of signals over long distances, and with imperfections of signal routing devices.

(a) Techniques for reducing interference pick-up[42]

Signals which appear at the output of a signal processing system in some form as to add or subtract with respect to the desired signal and which cannot be calibrated constitute 'noise' or interference. Those effects internal to the transducer and amplifier are examined in Sections 1.2.4 and 1.7.6. Here our concern will be with external effects in terms of the source of the noise. Of the external effects, electrostatic and electromagnetic fields, and currents through common impedances can be reduced by shielding and isolation, and will be discussed here. Noise associated with cable movement will also be discussed.

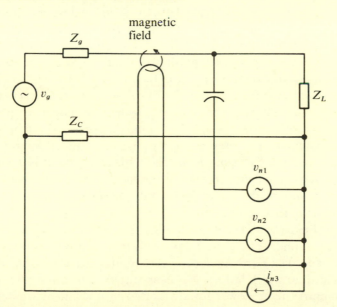

Fig. 3.38. Mechanisms of noise coupling; Z_g source impedance, Z_C cable impedance, v_{n1} capacitively coupled noise source, v_{n2} magnetically coupled noise source, i_{n3} current coupled noise source, z_L load impedance. Adapted from [42].

Figure 3.38 illustrates the mechanisms by which these undesired signals can be introduced into the signal paths. The nature of the impedances and noise sources depends on the portion of the frequency spectrum with which one is concerned, be it low-frequency interference (50, 60 or 400 Hz power typically), radio-frequency interference (RFI), or electromagnetic interference (EMI) which can encompass low to radio frequencies. Generally, at the low frequencies the impedances of concern are either resistive or capacitive, while at the higher frequencies the inductance of a short length of wire can become important as well.

Capacitively coupled interference is associated with the presence of a varying electrostatic field, or difference in potential, between two conductors coupled electrically by some stray capacitance. For example, two pins in a connector can be the 'plates' of the capacitor, while the mounting insulator be the 'dielectric'.

Magnetically coupled interference is associated with the presence of a varying magnetic field in the vicinity of the signal paths. For example, if one of the two pins considered above is conducting a varying current, there exists a varying magnetic field around it. The second pin, if it is part of an electrically closed circuit, can have a voltage induced by the varying field, resulting in noise.

Current coupled noise occurs where the signal and other currents use a common path, intentionally or otherwise. Any impedance in this path causes the interfering currents to develop extraneous signals, which the following devices cannot distinguish from data. For example, if a conductor were used both as the 'low' side of the signal path from a transducer to its associated amplifier, and as the 'low' side of an ac excitation for another device, the common resistance would develop a component of the ac that would appear as signal. Ground ('earth', or 'common') is used quite frequently as a reference for signal magnitude, polarity, shield terminations, etc. But what is ground? Ideally, ground is a reference plane of zero impedance so that no potential difference exists between any two points of the plane. Practically, however, zero impedance is unobtainable; thus, with currents injected and removed from the plane (or cable, bus bar, chassis, as the case may be), potentials do exist between points on ground. Thus, a ground point is often used as a reference, rather than a plane. The Earth, under the best conditions, is a less than perfect conductor into which large currents are injected by 50 or 60 Hz power distribution systems. Power ground is often an attachment to a water pipe (or rod driven into the earth) with its associated resistance of conductors, connectors, and corroded joints. Thus many facilities establish an instrument ground, which often takes the form of a large conductor or bus

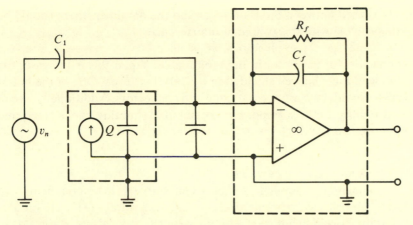

Fig. 3.39. Capacitive coupled noise source. Adapted from [42].

bar judiciously protected from the ac power system. Even this may not be sufficient; thus we attack the problems by shielding and isolating in a planned manner.

Shielding for capacitive coupling
To illustrate the effect of capacitively coupled interferences we choose the circuit of Fig. 3.39, showing an amplifying system consisting of a piezoelectric accelerometer and a charge amplifier. The input to a charge amplifier is particularly susceptible to electrostatic coupling because it responds directly to charge, i.e. coulombs input. Assume that the accelerometer has a sensitivity Q_a of 10 pC/g and that the vibration level which is to be measured is approximately 1 g. Assume also that there is an ac electrostatic voltage (v_n) present of 100 V (this is typical). The charge amplifier measures the total charge appearing at its input, which in this example would be

$$Q_{\text{total}} = Q_a \times g + v_n \times C_1 = 10 \text{ pC} + 100 C_1 \text{ pC},$$

where C_1 is the capacitance in picofarads between the interference source and the charge amplifier input.

This equation indicates that a capacitance of only 0.01 pF will give a 10% interference. In many cases it is desirable to make measurements to less than 0.1 pC input, and at these levels the interference problem becomes even greater. To insure rejection of the electrostatic voltage, the input lead of a charge amplifier must be entirely shielded, with this shield connected to the input common of the amplifier. In a system where the transducer case is connected to the shield of the cable and a single coaxial cable is used between the transducer and the amplifier, there should be little problem with this source of interference.

In some system applications, it is necessary to have an interface (connector, for example) in the input cable. When this is the case, the input lead must be well shielded in the interface area. One of the better ways to do this is to use coaxial cable and connectors. Additionally, use of good quality cable is important to stabilize shunting capacitance and minimize leakage (high insulation resistance).

Shielding for magnetic coupling

Electromagnetic coupling occurs when varying magnetic fields are present and physically oriented so as to induce a voltage in the circuit. It is generally more difficult to shield magnetically than electrostatically. In most good amplifiers, some provision is made for magnetic shielding (as well as minimizing internally generated fields). However, the instrumentation engineer must take precautions to insure that magnetic fields are minimized in the area of the amplifier so as to reduce electromagnetically coupled signals to acceptable levels. This requires electromagnetic field sources such as power transformers, solenoids, motors, and other electrical devices to be a sufficient distance away so that the induced voltage is negligible. Alternately or additionally, magnetic shielding and other field-reducing techniques can be employed around the source of the disturbance, to constrain the magnetic field sufficiently.

The designer should always bear in mind that any conductor carrying a current is surrounded by a magnetic field, whose strength is in proportion to the current. Thus, signal leads sharing a cable trough with power or other high-current conductors are subject to a high degree of electromagnetic coupling. To attack this problem, let us examine the circuit of Fig. 3.40(a), which illustrates a conductor carrying a current i lying next to the input signal lines. The flux produces voltages v_1 and v_2, which will have the polarities shown (at one particular instant). Their magnitudes are

$$v_1 = \frac{\mathrm{d}}{\mathrm{d}t}(N_1\Phi_1), \qquad v_2 = \frac{\mathrm{d}}{\mathrm{d}t}(N_2\Phi_2),$$

where N is the number of turns being coupled (essentially unity unless there is an excess of signal lead that has been neatly coiled!), and Φ is the number of lines of flux coupling the conductors – a function of the magnitude of i, how close the conductors are physically, over how great a length, and the effect of any magnetic shielding.

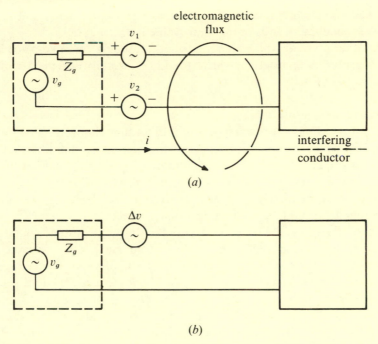

Fig. 3.40. (*a*) Electromagnetically coupled noise source. (*b*) Equivalent diagram. Adapted from [42].

Thus we can see that the induced voltages will be larger for higher frequencies (faster rate of change) but can be reduced by more direct signal cable runs (fewer loops and turns), greater separation between signal cable and source of undesired field, and minimizing the length of signal cable exposed to the field (where it cannot be completely avoided). An additional large improvement is obtained when the conductor carrying i is tightly twisted with the return conductor for i. This results from the cancellation effect of having essentially equal and opposite electromagnetic fields in close proximity, with the resultant reduction of Φ.

If v_1 and v_2 were exactly equal, there would be no effective error signal. However, this hardly ever occurs. Any difference in the number of flux linkages taking place, due to length or positions of the lines, will cause a slight difference in the voltages, shown as Δv in fig. 3.40(*b*). The magnitude of Δv can be reduced by tightly twisting the input signal lines, using small-diameter coaxial cable, or taking similar precautions to insure a high degree of balance or symmetry with respect to the source of the electromagnetic interference. Even with these precautions, typical cables will have a $\Delta v / v_2$ ratio of the order of 10^{-3}. Maintaining the symmetry through connectors, and insuring wide separation of signal and inter-ference conductors adds to the improvement obtainable.

Another point at which electromagnetic coupling can take place is in the transducers themselves, which should be carefully designed to mini-mize this possibility.

Isolation for current coupling

Current coupling occurs when a current other than that generated by the signal introduces an apparent signal into the transducer/amplifier system. Referring to Fig. 3.41(a), we see that there will be no current flow in the cable shield resistance R_C if there is no potential difference between instrument ground G_I and signal ground G_S. But, as previously mentioned, 'ground' tends to be less than perfect.

Referring to Fig. 3.41(b), we assume that 'ground' is some conductor with distributed resistance (wire, cable, bus bar, earth, etc.); even though the resistance may be small, the analysis will still be valid. The voltage v_1 represents a source such as the 50 or 60 Hz power system. Because the power system has a return path to ground G_I and/or ground G_S (unless specialized efforts are put forth to prevent this), the 'high' side of the line will electrostatically couple (via the stray distributed capacitance) a current into the ground conductor. This current develops a voltage between grounds G_I and G_S which also appears across R_C (which is part of the current path). The voltage source v_2 could also be part of the power system driving a motor (R_{L1}). The electromagnetic field around the conductor, passing through the ground, induces a voltage between grounds G_I and G_S – another signal across R_C. Voltage v_3 should never happen, but often does; it is supplying a load (R_{L2}) whose current return is through the ground, with a resultant voltage drop between grounds G_I and G_S – a third signal across R_C.

As an example, consider a transducer attached to a specimen mounted on a shaker. The coax shield is grounded to the transducer case, and is the low side of the signal path. The coax runs some distance to an amplifier mounted in an equipment rack, which is grounded to the power system ground. The specimen is grounded to the shaker which is grounded to the shaker drive system – also grounded to the power system ground. If we identify the transducer case as ground G_S and the low side of the amplifier output as ground G_I, we have the 'circuit' of Fig. 3.41(b). Into this circuit we have electrostatic and electromagnetic coupling from the lighting system, motors, heaters, switching circuits, and possible 'static' genera-tion due to cable motion. The shaker drive circuits, specimen heaters, and other electrical devices attached to the specimen could add conduction

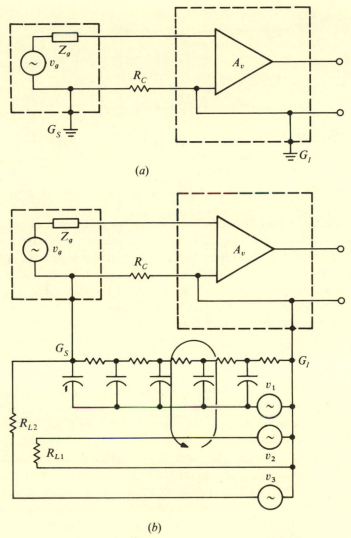

(a)

(b)

Fig 3.41. Current coupled noise sources. Adapted from [42].

currents to the specimen/shaker ground path. Thus we could have all the evils described previously.

Techniques for reducing ground-loop current and its effects
For didactic purposes, v_1, v_2 and v_3 of Fig. 3.41(b) can be reduced to an equivalent source v_{GL} (equivalent ground-loop voltage) and Z_{GL} (equivalent ground-loop source impedance) as shown in Fig. 3.42, which

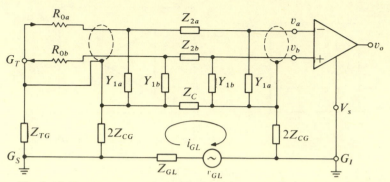

Fig. 3.42. Equivalent diagram for evaluation of ground-loop effects.

enables the interplay between the various impedances and the interferences reaching the amplifier to be examined. Z_{TG} denotes the impedance between transducer ground G_T and signal ground G_S, Z_{CG} the impedance between cable and ground, R_0 the cable termination* and Y_1, Z_2 the losses of the interconnecting twin lead coaxial cable. From this diagram the following conclusions can be drawn.

First, the interference signal reaching the amplifier may be CM and DM. Since all amplifiers have a finite CMRR, both terms will cause an interference signal which is indistinguishable at the amplifier output from the desirable signal emitted by the transducer. However, the DM signal due to interferences is potentially more damaging since it is fully amplified, whereas the CM signal is attenuated by a factor equal to the CMRR. This situation may be examined considering Fig. 3.42, with all cable losses lumped into an equivalent circuit. We shall first assume the two signal paths through the twin lead coaxial cable to be ideally balanced, i.e. $R_{0b} = R_{0a}$ and the equivalent lumped cable losses Y_{1a}, Y_{1b} and Z_{2a}, Z_{2b} to be identical to each other in both branches. Under this condition it can be seen by inspection that the voltage across Z_C, due to i_{GL}, will produce only a CM signal, whereas a DM signal will be created merely by imbalances between corresponding impedances, in the two branches. This shows the importance of using a uniform cable and, in particular, of employing equal resistors $R_{0b}R_{0a}$ in series with each cable lead as terminations.

Second, i_{GL} should be made as small as possible to minimize the secondary effects due to imbalances in the cable losses and the finite

---

* A termination at the transmitting end of an open-ended transmission line serves to absorb the reflections originating from the unterminated receiving end.

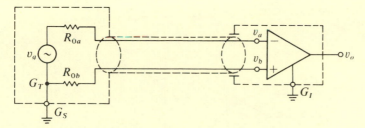

Fig. 3.43. Grounded transducer guarded amplifier system.

CMRR of the amplifier. To this end, the loop impedance $\{Z_{GL} + [Z_{TG}\|2Z_{CG}] + Z_C + 2Z_{CG}\}$. should be made as high as possible.

Third, it is essential to provide a low impedance between the cable screen and ground either at the transducer terminal ($Z_{TG} = 0$) or at the amplifier terminal ($2Z_{CG} = 0$).This avoids excessive CM signals reaching the amplifier, which would otherwise be produced across these impedances by i_{GL} (on no account should both sides of the cable screen be grounded, since this would result in an excessive increase of i_{GL}). This important point will be considered in greater detail.

One-sided grounding of the cable screen requires the use of the *guarding technique*. In this case, guarding relates to the technique of connecting the screen of the twin lead cable to such a voltage that the DM component due to i_{GL} at the output ($v_b - v_a$) equals zero and only a CM component ($v_b + v_a$) appears, due to i_{GL}.

This requirement can be satisfied (a) if $Y_{1a} = Y_{1b}$, $Z_{2a} = Z_{2b}$; and (b) if the cable screen is connected either at the transducer terminal to ground as shown in Fig. 3.43, or at the output terminal to a guarding voltage $v_G = (v_a + v_b)/2$, as shown in Fig. 3.44. The proof is easily obtained: bisect the equivalent cable circuit into two symmetrical sections by splitting Z_C and $2Z_{CG}$ into two equal impedances each of double value

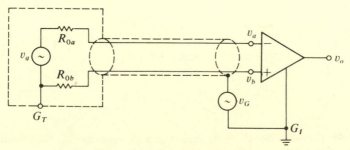

Fig. 3.44. Guarded transducer system.

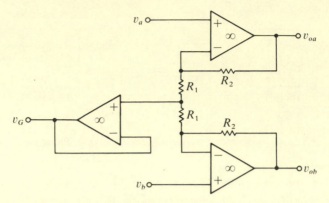

Fig. 3.45. Circuit for deriving guarding voltage v_G.

and show that potentials at corresponding points of the two split sections are the same. The two alternative methods of short-circuiting Z_{TG} to ground (at the transducer) or $2Z_{CG}$ to a voltage source (at the amplifier) are generally referred to as the *grounded transducer guarded amplifier system* (Fig. 3.43), and as the *guarded transducer system* (Fig. 3.44), respectively. They ensure that any signal output due to i_{GL} will be solely due to impedance asymmetry and due to a finite CMRR of the following amplifier – both second order effects. A suitable input circuit for deriving the guarding voltage is shown in Fig. 3.45.

Both the grounded transducer guarded amplifier and the guarded transducer systems require a differential amplifier with a high CMRR, a standard feature of voltage amplifiers. However, charge-sensitive amplifiers such as are commonly employed with capacitive transducers do not lend themselves easily to be designed as differential amplifiers. A practical alternative, shown in Fig. 3.46, is recommended by Endevco. A

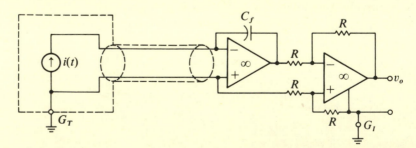

Fig. 3.46. Connection of single-ended charge amplifier to twin lead cable. Adapted from [42].

single-ended input charge amplifier is floating and shielded with respect to the instrument ground G_I. It is connected to a differential voltage amplifier providing a single-ended output, which is referenced to the instrument ground.

Ground-loop voltage rejection techniques may be essential not only for interconnections between transducer and amplifier, but also in some cases in the interconnection between two self-contained units, each complete with its own power supply. As an example, consider a self-contained amplifier connected to an A/D converter including the associated digital circuits. In practice, the two units will be interconnected by a single lead standard coaxial cable. In this case, strong ground-loop currents will flow between the two units due to their separate power supplies and may cause interferences. Here, too, interrupting the ground-loop by disconnecting the screen at one end may be desirable, but the sophistication of using a twin lead coaxial cable or optimal ground-loop rejection may be neither justified nor desirable, since signal connections are preferably made employing standardized cables and connectors. A simplified scheme applicable in this case is shown in Fig. 3.47. The cable connector at the receiving end is isolated from the chassis and the cable screen is connected to the second terminal of the differential input stage. The same method is applicable between transducer and amplifier in cases with less critical interference problems.

Effect of high frequencies
Thus far our discussions have been general in nature and applicable over a wide spectrum of frequencies. At higher frequencies (10 kHz and more), careful consideration of the impedances involved (and alternative paths) is important in the understanding and elimination of the noise errors. For instance, the proximity of interference sources that were no problem at 50 Hz can be serious at 500 kHz, since the impedance of a capacitively coupled path is 1/10 000 its previous value. Additionally, a conductor that was primarily a low-resistance path at 50 Hz can have a

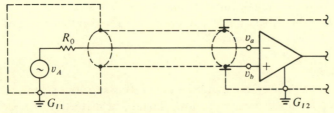

Fig. 3.47. Reduction of i_{GL} between two signal processing units.

significant inductive impedance at 500 kHz, thus changing the nature and magnitude of the noise sources and their paths.

At higher frequencies, amplifiers no longer fit the classical picture, both because the characteristics of the active elements come into play and the passive components and paths take on reactive characteristics. A well-designed amplifier is stable and predictable over the band of frequencies and magnitudes for which it is designed. When subjected to stimuli from extraneous sources, erroneous outputs can result. For instance, radio-frequency energy can be rectified at the emitter base junction of a transistor causing a dc error. Additionally, modulation of the desired signal by unwanted radio-frequency noise can cause spurious information to appear at the output.

The sources of high-frequency signals are as numerous as electrically operated devices. Some examples are: digital devices – computers, counters, time code generators; telemetry equipment and systems; radio-frequency transmitters; radar; power control systems – thyratron and SCR controls in particular; rotating machinery – electrostatic discharge, for example; relays and other power switching devices; welders, strobes and other energy discharge devices; fluorescent lights; office machines and many others. A point to consider with this type of interference is that the 'ground loop' with which one is dealing may be much larger than is apparent.

At high frequencies, the stray capacitance from primary to secondary of power transformers may be an 'easy path' for these signals so that one is dealing with conducted interference, not present at the lower frequencies. Such interference from the power line is one of the most obstinate sources of man-made noise. The attack on the problem is usually threefold:

(1) suppression of the signal at its source;
(2) highly shielded transformers to reduce the 'feedthrough' capacitance;
(3) power line filters to block (or bypass) the troublesome noise.

One technique of reducing high-frequency ground-loop currents consists of increasing the impedance of the cable screen (Z_C in Fig. 3.42) by winding the cable on a ferrite core, thereby creating a coil of greatly increased series inductance.[43]

Cable noise

The input cable itself is capable of generating spurious signals when it is subjected to mechanical vibration, flexure, or distortion. Experimental studies have shown that a major cause of noise generation is due to the

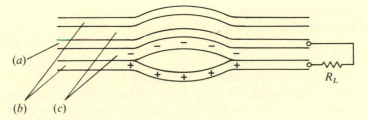

Fig. 3.48. The triboelectric effect on coaxial cables: (*a*) center conductor, (*b*) shield, (*c*) dielectric. Adapted from [42].

triboelectric effect (charge generated by friction) associated with the relative motion, or localized separation, between the cable dielectric and the outer shield around the dielectric.

When the outer shield of the cable separates from the cable dielectric as shown in Fig. 3.48, the steady-state charge distribution at the interface of the localized separation becomes unbalanced. The triboelectric charge in the dielectric is immobilized because of the low conductivity of the dielectric material; however, the charge in the shield is free to flow through an external conduction path, R_L, to neutralize the electric field between the inner and the outer conductors created by the trapped triboelectric charge in the cable dielectric. When the shield re-establishes contact at the point of separation, the process reverses and a flow of opposite polarity takes place through the conduction path.

Cable noise tends to be a low-frequency phenomenon with little noise generated above 200 Hz. Charge-sensitive amplifiers connected to a capacitive transducer through a long cable are especially sensitive to cable noise, owing to their low input impedance. In this case, the external conduction path is generally the input impedance of the amplifier; hence, the spurious charge flow in untreated cables can generate signals of greater amplitude than the transducer output, and only specially treated low-noise cables should be employed.

Alternative ground-loop suppression techniques
Excellent ground isolation can be achieved with isolation amplifiers, employing as a separation element LED coupled phototransistors, photoresistors, magnetoresistors, Hall-effect devices or thermal couplers. The non-linearity introduced by these devices can be corrected by applying feedack through a signal path having a matching non-linearity.[2,44] Alternatively, the non-linearity can be avoided employing digital encoding in conjunction with optical couplers, or transmitting a modulated carrier through an isolation transformer.

A variety of isolation amplifiers is commercially available, and relevant application notes are available from the manufacturers.

Utilization of remote amplifiers

The longer the distance between transducer and input amplifier of the signal processing system, the more serious will the interference problems become. When interference problems due to excessively long cable runs become insurmountable, a remote preamplifier mounted next to the tranducer will improve the SNR due to interferences by a factor equal to its amplification. Where the preamplifier is powered by its own power supply connected to the mains at the remote transducer location, the grounded transducer guarded amplifier system is preferable. Alternatively, the preamplifier may be connected to the dc power supply of the signal processing system, in which case the dc supply leads should be incorporated in the interconnecting cable. The guarded transducer system should then be used.

Filtering

Selective rejection of interfering frequencies by a notch filter may in some cases be cheaper than suppression at the source, or a last resort where other methods alone are insufficient. The conditions are (*a*) that the interfering frequency is stable and (*b*) that the attenuated frequency band does not contain essential signal frequencies.

An elegant and effective method for eliminating sinusoidal interference has been described by B. Widrow.[45] This method of adaptive noise cancelling makes possible the design of a notch filter which offers easy control of bandwidth, an infinite null, and the capability of adaptively tracking the exact frequency of interference, which therefore need not be stable.

Referencing of differential amplifiers

All efforts spent on reducing interferences can be wasted if the input differential amplifier is not referenced to signal zero. The sensing voltage V_s in Fig. 1.27 should be chosen accordingly, all consecutive signal processing such as amplification, shaping and A/D conversion should be referred to this voltage, and the layout and wiring should be carefully designed to prevent any kind of interference being added to V_s.

Problem

3.12. A transducer (v_g) is connected through a cable to a differential amplifier, as shown in Fig. 3.49(*a*). The ground of the transducer is at ground-loop potential v_{GL}. Fig. 3.49(*b*) shows a simplified equivalent diagram, C_a and C_b modelling the

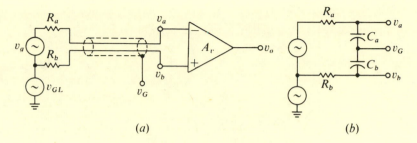

Fig. 3.49. Problem 3.12.

distributed cable capacitance. If A_v denotes the differential gain of the amplifier and $CMRR_A$ its CMRR, then

$$v_o = A_v(v_b - v_a) + \frac{A_v}{CMRR_A}(v_b + v_a)$$

Let $R_a C_a = \tau_a$, $R_b C_b = \tau_b$, and $R_a C_a - R_b C_b = \delta\tau$.

We define the overall CM and DM gains as $A_{CM} = v_o/v_{GL}(v_g = 0)$ and $A_{DM} = v_o/v_g(v_{GL} = 0)$.

(a) Assume that $CMRR_A$ is infinite, and the guard voltage $v_G = 0$. Show that the overall CMRR (which is defined as A_{DM}/A_{CM}) is

$$(1 + s\tau_b)/s\delta\tau.$$

(b) Show that the CMRR is infinite for infinite $CMRR_A$ and $v_G = (v_a + v_b)/2$.

(c) If $CMRR_A \gg 1$ and $v_G = (v_a + v_b)/2$, show that $CMRR \approx CMRR_A/2(1 + s\tau)$, where $\tau = (\tau_a + \tau_b)/2$.

(b) Gating and multiplexing

Linear gating

The linear gate is an important building block of analog signal processing systems. Operating as an analog switch, it performs functions such as elimination of undesirable signals in a single analog channel, or waveform sampling, to make possible time sharing of a single signal processing system by several transducers (time multiplexing).

We shall first turn our attention to the imperfections encountered in practical circuits. Consider Fig. 3.50, showing the symbol for a linear gate and waveforms of the input analog signal v_A, the logic gate control and the analog output v_o. The following imperfections (spurious signals) are observed: signal feedthrough if the gate control is off, and gating transients. Both are due to capacitive coupling between the analog input circuit or gate control input circuit, respectively, and the analog output. When the gate control is on, a pedestal appears at the output mainly due

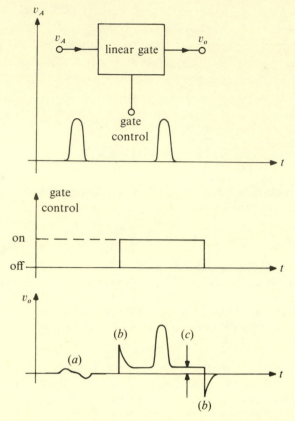

Fig. 3.50. Imperfections of linear gate: (*a*) feedthrough, (*b*) gating transient, (*c*) pedestal.

to dc offset of the analog transfer characteristic, but possibly also due to galvanic coupling between the gate control and the output circuit.

In Fig. 3.50, the amplitude of the feedthrough in relation to the signal has been exaggerated for the sake of clarity but in practice the amplitude of the spurious signals and of the pedestal is typically between two to three orders of magnitude below FS.

Specifications
The criteria specifying the performance of a linear gate with the gate control in the on position are practically the same as those defined for a linear amplifier.

Further imperfections to be specified are the rejection ratio and the acquisition time. The *rejection ratio* relates to the ratio between output and input signals when the gate is disconnected. This specification should include a definition of the input signal shape for which the rejection ratio is measured, since the amplitude of the signal feedthrough is a function of the signal derivative. The *acquisition time* is defined as the total time required for the output to settle to within a specified percentage of the final output response from the time the gate control on signal is applied.

Multiplexing

An analog multiplexer is essentially a combination of linear gates, connecting several analog signal sources sequentially to a centralized signal processing system. Multiplexers are available as units comprising n linear gates, with n ranging typically between 8 and 256. They will accept n single-ended or $n/2$ differential analog inputs, and usually feature self-contained binary channel address coding.

A circuit-based multiplexer has been designed by Gere and Miller,[46] consisting of n differential input stages which are alternatively connected to a common output stage. The complete circuit is feedback-stabilized for unity gain. The same basic idea was applied by van de Plassche[47] to a highly accurate instrumentation amplifier.

Imperfections of a multiplexer are the same as those relating to a linear gate. In addition, crosstalk between the on and the remaining off channels and maximum *throughput-rate* should be specified. The latter relates to the maximum sampling rate per second at the output of the multiplexer.

The number of channels multiplexed with a single unit can be increased by using several units either in the single mode, or in multitiered matrix multiplexer configurations.

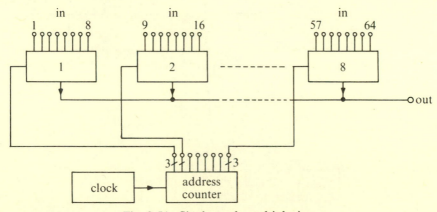

Fig. 3.51. Single mode multiplexing.

In the single mode shown in Fig. 3.51, the individual outputs of the different multiplexers are connected in parallel. As a result, the off channel shunt capacitances of all channels add at the common output line and are reflected to the input of the on channel, resulting in a reduction in switching speed and settling time. This limits the maximum number of multiplexers which can be parallel connected in this fashion.

Multitiered or second level multiplexing circumvents this limitation by connecting the individual outputs of the first level multiplexers to the inputs of a single second level multiplexer, as shown in Fig. 3.52 for 64 inputs. In general, 2^n inputs require $2^{n/2}+1$ multiplexers each accepting $2^{n/2}$ channels. This is preferable to the single mode, since it offers advantages with regard to output loading and simpler logic.

On the debit side, the multitiered multiplexer employs two cascaded linear gates in the signal channel.

Modular design of a multiplexer

Little and Capel have combined a multiplexer with an A/D converter,[48] replacing the analog switches by comparators. Fig. 3.53 shows the principle involved.

A number of analog channels are connected to the same number of comparators. The output from a single comparator is selected by the logic

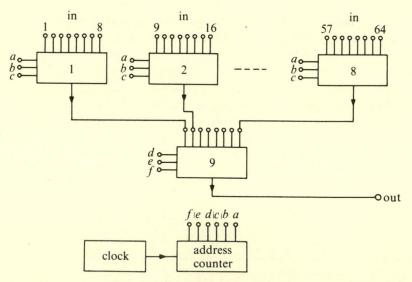

Fig. 3.52. Multitiered multiplexing.

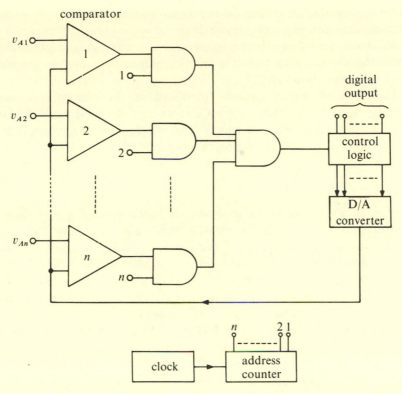

Fig. 3.53. Modular multiplexer. Adapted from [48].

and connected to the control logic of a closed-loop A/D converter (see Section 6.2), which converts the selected analog voltage into a digital number. Loading effects of the parallel connected comparators and possible interference pick-up by the analog feedback path should be considered in the design of such a system, which could be the preferred approach for certain applications.

3.2 Non-linear amplification and shaping

High reliability, improved accuracy due to a laser trimming, and increased versatility owing to novel processing techniques have led to a widespread use of ICs in non-linear circuit applications. Multipliers and dividers perform a variety of functions such as power monitoring, square rooting, generation of polynomial functions, and modulation, to mention just a few, and log and antilog circuits are employed to adapt the signal range to special needs. Most of these circuits operate on a variation of the transconductance principle, which is especially suitable for implementation by IC techniques owing to the superior matching characteristics available. Theoretical aspects and practical uses of these modules are well covered in the literature.[20,49]

Here we shall restrict ourselves to a discussion of comparators, a basic interfacing building block between the analog and digital world, and of baseline restorers, which are widely employed in nuclear electronics but have found only little application in other fields.

3.2.1 The comparator

The comparator serves as an interface between analog and digital instrumentation. Its output a_i may be 'high' (logic 1) or 'low' (logic 0), depending on whether the non-inverting input voltage is higher or lower than the inverting one (the indicated relationship between output level and logic state corresponds to positive true coding). The *threshold voltage* of a comparator is defined as the input voltage for which its logic output starts changing from high to low or vice versa.

Although in principle any amplifier could be employed as a comparator, the available output voltage swing is generally not directly compatible with the various logic families. Long propagation delay and slow recovery are two more drawbacks of most amplifiers in this application.

Because of this, the differential comparator has been developed and is today available as a standard component.

Specifications
Specifications for a typical high-speed comparator (AM 685 manufactured by Advanced Micro Devices) are given in Table 3.2.[50] This comparator is ECL compatible, i.e. the output swing is 800 mV.

The logic output of a comparator referred to its input is termed its *resolution*, because – disregarding speed effects – the threshold for input signals of positive or negative polarity will differ by this voltage difference. If we add input offset, temperature drift and the effect of power supply variations to this, we obtain the *region of uncertainty* as specified for the device under consideration – less than 3 mV for the AM 685.

Most comparators incorporate a strobe and/or latch function. A *strobe* forces the output of the comparator to one fixed state, independent of input signal conditions, whereas a *latch* locks the output in the logical

Table 3.2

Propagation delay	
(100 mV step, 5 mV overdrive)	7.5 ns max
Input offset voltage	±2.0 mV max
Average temperture coefficient	
of input offset voltage	10 μV/°C max
Input offset current	1.0 μA max
Input bias current	10 μA max
CM voltage range	±3.3 V min
CMRR	80 dB min
Supply voltage rejection ratio	70 dB min
Positive supply current	22 mA max
Negative supply current	26 mA max
Gain	$\geqslant$1600

Conditions: $\mathcal{T}_0 = 25$ °C, $V^+ = 6.0$ V,
$V^- = -5.2$ V, $R_L = 50\ \Omega$ to -2.0 V

Specifications for AM 685 comparator.

state it was in at the instant the latch was enabled. The latch can thus perform a sample-and-hold function, allowing short input signals to be detected and held for further processing. If the latch is designed to operate directly upon the input stage – so the signal does not suffer any additional delays through the comparator – signals only a few nanoseconds wide can be acquired and held. A latch, therefore, provides a more useful function than a strobe for very high speed processing.

A strobe holding the comparator output at zero may be employed to inactivate it for a selected time interval. Removal of the strobe in effect *arms* or activates the comparator, which will then respond to the next threshold crossing of the input signal.

The most difficult input signal for a comparator to respond to is a large amplitude pulse that just barely exceeds the input threshold. This forces the input stage of the comparator to swing from a full off (or on) state to a point somewhere near the center of its linear range. This exercises both the large-signal and small-signal responses of the stage. If the comparator has a specified delay under these stringent conditions, then it should be as fast or faster for any other circumstances. The industry standard measurement is with a 100 mV input pulse and an overdrive 5 mV above input theshold. Pulses larger than 100 mV might be used, but this would multiply measurement difficulties, since a percentage aberration of only a few tenths or ripple in the pulse generator waveform would be enough to affect the accuracy of the small overdrive seriously, and thus would give misleading results for the propagation delay.

Simplified analysis of performance

A simplified model simulates the small-signal response of a comparator reasonably well.

Assume the gain of the comparator to be zero below threshold and

$$A(s) = [A(0) \exp(-s\tau_D)]/[1 + sA(0)\tau_G] \qquad (3.71)$$

above threshold, where τ_D is the signal delay of the amplifier.

An input with a slope of $\dot{v}_g$ at threshold crossing yields for $t \geq \tau_D$ and $t \ll A(0)\tau_0$ a time response of

$$v_o(t) = \dot{v}_g(t - \tau_D)^2/\tau_0. \qquad (3.72)$$

Hence, the response time T_r required to reach a predetermined output level V_o, for a given derivative $\dot{v}_g$ of the input signal, equals

$$T_r = (V_o\tau_0/\dot{v}_g)^{1/2} + \tau_D. \qquad (3.73)$$

Equation (3.73) shows that the response time of a comparator is – apart from the signal delay – inversely proportional to $\dot{v}_g^{1/2}$.

Positive feedback is frequently applied to a comparator, and affects its performance in two ways: it introduces hysteresis, and improves its speed of response.

Hysteresis

Hysteresis is a form of non-linearity, in which the response of a circuit to a particular set of input conditions depends not only on the instantaneous values of those conditions, but also on the immediate past (recent history) of the input and output signals. Hysteretic behavior is characterized by an inability to 'retrace' exactly a particular locus of input/output conditions on the reverse swing.

Consider the circuit shown in Fig. 3.54, in which positive feedback is applied to a comparator's non-inverting input. Assuming $C = 0$, its response to a slowly rising and falling input voltage will be as shown in Fig. 3.55.

Although the large-signal gain of an amplifier is a highly non-linear function of the input voltage in addition to being frequency dependent, a simplified analysis is both easily comprehensible and adequate for a reasonable prediction of the actual performance. Noting that the voltage gain of a typical high-speed comparator is greater than 10^3, its resolution is better than 1 mV for an output voltage swing of 1 V. We shall presently assume the resolution to be infinite; hence, amplifier gain will be assumed to equal zero for $v_b \neq v_a$ and sufficiently high to initiate regenerative action (i.e. $A_v(0)R_1/(R_1 + R_2) > 1$), if $v_b = v_a$.

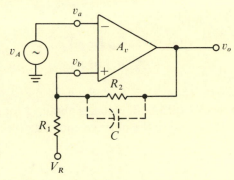

Fig. 3.54. Positive feedback applied to comparator.

Referring to Fig. 3.55, we observe that v_b may attain two distinct voltage levels, depending on the polarity of $(v_b - v_a)$ and the two saturation levels V_{o+} and V_{o-} of v_o:

If $v_b > v_a$, then $v_o = V_{o+}$ and

$$v_{b+} = (V_{o+}R_1 + V_R R_2)/(R_1 + R_2). \qquad (3.74)$$

If $v_b < v_a$, then $v_o = V_{o-}$ and

$$v_{b-} = (V_{o-}R_1 + V_R R_2)/(R_1 + R_2). \qquad (3.75)$$

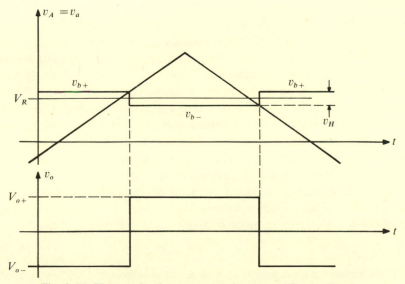

Fig. 3.55. Hysteresis of comparator due to positive feedback.

Practically $R_2 \gg R_1$, so that equations (3.74) and (3.75) can be simplified:

$$v_{b+} \simeq V_{o+}R_1/R_2 + V_R, \tag{3.76}$$

$$v_{b-} \simeq V_{o-}R_1/R_2 + V_R. \tag{3.77}$$

Finally, the *hysteresis voltage* v_H is defined as the voltage difference between the comparator threshold voltages obtained for an input signal, which crosses the threshold in the positive and negative direction. Hence,

$$v_H = v_{b+} - v_{b-} \simeq (V_{o+} - V_{o-})R_1/R_2. \tag{3.78}$$

The *reference voltage* V_R is the main factor determining the levels of comparison v_{b+} and v_{b-}, but accuracy of comparison between v_A and v_R also marginally depends on the stability of V_{o+}, V_{o-} and the ratio R_1/R_2.

Fig. 3.55 shows that, following a change in the logic output of the comparator, there is a certain dead region of width v_H which must be traversed by v_A before the logic output returns to its preceding state. This has important implications if noise is present in the input signal, which in absence of hysteresis can cause the output to change repeatedly between the two logic states during the time the signal is passing through the transition region of the comparator. This problem can be eliminated by hysteresis.

Note that, for $A_v(0)R_1/(R_1 + R_2) < 1$, v_H equals zero.

Speed-up by positive feedback
Positive feedback in addition to producing hysteresis, also increases the speed of response of a comparator. An accurate speed analysis should again take account of the highly non-linear characteristic of the amplifier in the threshold region, which requires a computer-aided analysis. Much insight into the typical behaviour of a regenerative comparator may, however, be gained employing a linear model.

Consider again the comparator shown in Fig. 3.54, with $C = 0$ and $A_v(s) = A(0)/[1 + sA(0)\tau_0]$, disregarding delay and parasitic poles. If the positive feedback factor is adjusted as $(R_1 + R_2)/R_1 = A(0)$, the comparator is on the threshold of instability and the transform of its dominant time responses becomes $A_f(s) = -1/s\tau_0$. Hence, the time response to a step of unit magnitude is

$$v_o(t) = -t/\tau_0. \tag{3.79}$$

If the speed-up capacitor C is connected as shown, the LT for $R_2 \gg R_1$ becomes $LT(s) \simeq +A(0)R_1(1 + sR_2C)/R_2[1 + sA(0)\tau_0](1 + sR_1C)$. The dominant pole at the frequency $1/A(0)\tau_0$ may be cancelled against the zero introduced by the feedback by making $R_2C = A(0)\tau_0$, and the gain can again be made infinite due to the positive feedback by choosing $(R_1 + R_2)/R_1 = A(0)$. Hence, for $A(0) \gg 1$, $R_1C \simeq \tau_0$ and the resulting transfer function of the positive feedback amplifier is, in Laplacean notation,

$$A_f(s) \simeq -A(0)(1 + s\tau_0)/s\tau_0(1 + s\tau),\qquad(3.80)$$

with $\tau = A(0)\tau_0$, and its time response to a step input of unit amplitude is

$$v_o(t) \simeq -A(0)\left[\left(\frac{\tau}{\tau_0} - 1\right)(e^{-t/\tau} - 1) + \frac{t}{\tau_0}\right]$$

which for $\tau/\tau_0 \gg 1$ can be simplified as

$$v_o(t) \simeq -A(0)\left(\frac{\tau}{\tau_0}e^{-t/\tau} + \frac{t - \tau}{\tau_0}\right) \simeq -\frac{t^2}{2\tau_0^2}.\qquad(3.81)$$

This time response is a parabola – a significant improvement compared with (3.79).

In conclusion it should again be stressed that the above treatment assumes that the amplifier is linear, and that passive circuit components are matched with rather ill-defined parameters such as $A(0)$ and τ. Still, the following conclusions are qualitatively valid in the practical case.

(a) If R and C are adjusted as described above, the output slope is significantly improved.

(b) In the above treatment, stray capacitance between the non-inverting amplifier input and ground has been disregarded. Its presence increases the value of C satisfying the above relationships, which can easily be taken care of by an empirical adjustment.

In practice, the feedback network is usually adjusted to make $(R_1 + R_2)/R_1 > A(0)$. This makes the circuit regenerative and further improves its speed. It also introduces hysteresis, a desirable feature in many cases.

Applications
Comparators are employed in A/D converters as decision elements, as core memory sense amplifiers, as single-ended or differential line receivers, to mention just a few applications. Monostable and astable multivibrators can be realized employing two crosscoupled comparators.

A useful property of comparators is their capability to perform AND and OR functions if their output is connected in parallel. With comparators whose output stage is an npn emitter follower, parallel connection of

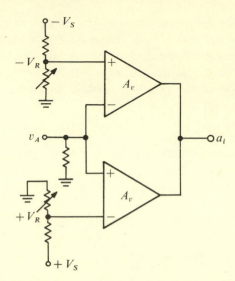

Fig. 3.56. Window comparator. Adapted from [51].

their outputs yields an OR function for positive true coding and an AND function for negative true coding. The opposite is true if the comparator output is obtained from the collector of an npn transistor.

This is utilized in the circuit of Fig. 3.56, which shows a window comparator,[51] capable of converting a sine wave into a train of narrow clock pulses. The corresponding waveforms are shown in Fig. 3.57. The output of the two ORed comparators is low only during the time the input signal is between the $-V_R$ and $+V_R$ voltages; otherwise the output is high. Note that the frequency of the resulting pulse train is twice the

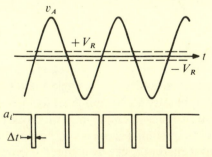

Fig. 3.57. Waveforms of window comparator. Adapted from [51].

frequency of the sine wave. The width of the pulse obtained, neglecting speed effects, is $\Delta t = V_R / V\pi f$, where V is the amplitude and f the frequency of the sinusoidal input. Equal spacing between the pulses depends critically on correct adjustment of the window by the reference voltages V_R. Different realizations of window comparators are described in reference [2].

The use of an operational amplifier as comparator
In spite of the availability of design-committed comparators, there are applications where the use of an IC operational amplifier is preferable because of the greater variety offered, lower cost and possibility of using a standard stock item.

One problem encountered in employing operational amplifiers as comparators is the thermal feedback from the output stage to the input differential transistor pair, which may cause considerable input offset drift if the comparator remains in one of the two overdriven states too long. This can be avoided in the circuit shown in Fig. 3.58(a), in which the full gain of the comparator is available near the level of comparison where $v_A = V_R$, but the output voltage swing is limited by feedback to less than 2 V peak-to-peak, due to the two diodes connected in the feedback path. Hence, the comparator remains permanently within the linear range of operation even for extreme signal amplitudes.

The feedback also allows us to compensate for equal input bias currents, by equalizing the resistance seen by each of the amplifier's inputs at cross-over, when both diodes are open-circuited. This kind of compensation cannot be applied to a comparator not employing negative feedback, since then the input stage is overdriven most of the time and hence one of the two input bias currents is zero. Consequently, the average bias current of the two inputs is in this case a function of the input signal shape, which is undesirable.

This circuit is quite satisfactory for relatively large inputs, but its response becomes sluggish if the signal derivative is low near cross-over. This is because the total capacitance C_T, due to the diodes in the feedback path, dominates the response when both diodes are near zero bias and limits the output derivative to

$$dv_o/dt = -(v_A - v_R)/R_g C_T. \tag{3.52}$$

The speed of response near cross-over can be significantly improved by the circuit shown in Fig. 3.58(b). From the equivalent diagram of Fig. 3.58(c), the output current i_o can be computed for the condition that the diodes are cut off and the total combined capacitance of each pair is again

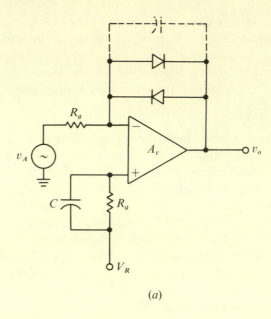

(a)

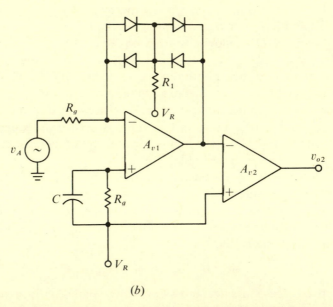

(b)

Fig. 3.58. Comparator whose output swing is limited by feedback: (a) basic circuit, (b) modification for improvement of switching speed, (c) equivalent diagram at cross-over, for $V_R = 0$.

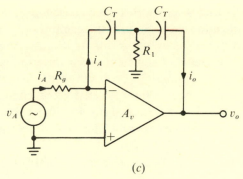

(c)

Fig. 3.58 (cont.)

C_T. The output derivative is now for $V_R = 0$

$$\mathrm{d}v_o/\mathrm{d}t = -(2v_A/R_gC_T)(1 + t/2R_1C_t), \qquad (3.83)$$

an improvement by a factor of $2 + t/R_1C_T$.

Assuming $C_T = 10^{-12}$ pF, $R_g = R_1 = 10^3\ \Omega$, $v_A = 1$ mV, the circuit shown in Fig. 3.58(a) yields a rise time of 1 V/μs. The improved circuit of Fig. 3.58(b) yields an initial rise time of 2 V/μs, which increases by a factor of $0.5 \times 10^3/\mu$s, indeed a significant improvement. In practice, the limit in improvement achievable by employing the circuit of Fig. 3.58(b) is set by the frequency response of the amplifier which has in the preceding treatment been assumed to have infinite gain.

A conventional design would limit the open-loop gain-bandwidth product of the amplifier to a value ensuring monotonic response for $B' = 1$, the latter being the case if one of the feedback diode pairs is heavily conducting. In practice, however, we do not require strict short-circuit monotonicity, since at cross-over the loop is actually open and closes only gradually as $|v_A - V_R|$ increases. Hence, it is good practice to adjust the gain-bandwidth product by trial and error. If the output swing of this comparator is too small, a second amplifier A_2 can be employed in cascade, as shown in Fig. 3.58(b).

A variety of comparator designs has been described by J. Graeme.[8]

Problem

3.13. Verify (3.83).

3.2.2 Baseline restoration

As its name implies, the baseline restorer restores the baseline of a signal to an arbitrary dc level. It is applicable in cases where the signal

amplitude, as measured from a well-defined base line, conveys information, but dc amplification throughout is undesirable.

The baseline restorer is, by nature of its dc restoring capability, a
differentiator. If a dc signal of sufficiently long duration is applied to its
input, the output must ultimately be restored to the reference level,
because absence of change is interpreted by the circuit as baseline.
Moreover, the differentiation is necessarily non-linear, since the average
output dc level – unlike in the case of a linear RC differentiator – can be
different from zero.

These properties make the baseline restorer suitable for the following
applications.

(a) If the signal is so small that the relative error introduced by the dc
drift of the following high-gain amplifier makes direct coupling prohibitive.

(b) If the dc level of the transducer output is not well defined. This is
the case with signals obtained from capacitive detectors, and with bioelectric signals.

(c) If the baseline is contaminated with low-frequency interferences
and/or low-frequency noise, the baseline restorer is capable of improving
the SNR, since by virtue of its differentiating capacity it attenuates the
low frequencies.

Open-loop baseline restorers
Consider Fig. 3.59, which shows a primitive baseline restorer, widely
employed in the era of vacuum tubes, known as a dc restorer. Assuming
an infinite C, an ideal diode D and a perfect baseline without undershoot,
the output dc level in absence of any signal is zero and remains at this level
if positive pulses are applied, since the charge which has accumulated
across C during the pulse immediately leaks away through D as soon as v_g
returns to zero. Deviation from an ideal characteristic mainly due to the
junction voltage of the germanium diodes available at that time was not
too serious in relation to the FS signal voltage of 100 V commonly
employed.

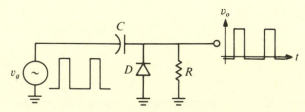

Fig. 3.59. The dc restorer.

The most significant step in the evolution of baseline restorers was the Robinson restorer,[52] which is shown in Fig. 3.60(a). This circuit has served as prototype for all further baseline restorer developments. We shall consider its response to an idealized signal v_g as shown in Fig. 3.60(b), which completely cuts off D_2 if $v_g = v_{g\,\text{max}}$. That response depends on the voltage droop $\Delta v_o = \tau_1(I_2 - I_1)/C$ whilst $v_g = v_{g\,\text{max}}$, as will be shown in the sequel.

The baseline restorer with C nearing infinity
If C is sufficiently large to make Δv_o negligibly small, we may disregard Δv_o and assume that C is infinite. We will compute the resulting baseline shift ΔV_C as shown in Fig. 3.60(b), assuming D_1 and D_2 to exhibit ideal diode characteristics, and $I_2 = 2I_1 = 2I_D$.

In the absence of any signal, with $v_g = 0$, each diode carries a current I_D, and hence the output voltage v_o equals V_R. If the pulse train $v_g(t)$, shown in Fig. 3.60(b), has been applied to the input long enough, the voltage ΔV_C across the capacitor C reaches steady state. For this condition, the following relationships are derived.

Whilst $v_g = v_{g\,\text{max}}$, the capacitor C discharges by a current $I_2 - I_1$ as shown in Fig. 3.60(c). After each termination of a pulse, v_o returns to the baseline and the capacitor C is recharged by a current ΔI as shown in Fig. 3.60(d).

Conditions for steady state require that the charge lost by C during the time interval $\tau_1(v_g = v_{g\,\text{max}})$ be equal to the charge recovered during the time interval $\tau_2(v_g = v_{g\,\text{min}})$: i.e.

$$(I_2 - I_1)\tau_1 = \Delta I \tau_2. \tag{3.84}$$

The current ΔI is created by ΔV_C, the voltage change across C, which in turn is caused by the discharge current $I_2 - I_1$ flowing when $v_g = v_{g\,\text{max}}$. Upon application of a pulse train, this discharge current gradually charges C until (3.84) is satisfied. The steady state relationship between ΔV_C and ΔI is obtained by noting that ΔI is the current created by the voltage drop ΔV_D across the dynamic resistance of the diodes D_1 and D_2. Since the dynamic resistance of a diode r_D equals V_T/I_D, where $V_T = k\mathcal{T}/q$, we obtain for $\Delta I_D/I_D \ll 1$

$$\Delta V_C = \Delta V_{D2} - \Delta V_{D1} \approx \Delta I_{D2} r_{D2} - \Delta I_{D1} r_{D1}$$

$$\approx \Delta I V_T\left(\frac{1}{I_2 - I_1} + \frac{1}{I_1}\right) = V_T\frac{I_2}{I_1}\frac{\Delta I}{(I_2 - I_1)}.$$

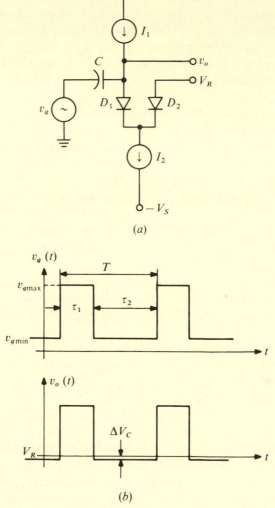

(a)

(b)

Fig. 3.60. The Robinson restorer: (a) circuit, (b) waveforms, (c) conditions for $v_g = v_{g\,\text{max}}$, (d) conditions for $v_g = v_{g\,\text{min}}$.

Finally, using (3.84),

$$\Delta V_C|_{\Delta I_D/I_D \ll 1} \simeq V_T \frac{I_2}{I_1} \frac{\tau_1}{\tau_2}. \tag{3.85}$$

Equation (3.85) shows that the baseline shift ΔV_C is a function of the ratios I_2/I_1 and τ_1/τ_2. The latter is preferably expressed in terms of the

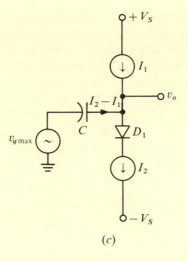

(c)

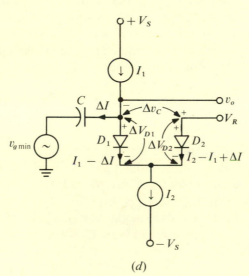

(d)

Fig. 3.60 (cont.)

duty ratio d, which is defined as the ratio between the signal duration and the period. Referring to Fig. 3.60(*b*),

$$d = \tau_1/(\tau_1 + \tau_2) = \tau_1/T; \tag{3.86}$$

hence

$$\Delta V_C \simeq V_T(I_2/I_1)d/(1-d). \tag{3.87}$$

Our previous assumption that $I_2 = 2I_1$ is important for the basic circuit shown in Fig. 3.60(a), since it eliminates any initial offset of v_0 in the absence of an input signal and makes the baseline shift equal to ΔV_C. If, however, ΔV_C is interpreted as the voltage difference between the baseline at the output in absence of any signal and in its presence, then (3.85) and (3.87) hold for any ratio I_2/I_1 as long as $\Delta I_D/I_D \ll 1$.

In view of its simplicity, the performance of the baseline restorer shown in Fig. 3.60(a), yielding for small duty ratios a baseline shift of the order of 10 mV, is remarkable. Its main drawback is the initial offset of the output, which is negligible only if $I_2 = 2I_1$. Circuits overcoming this restriction will be discussed at a later stage. Here we shall disregard this offset and consider the implications of the ratio I_2/I_1 on the performance of the basic circuit.

Limitations of the duty ratio
There is an intimate relationship between I_2/I_1 and the maximum applicable duty ratio d, because of the restriction that the amount of charge lost while the signal is present can never be greater than the maximum amount of charge which can be restored in the presence of the baseline. In effect, to prevent an undefined condition, it should be smaller.

Since ΔI can never be greater than I_1, it follows from (3.84) that

$$(I_2 - I_1)\tau_1 < I_1\tau_2, \quad \text{or} \quad d < I_1/I_2. \tag{3.88}$$

Hence, the ratio I_1/I_2 can always be suitably adjusted to accommodate any duty ratio as long as a well-defined baseline is still available.

This basic relationship holds for all variations of the Robinson restorer. Assuming, for example, $I_1/I_2 < \tau_1/(\tau_1 + \tau_2)$, the restorer in Fig. 3.60 will recognize $v_{g\,max}$ as baseline and the negative square pulse of duration τ_2 as signal. Hence $I_1/I_2 > \tau_2/(\tau_1 + \tau_2)$ satisfies (3.88) since for this case the duty ratio is defined as $\tau_2/(\tau_1 + \tau_2)$.

The fast baseline restorer
The preceding considerations relate to the case in which the capacitor C is sufficiently large to make its steady state voltage practically constant. Here we shall consider the case in which the capacitor C is sufficiently small to let the circuit return to its quiescent state after each application of a signal pulse. Fig. 3.61 shows the resulting output voltage, due to the same input as shown in Fig. 3.60(b), assuming ideal diodes D_1 and D_2.

For the time interval $t_a \leqslant t \leqslant t_b$, v_o droops by

$$\Delta v_o = -\tau_1(dv_o/dt) = \tau_1(I_2 - I_1)/C. \tag{3.89}$$

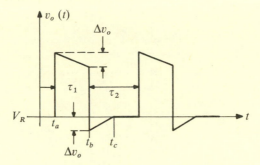

Fig. 3.61. Waveforms for fast baseline restorer.

At $t = t_b^+$, $v_o = -\Delta v_o$ and returns to V_R with a slope I_1/C. Hence,

$$t_c - t_b = \Delta v_o C/I_1 = \tau_1(I_2/I_1 - 1). \tag{3.90}$$

$t_c - t_b$ represents the dead time of the fast baseline restorer. For $I_2/I_1 = 2$, $t_c - t_b = \tau_1$ and the dead time following a signal pulse equals the pulse's duration. If the interval between successive pulses is greater than $t_c - t_b$, each pulse will rise from the baseline. Hence, neglecting secondary effects, baseline shift has been eliminated, at the cost of introducing dead time after each pulse.

The fast baseline restorer can be compared to an RC differentiator, with the exponential decay being replaced by a linear one and the possibility of choosing a different positive and negative slope, respectively. Hence the fast baseline restorer functions as a non-linear high-pass filter capable of suppressing slow baseline fluctuations, whose slope is smaller than that due to the baseline restorer.

Although the circuit shown in Fig. 3.60(a) serves well to demonstrate the basic properties and possible variations of the baseline restorer, it is in practice very little used because of the critical dependence of its offset upon the ratio I_2/I_1, and upon the identical characteristics of the diodes.

The active baseline restorer
The active baseline restorer has been developed in order to overcome the basic limitations of the passive circuit. In its simplest form[53] it consists of an operational amplifier suitably interconnected with diodes and current sources, as shown in Fig. 3.62. Placing the diodes D_1 and D_2 in the feedback path of the amplifier results in nearly ideal diode characteristics, since a very small signal is sufficient to completely cut off one of the two diodes. Furthermore, any offset due to mismatch between the diodes and the difference in the quiescent diode currents due to $I_2/I_1 \neq 2$ appears at

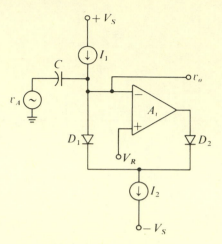

Fig. 3.62. Active baseline restorer.

the output of the amplifier, and the offset of the baseline equals the offset of the amplifier itself. With these modifications in mind, the response and restrictions of the basic Robinson restorer can be directly applied to the circuit of Fig. 3.62. Moreover, if the input is completely free from undershoots, the undershoot at the output of the active baseline restorer can be significantly shortened by replacing D_1 by a short circuit.

Closed-loop baseline restorers
Closed-loop baseline restorers delegate the restoring function to the feedback path. Clearly, since the baseline restorer acts as a non-linear differentiator, we expect the corresponding feedback counterpart to have a non-linear integrator in the feedback path.

A typical feedback baseline restorer is shown in Fig. 3.63(a) employing a basic transadmittance amplifier feeding a capacitor C as a non-linear integrator. By action of the feedback loop, the baseline is clamped to the reference voltage V_R. An input signal of several millivolts drives the transadmittance amplifier into saturation, charging or discharging C with a slope equal to $I_{\max}^+/C$ or $I_{\max}^-/C$, where $I_{\max}^+$ and $I_{\max}^-$ are the positive and negative saturation currents, respectively, of the transadmittance amplifier. Multiplication by $-R_f/R_1$ yields the corresponding slope appearing at the output of the amplifier. If the positive and negative saturation currents of the transadmittance amplifier are equal, its use for driving the capacitor C results in a symmetrical baseline restorer. An asymmetrical baseline restorer may be obtained if the transadmittance

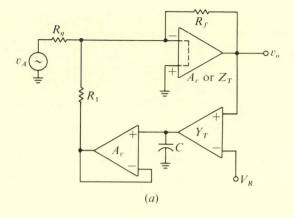

(a)

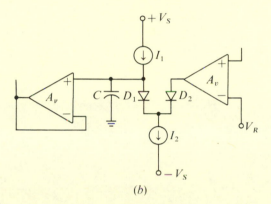

(b)

Fig. 3.63. Closed-loop baseline restorer: (a) using a transadmittance amplifier; (b) controlled asymmetry by voltage amplifier.

amplifier is replaced by a voltage amplifier, a diode pair and two current sources as shown in Fig. 3.63(b), with $I_2 \neq 2I_1$.

Note that the basic signal amplifier is not required to exhibit any degree of dc stability, since dc stabilization is performed by the baseline restorer. Hence, the choice of its realization as a voltage or transimpedance amplifier can be guided with complete disregard of its dc properties.

A recent medical instrumentation system has used the baseline restorer as an adaptive filter[54] for improving the SNR of an EKG. In the circuit shown in Fig. 3.64(a), the slope of differentiation is controlled by the baseline restorer control signal in proportion to the amplitude of the preceding R-level of the QRS wave. Since successive R-levels are correlated, this amounts to controlling the baseline restorer slope in

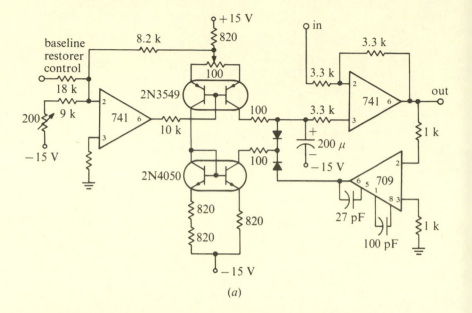

(a)

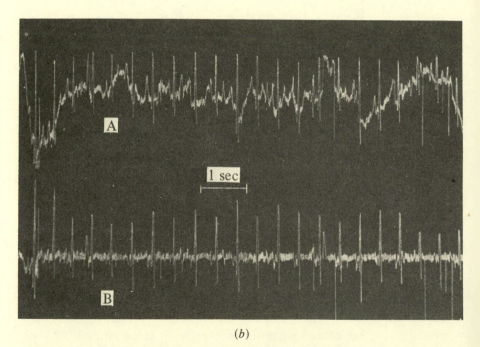

(b)

Fig. 3.64. (a) Controlled slope baseline restorer; (b) EKG (A) before and (B) after processing by controlled slope baseline restorer. Adapted from [54].

proportion to the signal amplitude and results in a better SNR than would be obtained with a constant slope baseline restorer. Fig. 3.64(*b*) shows the EKG before and after processing by baseline restorer.

The effect on the SNR
Baseline restoration necessarily modifies the transfer function, and therefore affects the SNR. This has been investigated by Radeka[55] for various types of baseline restorers.

4

Analog signal measurement in the time domain

4.1 Amplitude measurement

The information contained in a signal could be in its shape, or in its extreme values. A sample-hold module is applicable in the first case, and provides analog samples of the signal voltage at predetermined instants. A peak detector, applicable in the second case, measures the first maximum or minimum signal value occurring after it has been activated by a suitable mode control.

4.1.1 Sample-hold modules[1]

A *sample-hold module* is a device having a signal input, an output, and a control input. It has two steady-state operating modes: *sample* (or *track*), in which it acquires the input signal as rapidly as possible and tracks it faithfully until commanded to *hold*, at which time it retains the last value of input signal that it had at the time the control signal called for a mode change.

Sample-holds usually have unity gain and are non-inverting. The control inputs are operated by 'standard' logic levels. Logic 1 is usually the sample command and logic 0 the hold command.

In data acquisition systems, sample-holds are used either to 'freeze' fast-changing signals during conversion or to store multiplexer outputs while the last signal is being converted and the multiplexer is seeking the next signal to be converted. In analog data reduction, they may facilitate analog computations involving signals obtained at different instants of time. In data distribution systems, sample-holds are used to hold converted data between updates.

Imperfections of sample-holds
In the ideal sample-hold shown in Fig. 4.1, tracking is error free, acquisition and release occur instantaneously, settling times are zero, and hold is infinite. Commercially available units are specified by their departures from the ideal. We shall deal with these imperfections in the order they occur during a complete cycle of operation.

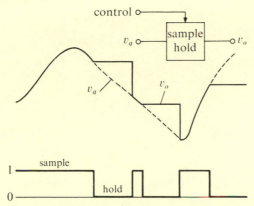

Fig. 4.1. Typical waveforms of an ideal sample-hold. Adapted from [1].

(*a*) *Tracking errors during sample.* In the sample mode, the module behaves as an amplifier or unity-gain follower. Thus, typical specifications are non-linearity, gain error, offset, frequency response, settling time, slewing rate, noise performance etc.

(*b*) *Sample-to-hold imperfections* (*see Fig. 4.2*). Switching action from sample to hold is not instantaneous. A measure of the speed of the sample-hold module is the *aperture time* τ_a, often rather loosely defined as the time required to switch from sample to hold. It is measured from the 50% point of the mode control transition (its nominal time of arrival) to the time at which the output stops tracking the input.

A finite aperture time introduces an *aperture error*, defined for a sample-hold module as the residual voltage difference accumulating

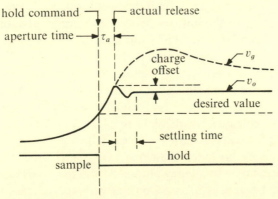

Fig. 4.2. Sample-to-hold imperfections. Adapted from [1].

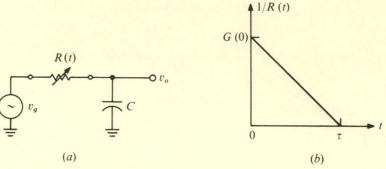

(a)   (b)

Fig. 4.3. (a) Simplified equivalent diagram for sample-hold. (b) Linear time trajectory for $G(t)$.

across the storage capacitor after arrival of the mode control, due to the slowness of the disconnect switch.

Goldstone and Hertz[2] have modelled a sample-hold module by the simplified diagram shown in Fig. 4.3(a), assuming either a linearly decreasing conductance as shown in Fig. 4.3(b), or an exponentially decreasing one.

Assuming a constant signal derivative $\dot{v}_g$ during switching time and making the reasonable assumption that $\tau \ll R(0)C$, the aperture error Δv can be evaluated for a linearly decreasing conductance as

$$\Delta v = \dot{v}_g \tau^2 / 6R(0)C \qquad (4.1)$$

and for an exponentially decreasing one as

$$\Delta v = \dot{v}_g \tau^2 / R(0)C, \qquad (4.2)$$

where $R(0)C$ is the charging circuit time constant, and τ the cut-off time of the linear decay or the time constant of the exponential one. The rms aperture error can be evaluated from (4.1) or (4.2), if the probability density function of the input signal derivative is known.

Usually manufacturers specify neither the time trajectory of the disconnect switch, nor the charging circuit time constant. The aperture error is frequently evaluated as $\Delta v = \dot{v}_g \tau_a$, which in itself is a rather pessimistic estimate when compared with (4.1) or (4.2). Furthermore, it is often stipulated that the aperture time must be small enough to permit the input data to change by no more than one LSB (*least significant bit*). The maximum permissible frequency obtained from this analysis is an unrealistic estimate, because it is based on the very special case in which all samples are taken near the zero crossing of a maximum amplitude sine wave. Actually, data are sampled continuously and only few will fall within the range of maximum slope, and when they are restored to a continuous signal in a data reconstruction device, the final error is much less than that inferred from such a worst-case analysis.[3] Evaluation of the rms error and of the error distribution function yields a much better estimate, but requires knowledge of the statistical signal properties and needs computational facilities which are not always available.

Signal processing assumes a known spacing between samples. Limitations in the timing accuracy create a jitter in the sampling period, the *aperture jitter*. In a sampled data system, a time error has an effect similar to that of an amplitude error, since in reconstructing the data, equal sampling intervals are assumed. Hence, the aperture jitter shows up as distortion or noise, depending on its statistical properties. The amplitude error due to jitter depends on the instantaneous signal derivative at the time of sampling, and equals the product of the two.

At the instant of actual release, a switching transient will occur at the output mainly due to capacitive feedthrough from the control circuit. This transient requires a *settling time* to be specified, measured from the instant of actual release and defined as the interval required for the output to attain its final value within a specified fraction of FS.

A *sample-to-hold offset* or *charge offset* occurs at the initiation of the hold mode, caused by 'dumping' of charge into the storage capacitor via the capacitance between the control circuit and the capacitor side of the switch. It can be partially compensated by coupling an out of phase signal through a compensating capacitor, but usually only under a given set of tightly controlled conditions.

(*c*) *Hold imperfections* (*see Fig. 4.4*). The *droop* during the *hold* period is defined as the drift of the output at an approximately constant rate caused by the flow of current through the storage capacitor: $dV/dt = I/C$. The current I is the sum of the leakage across the switch, the bias current of

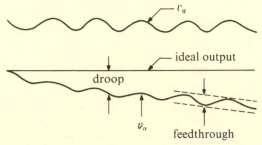

Fig. 4.4. Errors in hold. Adapted from [1].

the output stage, and leakage to the power supplies and to ground. In a well-designed unit, only the first is of any consequence.

A rough figure of merit for analog sample-holds is the ratio of droop time to settling time for the same percentage (FS) error. For example, a unit having settling time of 5 μs to 0.01% (1 mV for 10 V FS) and droop rate of 50 mV/s (0.02 s to 1 mV) would have a figure of merit of $20\,000/5 = 4000$ at 0.01% accuracy.

Another way of specifying droop is the charge-to-droop current ratio, which relates the slewing rate to the droop (the capacitor value cancels out). If I_{max} is the maximum charging current, C the value of the storage capacitor, and I_l the leakage current causing the droop, then $(I_{max}/C)/(I_l/C)$ is the ratio between slewing rate and droop rate and I_{max}/I_l the charge-to-droop current ratio.

This figure of merit does not require the settling accuracy to be specified. It is based on the facts (a) that settling time and rise time are related, and (b) that the rise time of the FS step response, for a well-designed sample-hold, should not be slewing rate limited, in order to prevent opening of the feedback loop. Charge-to-droop current ratios of 10^9 have been reported.[4]

Feedthrough is defined as the fraction of input signal that appears at the output in hold, caused primarily by capacitance across the switch. It is usually measured by applying an FS sinusoidal input at a fixed frequency (e.g. 20 Vpp at 10 kHz), and observing the output.

Dielectric absorption (soaking effect)[5] is the tendency of charges within a capacitor to redistribute themselves over a period of time, resulting in a

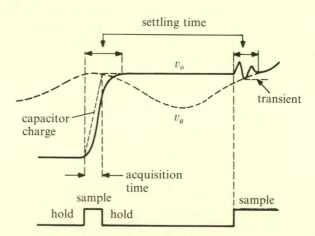

Fig. 4.5. Hold-to-sample errors shown qualitatively. Adapted from [1].

'creep' to a new level when allowed to rest after large fast changes. This is less than 0.01% of FS for good polystyrene and Teflon capacitors, but can be greater-than 1% for ceramic and mylar capacitors.

(*d*) *Hold-to-sample imperfections* (*see Fig. 4.5*). Of particular significance is the *acquisition time* of a sample-hold module, defined as the time required for the capacitor voltage to follow the input within a given error band, after the mode control is switched from hold to sample. For sample-hold units employing a single feedback loop comprising the capacitor charging circuits and the output stage (slow- and medium-speed units), acquisition time and settling time are identical, since both capacitor and output reach steady state simultaneously. High-speed units, however, employ a separate stage for charging the capacitor, and another as buffer between capacitor and output terminal, providing a low output impedance. For such units, the capacitor starts following the input earlier than the output; hence, the sample command can be terminated before the output reaches steady state, because the acquisition time is shorter than the settling time. This situation is shown in Fig. 4.5.

Hold-to-sample transients (shown only at the second transition from hold to sample) occur between the sample command and final settling. They are not too important for large charges, but can be crucially important in some applications if the spikes are large compared to the actual change (e.g. at constant input). Such glitch-like spikes may be due to limiting and other sources of dynamic disequilibria within the sample-hold circuit, or due to capacitive feedthrough from the control signals.

In addition to the above imperfections, a complete data sheet should specify input and output impedances, power supply requirements, and absolute maximum values for the input and operating temperature.

Typical designs
The choice of storage element divides sample-holds into two major classes. The more conventional, popular, and cheaper employ a capacitor for storage (analog storage). The other technique uses an A/D converter and a register for storage, and reads out via a D/A converter. This type of sample-hold is slower, more complex, and costly (especially where high accuracy or fast sampling are necessary), but it has the undisputed advantage of an arbitrary and essentially 'infinite' hold time.

4.1.2 Peak detection

Wherever peak amplitude is the parameter of interest in signal analysis, a peak detector is required. Examples are kinetic energy measurement by

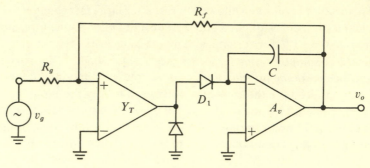

Fig. 4.6. Peak detector (simplified).

mechanical impact on a pressure transducer, or energy measurement in nuclear spectrometry.

Sample-hold circuits and peak detectors are closely related. Both employ a capacitor as memory element. But, whereas in the sample-hold circuit the capacitor is connected to the amplifier through an electronic bidirectional switch, in the peak detector it is charged through a diode – a unidirectional switch.

Fig. 4.6 represents a typical circuit.[6] Since the voltage across the capacitor C is proportional to the signal, the feedback-generated current through D_1 is necessarily proportional to the signal derivative. Hence, disregarding secondary effects such as charge storage in the diode and frequency dependence of the amplifier gain, the diode D_1 acts as a unidirectional switch which disconnects the capacitor upon reversal of the charging current, and the resulting voltage across the capacitor corresponds to the peak voltage for which the signal derivative equals zero.

A complete peak detector includes means (not shown in Fig. 4.6) for preventing the capacitor C to be further charged by successive signals after a peak amplitude has been stored, and for discharging it before the next peak detection cycle.

A survey of peak detection techniques has been published.[7]

4.2 Time measurement

In various signal processing systems such as nuclear spectrometers or radar, the time relationship between the occurrence of two signals is to be measured. A/D conversion based upon amplitude-to-time conversion is another example for the need of accurate time measurement. In these and related applications, a comparator serves to indicate the instant at which the signal voltage crosses a certain level of reference.

Errors in time measurement[8]

Let us first consider the errors occurring in the time measurement of a single level crossing. Naturally, any variation in the delay between the theoretically correct instant of level crossing and the actually obtained comparator output limits the ultimate accuracy of the time measurement. The two factors involved are *time jitter* due to noise, and a variable delay termed *walk* or *time slewing*.

Referring to Fig. 4.7 the standard deviation $\sigma(t_z)$ of the *time jitter* is related to the rms noise $\bar{v}_n$ of the signal channel as follows:

$$\sigma(t_z) = \bar{v}_n / \dot{v}_g(t_z), \qquad (4.3)$$

i.e. inversely proportional to the signal derivative $\dot{v}_g$ at the time t_z of level crossing.

Walk is defined to be the error in time measurement as a function of signal amplitude. It is also a function of the input waveform and the speed of response of the comparator. Fig. 4.8 demonstrates the response of a comparator to two identically shaped signals of different amplitude, from which two fundamental sources of walk may be recognized.

(*a*) Although the two signals shown were caused by events occurring at the same time t_0, the larger signal (v_{g1}) crosses the threshold V_R before the smaller one (v_{g2}). The resulting walk equals $t_{z2} - t_{z1}$.

(*b*) The second walk effect is, according to Section 3.2.1 a function of the signal derivative at the time t_z of level crossing, the gain-bandwidth product $1/\tau_0$ of the comparator, and the nominal output amplitude V_o of the comparator:

$$\Delta t = [V_o \tau_0 / \dot{v}_g(t_z)]^{1/2} \qquad (4.4)$$

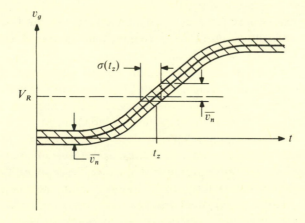

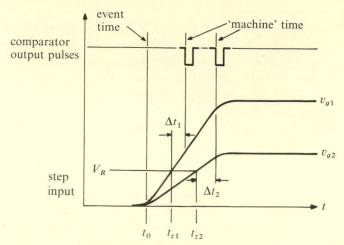

Fig. 4.8. Walk effect of leading edge timing. Adapted from [8].

In measuring the time interval between two events the contribution of these two factors must be carefully evaluated. The time jitter of the measured interval, due to noise, is a function of the rms noise of each level crossing signal, and of the crosscorrelation between the two. In this particular case, correlation will actually reduce the standard deviation of this measurement. Chapter 11 deals with this aspect in greater detail. Similarly, the error in the measured time difference due to walk of each level crossing will depend on the correlation between the two signal amr iitudes.

Timing methods for pulse signals[9]
In measurements where the event to be timed is represented by a signal pulse, three basically different measuring techniques may be employed, the choice depending on the signal properties.

The simplest but least accurate technique is *leading edge timing*. It employs a low-level comparator, adjusted for the maximum sensitivity at which its response to noise pulses is still negligible. Its response to pulses of constant shape but variable amplitude is shown in Fig. 4.8, and has been discussed in the preceding section. Clearly, accuracy of leading edge timing increases (since total walk decreases) as the level of comparison is lowered. The limit to this improvement is set by the probability of level crossings due to noise. In a more sophisticated scheme, the comparator is normally inactive and activated only shortly before the arrival of the signal.[10] In this case the probability of responding to noise pulses is

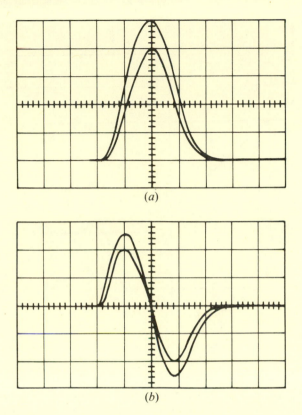

Fig. 4.9. Elimination of walk due to different signal amplitudes by zero cross timing of differentiated signals: (*a*) input, (*b*) differentiated signals.

reduced in proportion to the ratio between the *on* time and the average duty cycle of the comparator; thus a lower level of comparison can be used, resulting in reduced walk.

Whereas leading edge timing yields reasonably accurate results if the dynamic range of the signal is narrow (i.e. if the signal amplitude changes are slight), *zero cross timing* should be used with large dynamic pulse amplitude ranges provided the pulse shape is not a function of amplitude. In this technique, the signal pulses are differentiated as shown in Fig. 4.9 to produce a bipolar pulse with zero crossing. Since the instant of zero crossing represents the same phase point for all pulses irrespective of amplitude, the walk effect due to varying pulse amplitudes is eliminated. The walk that remains is due merely to the different signal slope, denoted Δt in Fig. 4.8. The problem of response to noise at zero signal level sensitivity is solved by employing a normally inactive comparator, and

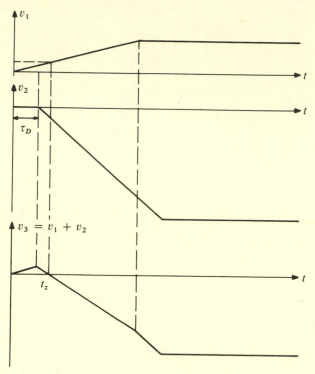

Fig. 4.10. Constant fraction timing method, idealized waveforms.

using the positive lobe of the differentiated pulse preceding the zero cross
for arming it.

The third and final timing method to be discussed is that of *constant
fraction*, applicable to pulses of variable rise time.[11] The method, shown
in Fig. 4.10 for a simplified pulse shape, adds the input pulse $v_1 = V_o t / t_1$ to
its own delayed, inverted and amplified version $v_2 = -A V_o (t - \tau_D)/\tau_1$.
The resulting sum pulse $v_3 = v_1 + v_2$ crosses zero at a delay $t_z =
A\tau_D/(A-1)$ from the extrapolated origin of the pulse, which is not a
function of the signal slope, but depends merely on the choice of the delay
τ_D and the gain A by which the inverted pulse is amplified.

4.3 Time-to-amplitude conversion[9]

TACs (*Time-to-amplitude converters*) are classic examples of judicious
conditioning of a signal before its A/D conversion. They measure time
differences by first converting them into amplitudes, which are
subsequently digitized by an A/D converter.

The operating principle of TACs is as follows. A capacitor is charged from a constant current source during the time interval Δt to be measured. The resulting increase in the capacitor voltage $\Delta v = I\,\Delta t/C$ is proportional to the time interval; this is stored by the capacitor for subsequent A/D conversion. After conversion is completed, the capacitor is discharged and ready for another measurement.

Commercially available TACs measure time differences up to 10^{-3} s FS. Their inherent noise contribution is less than 10^{-11} s rms, indeed a remarkable performance.

Problem

4.1. A signal of rise time τ_{RS} plus white noise of mean square value $\bar{v}_n^2$ are fed into a filter whose half-power frequency ω_h is to be adjusted (for accurate timing) such that $\bar{v}_{n_o}^2 \tau_{R_o}^2$ at its output is minimized, where τ_{R_o} is the rise time at the output of the filter. Assume that the ENB of the filter is $\omega_h/4$.

- (a) With τ_{RS} constant, find the half-power frequency of the filter satisfying the minimum defined above.
- (b) What is the physical significance of $\overline{v_{n_o}^2 \tau_{R_o}^2}$?
- (c) The conditions described above relate to the case in which a signal of rise time τ_{RS} is viewed on a scope, and the noise of the vertical amplifier is dominant; justify this statement.

5

Digital-to-analog converters

5.1 Introduction [1]

As their name implies, D/A converters convert digital information into analog form. They can be designed to accept a digital input in parallel or series form (including rate) and deliver an analog voltage or current whose relationship to the digital input may follow a linear or non-linear function. A D/A converter can also be used as feedback element in a closed-loop A/D converter, yielding a conversion characteristic which is the inverse of the D/A transfer function. This application will be discussed in Section 6.3.

Perhaps the most fruitful way of indicating the relationship between the analog and digital quantities involved is to plot a graph. Fig. 5.1 shows such a graph for an ideal three-bit D/A converter. A three-bit converter has eight discrete coded levels, thus a total of eight different inputs and eight corresponding outputs, ranging from zero to $\frac{7}{8}$ of FS. Since no other levels can exist with this coding, it is plotted as a bar graph. Note that the FS analog output is never reached, since the corresponding input code would require an additional bit.

5.2 Principles of digital-to-analog conversion

Fig. 5.2 demonstrates the basic principle of a four-bit binary coded D/A converter, with the input consisting of the binary coded word $N = a_1 a_2 \cdots a_n$, where a_1 is the MSB (*most significant bit*), a_n is the LSB (*least significant bit*), $a_j = 1$ or 0 signifies a closed or open switch, and V_R is the analog reference voltage. The output voltage is

$$v_o = V_R \sum_{}^{n} a_j 2^{-j}. \tag{5.1}$$

As an illustration of the voltage levels involved, assume $V_R = 10.24$ V, $n = 10$; this yields an LSB of 10 mV (assuming a one-to-one correspondence between V_R and FS).

The D/A converter shown in Fig. 5.2 employs *voltage switching*, in which the weighting resistors are connected either to V_R or to ground.

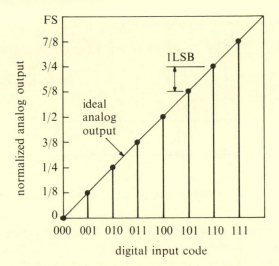

Fig. 5.1. Conversion relationships in three-bit D/A converter. Adapted from [1].

Because of the substantial voltage changes across the resistors, voltage switching poses a time-constant problem for output settling and hence affects adversely the settling time of the converter. A preferable arrangement is shown in Fig. 5.3, in which the switches operate between ground and the virtual ground of the operational amplifier. In this *current switching* arrangement, the current in the weighting resistors remains constant and therefore no parasitic capacitors need to be charged or discharged during switching. Furthermore, switch design is simplified, since all switch points are effectively maintained at the same potential.

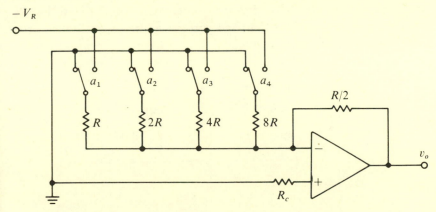

Fig. 5.2. Voltage switching binary ladder network. Adapted from [3].

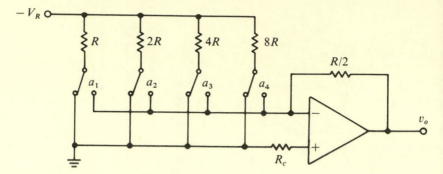

Fig. 5.3. Current switching binary ladder network. Adapted from [3].

The converters shown in Figs. 5.2 and 5.3 are impractical because of the large ratio in resistance values; temperature tracking between resistors deteriorates as their ratio increases. For a ten-bit converter, the largest resistor ratio would be $1:500$, in which case reasonable temperature tracking becomes impracticable, quite apart from the fact that precision resistors in the megohm range are very difficult to produce.

The practical answer to this problem is the $R-2R$ network, shown in Fig. 5.4 for the current switching variety. The $2R$ resistor between $-V_R$ and the network is not essential, but ensures that all switched resistors 'see' the same impedance R. This has a beneficial effect on the transient response of the D/A converter, but reduces the effective reference voltage by a factor of three.

Another current switching scheme is shown in Fig. 5.5. The voltage amplifier adjusts the collector current of the associated transistor to V_R/R_1. The excellent matching between transistors available by IC techniques ensures that all collector currents closely approximate V_R/R_1, yielding closer tolerances than the corresponding *on* resistance error encountered in voltage switching.

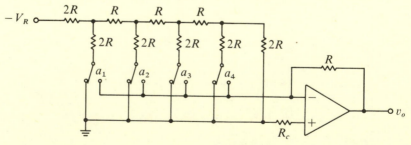

Fig. 5.4. $R-2R$ ladder network employing current switching. Adapted from [3].

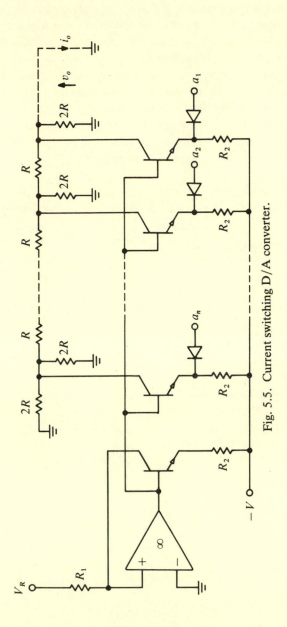

Fig. 5.5. Current switching D/A converter.

The network shown in Fig. 5.5 yields either voltage (output open-circuited, connected to a voltage follower), or current (output connected to the virtual ground of an operational amplifier). For highest speed, the voltage output is preferable, since the response of a voltage follower is faster than that of an inverting amplifier.

An important property of this network is its constant output impedance, which has a beneficial effect on the dc drift due to the offset current of the output amplifier. If the output impedance is a function of the digital input to the resistor network, as is the case for the network shown in Fig. 5.4, then the resulting dc offset will be also be a function of the digital input – clearly an undesirable property.

Integrated D/A converters

Conventional $R-2R$ techniques are unsuitable for realization of fully integrated D/A converters. Charge is a more suitable signal parameter than current, particularly if implemented by MOS (*metal oxide silicon*) technology, in which capacitors may be realized at accurate ratios and the MOS devices, serving as charge switches, have inherently zero offset voltages.

The *charge redistribution* technique previously used with discrete devices[4] has now been employed in the design of fully integrated A/D and D/A converters. Two related schemes have been proposed, both based on the principle of charge redistribution: the first uses $(n+1)$ binary weighted capacitors and the same number of switches for n-bit conversion,[5] whereas the second uses only two capacitors of nominally equal value and three switches.[6]

Conversion is inherently monotonic for the second scheme, even if the capacitance ratio differs from unity. Fig. 5.6 shows its basic diagram, and Table 5.1 explains its operation. The decoding process consists of two

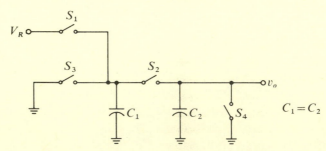

Fig. 5.6. Charge redistribution D/A conversion.

Table 5.1

j	a_j	t	S_1	S_2	S_3	S_4	Q_{C1}	Q_{C2}
n	1	t_1	**1**	0	0	**1**	$1 \times V_R C$	0
		t_2	0	**1**	0	0	$1 \times V_R C/2$	$QC_1(t_2)$
$n-1$	0	t_3	0	0	**1**	0	$0 \times V_R C$	$QC_1(t_2)$
		t_4	0	**1**	0	0	$0 \times V_R C/2 + 1 \times V_R C/4$	$QC_1(t_4)$
$n-2$	1	t_5	**1**	0	0	0	$1 \times V_R C$	$QC_1(t_4)$
		t_6	0	**1**	0	0	$1 \times V_R C/2 + 0 \times V_R C/4 + 1 \times V_R C/8$	$QC_1(t_6)$

steps per bit and starts with the LSB. The table demonstrates the conversion of 101 into an analog voltage V_o and is self-explanatory.

Both kinds of D/A converters require n redistributions of two steps in decoding a digital number to an analog voltage, and can be used in an A/D converter with the addition of a voltage comparator and sequencing logic. The two-capacitor scheme, however, requires $n(n+1)$ redistributions for n-bit A/D conversion, since, in contrast to D/A conversion, the MSB must be determined first, and for the A/D conversion of each kth bit the D/A converter must go through a complete cycle consisting of k redistributions.

A different approach has been chosen by Van de Plassche,[7] with the high accuracy $R-2R$ ladder network replaced by a time division concept which requires no trimming and is insensitive to element aging. An experimental twelve-bit D/A network was built on this basis and higher accuracies are envisaged in the future.

5.3 Errors in digital-to-analog converters

5.3.1 Conversion errors

Fig. 5.7 demonstrates the *scale factor* or *gain error*, due to deviation of the reference voltage V_R from its nominal value.

Fig. 5.8 demonstrates the *offset error* (also called *zero intercept* or *zero displacement error*), caused mainly by the offset voltage and current of the amplifier following the converter. One should distinguish between an offset error and purposely used level splitting, which is usually employed with converters designed for offset binary or two's complement codes (see Section 6.5).

Fig. 5.9 demonstrates the effect of *non-linearity* on the transfer characteristic of a D/A converter. This may occur due to mismatch of the weighting resistors and switch imperfections, or current mismatch in the

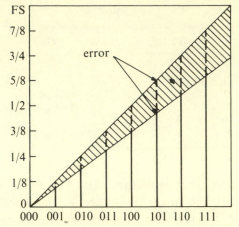

Fig. 5.7. Scale factor (gain) error. Adapted from [1].

case of the D/A converter shown in Fig. 5.5. Non-linearity due to the output amplifier should be negligible in a well-designed converter.

Deviation from linearity can be specified as integral or differential linearity. *Integral linearity* is usually defined as maximum deviation of the actual input–output characteristic from an ideal straight line, expressed as a percentage of FS. *Differential linearity* is measured by the maximum deviation of the actual quantum values from the average quantum value, the latter being defined as the specified (but never reached) FS analog output divided by 2^n.

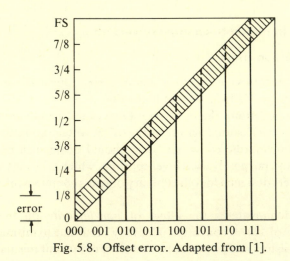

Fig. 5.8. Offset error. Adapted from [1].

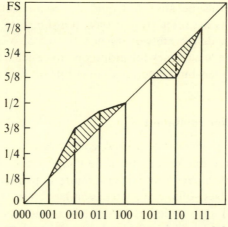

Fig. 5.9. Non-linearity. Adapted from [1].

A *monotonic* D/A converter relates an increasing analog output to every increasing bit of the input code. With a non-monotonic converter, as shown in Fig. 5.10, analog output decreases at some points with increasing input code. A D/A converter is potentially non-monotonic if the sum of errors for any given number of successive lesser significant bits is greater than one quantum level minus the maximum error produced by the next most significant bit. Hence, a non-monotonic response is most likely to occur when the MSB changes from 0 to 1 and all others from 1 to 0. In a closed-loop control system, a non-monotonic D/A converter

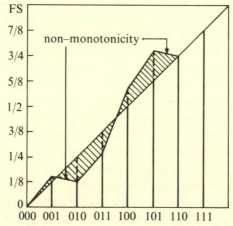

Fig. 5.10. Non-monotonicity. Adapted from [1].

could be disastrous because it would introduce 180° of phaseshift. Non-monotonicity is caused by excessive non-linearity.

Some designers deliberately overspace the analog weight at the major carry points to trade accuracy for monotonicity. Such a trick may reduce the accuracy of a ten-bit converter to nine bits or even fewer.

5.3.2 Temperature coefficient

The *temperature coefficient* of a D/A converter is a function of several factors such as the degree to which the temperature coefficients of the resistors differ from each other, temperature dependence of on and off resistance and offset voltage of switches and, of course, any temperature dependence of V_R, to which v_o is directly proportional. This proportionality may, however, also be a blessing in disguise, since it is the basis of the multiplying feature of D/As (see Section 7.2). The effect of power supply variations on the amplifier is generally sufficiently reduced by the CMRR to pose no problem.

The temperature coefficient of a D/A converter will be specified by the manufacturer. It must be sufficiently low to keep the total variation of v_o well below half an LSB over the full range of operating temperatures.

5.3.3 Speed errors

The speed of a D/A converter is a function of the *switching transients* or *glitches*, and of the output amplifier *settling time*. Glitches are analog spikes which occur at code transitions due to switch timing discrepancies. With natural binary coding, the largest glitches occur when the MSB turns on and all the others turn off. They affect adversely the settling time of a D/A converter, and when the converter is a component in automatic test equipment, ICs under test might give a response which could easily be interpreted as malfunctioning, or might even be damaged by glitches.

The width of glitches can be reduced by improving the symmetry of the switching transients. But the lack of switch perfection (stored charge and difference in the time constant of the on and off transients) causes them invariably to turn faster on than off. Moreover, the logic which generates the input code also has transition assymmetry problems, which it passes on to the D/A converter, thereby increasing the duration of the glitches.

A radical solution to this problem is the connection of a gated sample-hold circuit to the analog output of the D/A converter, at the price of increasing the conversion time. This principle is employed in 'deglitched' converters, such as the Model 4002 fourteen-bit D/A converter manufacturered by Teledyne Philbrick.

Manufacturer's specifications for D/A settling time do not, in general, take account of output loading effects: capacitive loading of the D/A output amplifier will add another time constant in conjunction with the output resistance of the amplifier, and may even render the amplifier response non-monotonic.

5.3.4 Errors in the recovered analog signal

An analog signal derived from a D/A converter contains all errors and false information, due to aliases introduced by analog signal processing and the following A/D conversion from which the digital signal has been originally derived, plus those due to the D/A conversion itself. This may seem a superfluous statement, but is frequently overlooked.

An ideal D/A converter should deliver at its output a train of impulses, each possessing an area proportional to the digital number it represents. A practical D/A converter creates a staircase-like waveform. Hence, it may be considered as an ideal impulse-producing converter followed by a linear filter whose impulse response is a rectangular pulse of amplitude proportional to the area of the impulse and width equal to $\tau = 1/f_c$, the inverse of the clock pulse frequency of the D/A converter. Such a filter is termed a *zero order hold* or *zero order recovery filter*, and the Laplace transform of its normalized impulse response is

$$H(s) = (e^{s\tau/2} - e^{-s\tau/2})/s\tau. \qquad (5.2)$$

The spectrum of (5.2), commonly known as the sampling function, is obtained by substituting $j\omega$ for s, yielding the Fourier transform of a rectangular pulse:

$$|H(j\omega)| = (\sin \pi f \tau)/\pi f \tau. \qquad (5.3)$$

Equation (5.3) is plotted in Fig. 5.11. Accordingly, a zero order hold acts as a low-pass filter, whose first zero occurs at $f = 1/\tau$ with relatively little energy being transmitted beyond. At frequencies $f \ll 1/\tau$, the relative amplitude error can be approximated by

$$\varepsilon = (\pi f/f_c)^2/6. \qquad (5.4)$$

The characteristic of the recovery filter is affected by glitches, which modify the shape of the staircase-forming rectangular pulses. An analytical treatment of this effect is impractical, but its quantitative evaluation can be based either on actual measurement or on a computer simulation.

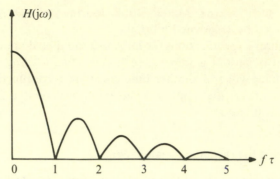

Fig. 5.11. Frequency attenuation due to linear gate of width τ, for $f \geq 0$.

Recovery filters

Glitches can be avoided by sampling the D/A output after it has reached steady state and holding it until the next updating time. The 'diglitched' D/A converter mentioned in the preceding section utilizes this principle. Such a sample-hold circuit is in effect a zero order recovery filter, also termed a *boxcar detector* due to its bumpy response. Higher order recovery filters provide greater accuracy of the recovered analog signal. Such filters predict the output resulting from the convolution of a series of inputs: a first order filter extrapolates the rate of change between the two preceding samples into the next sampling period, a second order filter makes use of three samples to predict also the second derivative, and so forth. The limit in the order of the recovery filter is set by the correlation time of the signal spectrum, which restricts the samples utilized in the prediction to those for which some degree of correlation still exists.

In many cases, the recovery filter is realized digitally, i.e. it is placed in front of the D/A converter. Realization of recovery filters or *data extrapolators* by open-loop and closed-loop techniques has been treated by Monroe.[8]

Problems

5.1. What is the purpose of the resistor R_c in Figs. 5.2, 5.3 and 5.4?

5.2. The value of R_c in Figs. 5.3 and 5.4 is not unique. Suggest a modification of the circuit shown in Fig. 5.4 to make it unique.

5.3. Assume that a voltage follower is connected to the output of the R–$2R$ network shown in Fig. 5.5. Draw a diagram including a resistor R_c and indicate its value.

5.4. Assume the D/A converter shown in Fig. 5.5 to exhibit three bits. Assign to the binary ladder resistors of value R, starting at the MSB, indices a and b, and those of value $2R$, again starting at the MSB, indices c, d, e and f. The network is connected to the virtual ground of an inverting amplifier, whose feedback resistor is R_g.

The nominal resistor values are $2R_a = 2R_b = R_c = R_d = 2R_g = 2R_h$, where R_h replaces the two resistors R_e and R_f. The actual resistor values are $R_j = R_{j0}(1 + \delta_{R_j} + \delta_{\mathcal{T}_j} \Delta \mathcal{T})$, $j = a, b, \ldots, h$, where δ_{R_j} denotes the relative resistor tolerance and $\delta_{\mathcal{T}_j}$ the temperature coefficient of resistor j. The current from all current sources equals exactly I. All binary weights $a_j = 1$, $j = 1, 2, 3$.

(a) Assume $\Delta \mathcal{T} = 0$, and all resistors to deviate from their nominal value by $+200 \pm 10$ ppm (i.e. $\delta_{R_j} = 2 \times 10^{-4} \pm 10^{-5}$). The errors of ± 10 ppm in the resistor values are uncorrelated. For these conditions find the total error in the output voltage. Note that uncorrelated errors add as $(\sum_j \varepsilon_j^2)^{1/2}$.

(b) Assume $\delta_{R_j} = 0$, and $\delta_{\mathcal{T}_j} = -10^{-5} \pm 10^{-6}$. Deviations from the nominal temperature coefficient are uncorrelated for the various resistors. Find the total error in the output voltage as a function of $\Delta \mathcal{T}$.

6

Analog-to-digital conversion techniques

6.1 Introduction[1]

We live in a truly analog world. Data taken from anything tested or measured are usually in analog form and are thus difficult to handle, process, and store for later use without introducing considerable error. If taken from a large number of sources, data accumulate at such a rate that they become a burden and a major problem to the laboratory running the test. A digital computer can process such data at rates comparable to those at which they were produced; however, the data must first be converted into a form compatible with digital processing. An A/D converter fulfills this need.

Although a pure analog system is basically more accurate than an analog–digital system because of the continuity of analog signal processing, this accuracy is rarely realizable. Furthermore, results are presented in a form that cannot be easily read, recorded, or interpreted with high accuracy. Digital data, however, are readily presented in numerical form regardless of the number of bits, and are just as easily manipulated, processed, and stored. Once data are converted into digital form they may be processed mathematically, sorted, analyzed, and used for control much more accurately than analog data. If data must be handled much after they are acquired, it is safer to digitize them because there is little chance of error accumulation in successive manipulation. Further, digital storage can be safer since it may be in non-volatile form (e.g. core memory, magnetic tape).

The applications of an A/D converter are almost unlimited. As the state-of-the-art in semiconductor technology advances, the cost of an A/D conversion system will continue to drop, and more system designers will be able to use A/D converters economically in places where they were formerly impractical. A few recent uses include: space telemetry systems, all digital voltmeters, voice security systems, closed-loop process control systems (i.e. chemical plants, steel mills, etc.), and inflight check-out systems (to code the output of sensors so that a small computer on board can process the information). Hybrid computers use A/D and D/A

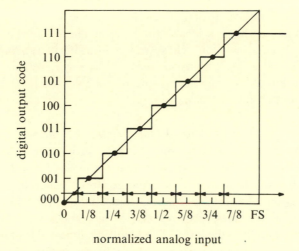

Fig. 6.1. Ideal conversion relationship in three-bit A/D converter. Adapted from [2].

converters as a means of interfacing analog and digital computers to solve large-system simulation problems. These applications indicate the versatility but represent only a small portion of the uses of A/D converters.

Basic principle of A/D conversion
An ideal three-bit A/D conversion relationship is shown in Fig. 6.1[2] for the *level splitting* version of encoding in which transitions occur in the middle of the nominal quantization levels. Since all values of the analog input are presumed to exist, they must be quantized by partitioning the continuum into eight discrete ranges. All analog values within a given range are represented by the same digital code, which corresponds to the nominal midrange value. These midrange values correspond to the bar heights of the D/A conversion chart shown in Fig. 5.1.

There is, therefore, in the A/D conversion process, an inherent quantization uncertainty of $\pm\frac{1}{2}$LSB in addition to the conversion errors analogous to those existing for the D/A converter. The only way to reduce this quantization uncertainty is to increase the number of bits. Note that $0\pm\frac{1}{2}$ of a quantum level corresponds to code 000 (non-splitting encoding), and that FS analog input (i.e. 8/8) is an overrange signal.

Design considerations
The decision when to convert from analog to digital signal processing requires a thorough understanding of analog, digital and

conversion techniques available, their limitations regarding speed, accuracy and – last but not least – their price. There are many ways to accomplish A/D conversion. These range from very slow inexpensive techniques to ultra-fast expensive ones.[3,4] The variety of conversion techniques is limited only by the ingenuity of the designer. Here we shall discuss the more popular techniques.

Two choices: open-loop and closed-loop (feedback) converters
A/D converters can be divided into two basic groups: open-loop types and feedback types. The open-loop converter generates a digital code directly upon application of an input voltage, and is generally an asynchronous operation. Most feedback converters, on the other hand, generate a sequence of digital codes, reconvert (D/A) each one in turn to an analog value, and compare it to the input. The resulting digital output will be the closest value of the reconstructed analog voltage compared to the real analog voltage.

6.2 Open-loop techniques

6.2.1 Voltage-to-frequency converters employing a voltage controlled oscillator

A basic diagram of an A/D converter employing a VCO (*voltage controlled oscillator*) is shown in Fig. 6.2. The VCO produces a signal whose frequency is ideally a precise linear function of the input analog voltage.

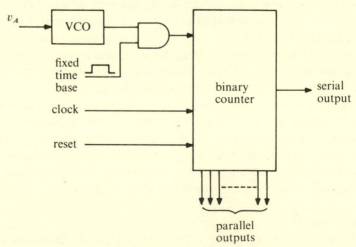

Fig. 6.2. V/F encoder. Adapted from [1].

VCOs employing feedback for their own linearization are available as self-contained ICs (see Chapter 7). The fixed time base input is a digital pulse whose on time is known and very precise. An accurate delay circuit will produce a pulse of this type. The leading edge of the fixed time base input will gate the VCO frequency into the binary counter. The output from the VCO must be properly shaped in the form of a pulse train. The counter is initially reset and could be one of a large number of designs, depending on the particular problem at hand. After the counter has had the oscillator frequency applied to it for the duration of the fixed time base input, the trailing edge of this signal will inhibit (gate off) the counter, and its contents will be a digital representation of the analog input voltage.

Because of the fixed time base, this technique has a predetermined conversion time, which is equal to its aperture time. The output code will represent the average of the input voltage during the aperture time. Parallel digital and discrete pulse serial outputs are readily available, but a binary coded serial output will require a parallel-to-serial conversion to be done after the V/F conversion is finished.

Accuracy limitations in this technique are due to non-linearities in the VCO over the entire analog input range, and the precision required in the time base signal pulsewidth. Neither of these are trivial problems, especially if the system is to operate in other than a controlled environment.

6.2.2 Converters employing a voltage-to-pulsewidth element

The constant and variable items in the previous method are interchanged in this method. There is now a fixed precise compensated clock and the system must generate a pulsewidth that is a linear function of the input analog voltage. This allows the clock pulses to be counted for a time proportional to the input analog voltage.

To obtain this variable linear pulsewidth, one can design a voltage controlled monostable and count pulses derived from a fixed frequency clock by a binary counter for the variable time thus created, as shown in Fig. 6.3. At the end of the gating pulse, the binary counter has accumulated the binary coded word N as defined in Section 5.2, and satisfies in accordance with (5.1) the relationship implicitly given by

$$v_A = V_{max} \sum_{j=1}^{n} a_j 2^{-j}, \tag{6.1}$$

where v_A is the unknown analog voltage, and V_{max} is the FS analog input voltage.

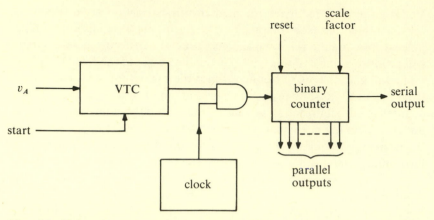

Fig. 6.3. Voltage-to-pulsewidth encoder. VTC: voltage-to-time (pulsewidth) converter. Adapted from [1].

The aperture time for this technique is equal to the clock period, and the conversion time is proportional to the input (v_A) magnitude and the clock period, but increases exponentially with the number of bits. The number of bits may have to be limited to prevent the conversion time from becoming too long (alternatively, the clock frequency can be increased to reduce conversion time). One obvious limitation in the accuracy of this approach is the linearity of the analog portion.

De Lotto *et al.*[5] have described a highly linear voltage-to-pulsewidth encoder converting signal amplitudes between 0 and 5 V into pulses of width between 0.4 and 1.2 μs. These time intervals are measured by a time sorter yielding 4000 quantization levels for the total variable range of 800 ns, delivered by the encoder (200 ps resolution).

Another encoder falling within this category is the *Wilkinson ramp* A/D converter,[6] developed in 1950 for use in nuclear spectroscopy. This converter measures pulse amplitudes by charging a capacitor to their peak and measuring the discharge time due to a constant current with a digital clock. Maximum conversion time is proportional to the number of quantization levels, yielding approximately 10 μs for a 100 MHz clock frequency and ten-bit accuracy.

A related method is employed in the *sawtooth ramp* encoder whose basic block diagram is shown in Fig. 6.4. Initially, the binary counter is cleared before arrival of the start signal at time t_1. The start signal causes the ramp generator (a linear sweep generator) to produce an increasing output and sets the *R–S* flip-flop which causes the AND gate to pass the clock pulses. The sweep generator starts at 0 V and increases at some

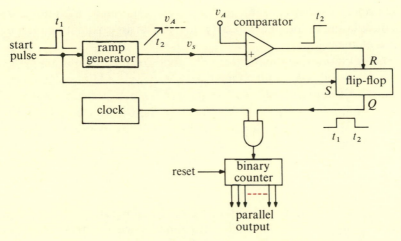

Fig. 6.4. Sawtooth ramp encoder. Adapted from [1].

precise known rate. While it is increasing, the binary counter counts the number of clock pulses. At some time t_2 $(t_2 > t_1)$ the sweep generator output voltage (v_s) crosses the level of the input analog voltage (v_A), causing the comparator to switch states, which in turn resets the R–S flip-flop causing the clock to be inhibited from the counter. The output digital count is proportional to v_A.

Limiting factors in these two methods are: (a) linearity of the ramp generator; (b) input offset in the voltage comparator; (c) output slew-rate of the comparator; (d) frequency limitations in the logic devices; (e) accuracy in clock generator frequency.

For a medium-range medium-accuracy slow-speed system, these techniques prove to be quite inexpensive. Four-bit and six-bit systems of these types have found applications in space-type vehicles and have performed very well. In nuclear spectroscopy, the Wilkinson ramp converter has been brought to a high level of sophistication, yielding twelve-bit accuracy and 1% differential linearity.

One advantage these encoders have over related counter type feed-back converters is that no D/A converter is required to generate the voltage to be compared to the input analog voltage. From an economic viewpoint, this is a very important aspect. Another advantage is the fact that a linearly increasing voltage is applied to the comparator's inputs which puts less of a strain on the comparator circuit than, say, the successive approximation technique where the comparator is receiving large input voltage changes in different directions and asked to resolve small differences. In general, all smooth conversion techniques operate at a considerably shorter time per step, are inherently monotonic and produce better differential linearity than digital approximation methods, but are much slower.

The sawtooth ramp method is very good for high-resolution systems since, as the number of bits is increased, very little additional circuitry is needed. However, N-bits will require 2^N clock periods for one FS conversion, although average conversion time will be closer to half this number. As more resolution is required, the conversion time increases very rapidly.

6.2.3 Integrating converters

As their name implies, integrating converters integrate the analog signal and thereby convert amplitude into time. Since the resulting digital information represents the value of the analog signal averaged over the interval of integration, the main application of integrating converters is for relatively slowly changing signals, which are quasi-static during each conversion. Practical examples of such signals are temperature, pressure, humidity and voltage or current as measured by panel meters. Apart from their simplicity and accuracy, integrating converters have the following properties.

(a) Conversion is inherently monotonic.

(b) Rapid signal fluctuations due to noise and interferences are averaged out and thereby attenuated.

(c) Since the integral of a sine wave vanishes over an integral number of periods, response to frequencies $f = n/\tau_i$ is zero (n is an integer and τ_i denotes the integrating period which also equals the aperture time in the case of an integrating converter). This is important for integrating converters with constant aperture time, which may be made insensitive to interference pick-up by synchronization with the interfering frequency (usually the mains frequency). Rejection ratios of 40 dB will be obtained if synchronization is accurate within a tolerance of 1%, usually achievable with reasonably simple techniques. Higher rejection ratios are possible using more sophisticated techniques for synchronization.

(a) *V/F converters*

A simple form of integrating converter is shown in Fig. 6.5. The analog input v_A is integrated, and the resulting ramp is reset to zero by S_1 each time it reaches the reference voltage V_R. Compared with the linear ramp analog-to-pulsewidth converter, V_R and v_A are exchanged.

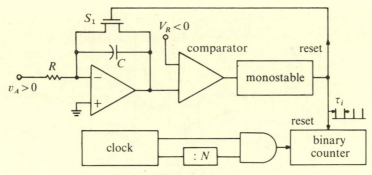

Fig. 6.5. Integrating converter.

Each conversion yields a time interval τ_i which is inversely proportional to $\overline{v_A}$, the analog input voltage averaged over τ_i. Two possible
methods may be applied to obtain a digital read-out.

The time interval τ_i may be measured by a digital clock, and $\overline{v_A}$ can be
computed as

$$\overline{v_A}/V_R = RC/\tau_i \qquad (6.2)$$

which is not a convenient solution.

Equation (6.2) indicates, however, that $\overline{v_A}$ is proportional to the
frequency of the resetting pulses, which points the way to a simpler
although much slower method of measuring it – namely to count the
resetting pulses during a constant time interval τ_a. The resulting output
code is proportional to v_A averaged over the conversion time, the latter
being constant and easily adaptable to suit various conditions.

This V/F converter is unsurpassed in simplicity. Its accuracy depends
on the stability of the time constant RC and the clock frequency.
Ultimate accuracy, however, is limited by the time Δt lost during resetting
the ramp to zero, which introduces an error into the measured frequency
that becomes intolerable for $\Delta t/\tau_i$ approaching the desired relative
accuracy. Note also that, since the aperture time is equal to τ_i and
therefore varies, interfering frequencies cannot be averaged out in order
to desensitize the converter. The same limitations apply also to the kind
of converters described in Section 6.2.2. V/F converters based on the
charge balancing technique are discussed in Section 7.1.

(b) Dual slope converters
A simple and elegant solution avoiding most drawbacks common to the
sawtooth ramp and the V/F converter described above is the *dual slope*
converter. It is an integrating converter and can be realized in two

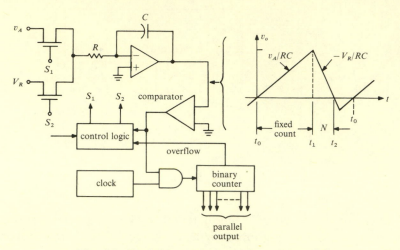

Fig. 6.6. Dual slope encoder. Adapted from [1].

versions, yielding either a digital number or a frequency proportional to v_A. In it, the abrupt reset of the sawtooth ramp is avoided, and the result is insensitive to the value of the integrating capacitor, to the clock frequency and to the offset and drift of the comparator employed. Moreover, the aperture time τ_i is constant in the first version, making synchronization with an interfering frequency possible. The basic idea is to integrate both the unknown input voltage v_A *and* the reference voltage V_R, and compare their respective slopes.

Fig. 6.6 shows a dual slope converter which computes directly the digital value of v_A, with $v_A \leqslant 0$. Initially, the output voltage from the integrator is below zero, S_1 is closed and v_A is integrated. When v_o crosses zero, the (initially reset) counter starts counting the clock pulses until the overflow opens S_1 and closes S_2, thereby initiating a discharge of the integrating capacitor C by the reference source V_R, since $V_R > 0$. During this 'ramp down' time, the counter counts the clock pulses starting again from zero (following the overflow), the comparator stops the count when the capacitor voltage crosses zero as before but in the opposite direction. At that time the count N in the counter will be proportional to the input voltage v_A:

$$v_A = (1 + \delta_1)NV_R/(1 + \delta_2)2^n, \qquad (6.3)$$

where n is the number of binary stages in the counter, $\delta_1 = r_{on1}/R$, $\delta_2 = r_{on2}/R$, and r_{on1}, r_{on2} are the *on* resistances of S_1 and S_2 respectively. Note that v_A is independent of C and the clock frequency. The proof of (6.3) is left to the reader as a problem.

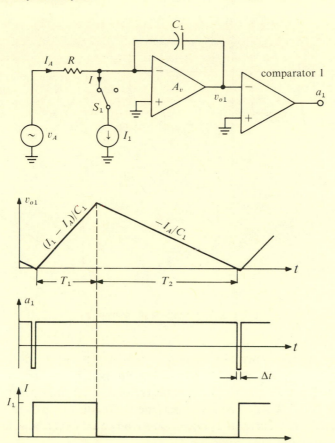

Fig. 6.7. Dual slope V/F converter.

Dual slope V/F converter. A dual slope V/F converter is shown in Fig. 6.7. It employs current switching for the reference source and leaves the signal source permanently connected, which is simpler to implement than the voltage switching used in Fig. 6.6. Assuming the dead time Δt of the converter to be negligibly small compared to the period $T_1 + T_2$, the resulting frequency can be shown to be proportional to v_A:

$$(T_1 + T_2)^{-1} = f = v_A / R I_1 T_1.$$

Stable operation of the V/F converter can be obtained by stabilizing I_1 and T_1 separately which leads to a rather complex design. An elegant solution is shown in the circuit diagram of Fig. 6.8, which stabilizes the charge $I_1 T_1$ rather than each parameter separately. This is accomplished by measuring the time T_1 required to charge C_2 by I_1 to a voltage V_R and

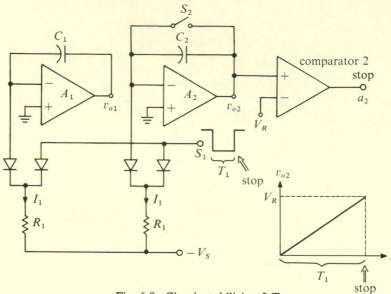

Fig. 6.8. Circuit stabilizing $I_1 T_1$.

gating an identical current I_1 in such a way that it charges C_1 during T_1. To this end, S_2 opens at the start of the time interval T_1, and the stop signal obtained from comparator 2 indicates the end of the time interval T_1. As a result, $I_1 T_1 = V_R C_2$. The stable parameters required are C_1, C_2 and V_R, whereas the two currents I_1 need merely be equal – their absolute value being irrelevant (see problem 6.1).

In comparison with the converter shown in Fig. 6.6, we have exchanged the inaccuracy due to differences in v_{on} of the series switches by imbalances between the two circuits injecting the current I_1 (R_1 in series with one of a balanced diode pair). Furthermore, we have added A_2, an accurate capacitor C_2 and a second comparator. This is the price for obtaining the digital output as a frequency instead of a digital number, which is an advantage in some applications (see Chapter 7).

6.2.4 Simultaneous conversion (flash) method

In the *simultaneous conversion* (or parallel) method,[7] all bits of the digital representation are determined simultaneously. It employs a parallel bank of voltage comparators, each responding to a different level of input voltage. An example of this type of encoder, also called a *flash encoder*, is shown in Fig. 6.9.

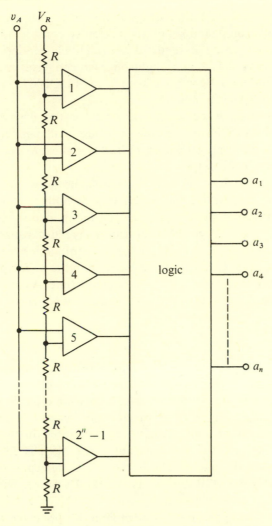

Fig. 6.9. Parallel comparator flash encoder. Adapted from [1].

It can be shown that for n-bits of binary information the system requires $2^n - 1$ comparators; that is, each comparator determines only one unique quantum level. Another disadvantage of this technique is that the output from the comparator bank is not directly usable information. These $2^n - 1$ outputs must be converted to n bits of binary information in some binary coded form. For very large values of n, the large quantity of comparators required and the massiveness of the conversion logic are prohibitive unless IC techniques are employed. This is economically more feasible now[8] and should be seriously considered where ultra-high-speed conversion is required. A relatively moderate speed LSI (*large-scale integrated circuit*) eight-bit flash converter is manufactured by TRW, Redondo Beach, California, as TDC1007J. Its conversion frequency is 20 MHz minimum.

The Gray code is eminently suitable for the flash converter: all comparators above the analog input level are 0; all comparators below it are 1. Multi-input logic gates are required to make the decisions to obtain a parallel Gray code output.

The parallel method illustrated is asynchronous by nature of its construction, and can be used effectively in both the sampled or the continuous tracking mode. Speed requirements determine the type of comparator needed. A commercial transient recorder manufactured by Biomation employs 128 comparators in conjunction with an analog folding circuit (which doubles the effective number of comparators) as an eight-bit flash converter with 10 ns conversion time. A conversion time of 2 ns has been specified for a more recent six-bit converter.

6.2.5 Successive approximation

The *successive approximation* converter translates from an analog input to a digital coded output by a sequence of internally generated inter-dependent events. Each event leads to a decision regarding the value of one binary digit. Fig. 6.10 illustrates the successive approximation method where three binary digits are generated in three sequential operations. Referring to the illustration, the MSB is determined first, then the next MSB, and so on until the LSB is determined. In a voltage A/D converter the MSB represents half the FS range. The next MSB is one fourth of the FS range, and so on.

The successive approximation technique is applicable to both open- and closed-loop converters, although the latter version is more widely employed. An n-bit open-loop successive approximation converter consists of n cascaded one-bit encoders as shown in Fig. 6.11.

A comparator and subtractor make up each decoding stage. The comparator compares the input voltage with a reference voltage, the value of which is determined by the serial position of that stage. If the input voltage ($v_{(j)}$, $v_{(j+1)}$, ...) to a coding stage is greater than or equal to the reference voltage ($V_{R(j)}$, $V_{R(j+1)}$, ...) for that stage, logic 1 is generated by the comparator, and the reference voltage is supplied to the subtraction circuit where it is subtracted from the stage input voltage,

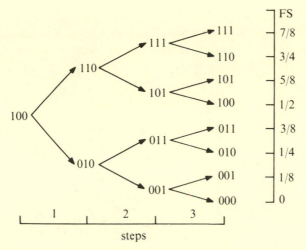

Fig. 6.10. Three-bit successive approximation conversion technique. Adapted from [1].

thus forming the input voltage for the next stage. When the input voltage is less than the reference voltage, logic 0 is generated, and zero voltage is applied to the negative input terminal of the subtractor. In this case, the same input will be passed on to the next coding stage. The input voltage to each succeeding stage, after the first, is often called the *quantization error*, which is defined as the difference between the stage input voltage and the decoded value of the binary digit produced by that stage.

For a cascaded-stage successive approximation encoder with a quantization level of X volts, the least significant encoder stage (stage n) uses a reference voltage of X volts ($V_{R(n)} = X$). Each stage preceding this uses a reference voltage of twice that of the following stage ($V_{R(j)} = 2 V_{R(j+1)}$). The output of this encoder appears in parallel form with the MSB appearing at the output of stage 1.

A modification of this technique allows the reference voltages of all stages to be equal. In this approach, each decoding stage multiplies its quantization error by two (hence the name *voltage doubling*), allowing the magnitude of the reference voltage to be 2^{n-1} times the quantization level (X volts). Two stages of such a converter are shown in Fig. 6.12. The analog polarity inverter (negative unity gain) is required to make each stage non-inverting. All conversion stages are now identical, resulting in a simplification of the design of the cascaded stage encoder.

The transfer function for the *j*th stage may be written as

$$v_{(j+1)} = 2(v_{(j)} - a_{(j)} V_R) \tag{6.4}$$

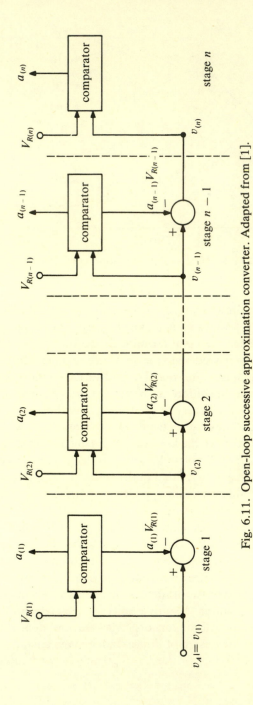

Fig. 6.11. Open-loop successive approximation converter. Adapted from [1].

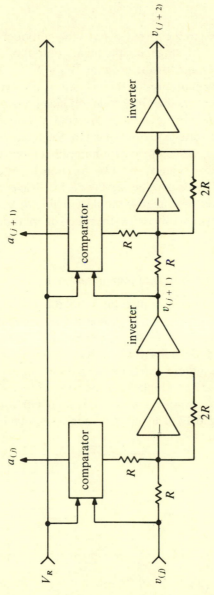

Fig. 6.12. Two stages of voltage doubling successive approximation converter.

with

$$a_{(j)} = \begin{cases} 0 & \text{if } v_{(j)} < V_R, \\ 1 & \text{if } v_{(j)} \geq V_R. \end{cases} \tag{6.5}$$

where $v_{(j)}$ is the input to the jth stage, $v_{(j+1)}$ the output of the jth stage and the input to the $(j+1)$th stage, $a_{(j)}$ the binary output of the jth stage, and V_R the reference voltage for the encoder. $V_R = 2^{n-1}X$.

A practical implementation of this scheme is the cyclic converter by Waldhauer.[9] It employs the Gray code, replacing the sawtooth transfer function of each stage by a smooth inverted-V function and thereby avoiding the abrupt voltage jumps typical for the binary code. Each stage operates in push–pull. It comprises four amplifiers, two for non-inverted and inverted positive signals and two for non-inverted and inverted negative signals. These amplifiers provide full wave rectification and subtraction of reference voltage without switching by external logic. The absence of analog switches increases the speed of operation and simplifies the design.

6.2.6 Multiple comparison subranging methods

In this technique, a compromise between the simultaneous and the successive approximation methods, k bits are determined simultaneously and the entire n bits are obtained in n/k steps. If $k = 1$ then the successive approximation method results. If $k = n$ then the flash encoding technique is obtained. Between these extremes the subranging method achieves a higher conversion rate at a higher cost because of increased complexity.

To illustrate this technique, consider a system with four subranges per step as shown in Fig. 6.13. The input voltage is simultaneously compared

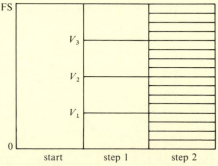

Fig. 6.13. Two-stage multiple comparison subranging converter with four subranges per stage. Adapted from [1].

with the four subrange boundaries 0, V_1, V_2 and V_3 to determine whether it is higher or lower than each. If the input signal is lower than the upper three boundaries, it must fall in the lowest range, and so forth. Once this information is determined, the selected subrange can be divided into four more subranges, and the process repeated.

If there are m subranges per step and s steps, the total resolution of this conversion will be $(1/m)^s$. For example, a twelve-bit system requiring a total resolution of 1/4096 can be implemented in six steps [$(1/4)^6 =$ 1/4096], in four steps [$(1/8)^4 = 1/4096$] or in three steps [$(1/16)^3 =$ 1/4096]. The step resolution does not have to be an integer power of 2. However, except in a BCD (*binary coded decimal*) where it useful to make $m = 10$, the complexity in control circuitry is usually sufficient to justify increasing the step resolution to the next power of 2.

A versatile realization of an open-loop multiple comparison subranging method is due to Arbel and Kurz.[10,11] The block diagram in Fig. 6.14 shows the basic principle. The input is divided into m subranges by the flash A/D converter connected to the input. Each stage consists of a folding circuit which folds the analog input $m/2$ times, i.e. once for every two subranges at the input, and a parallel A/D converter which selects the subrange for this stage. Relevant waveforms for a ramp input and $m = 4$ are shown in Fig. 6.15.

The inverted-V folding transfer characteristic of the converter stages yields a Gray code output, classifying it as a cyclic converter. Various combinations between m and the number of folding stages satisfy different speed requirements for a given n, and several circuit alternatives for the realization of the folding stages add to the versatility of this converter.[11]

6.3 Closed-loop (feedback type) techniques

Closed-loop converters compare the analog input signal with the analog output of a D/A converter, which is 'servoed' until the resulting error is minimized. A block diagram of the basic principle is shown in Fig. 6.16. The heart of this type of A/D system is the comparator circuit that compares an unknown analog voltage (v_A) with an internally generated analog voltage (v_D) and indicates which of the two is larger. This decision is then fed into the control logic, which determines the method of generation of the digital word that represents the input voltage by switching at each step one of two potentials to respective input legs of the ladder network. The output of the ladder network is the analog voltage v_D.

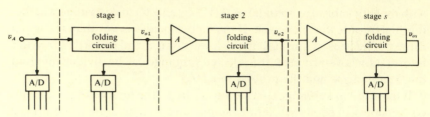

Fig. 6.14. Block diagram of Arbel–Kurz multiple comparison subranging converter. Adapted from [11].

The A/D transfer function will be determined by the D/A converter serving as feedback element. Most of the feedback systems use resistive ladder networks to perform a D/A conversion, which generates an analog voltage for comparison with the unknown input (v_A). Most feedback systems are also synchronous in that the control logic portion is operated by a system clock.

6.3.1 Counter techniques

There are two types of counter techniques. The first to be discussed is the counter equivalent of the sawtooth ramp method where an n-bit binary counter is used to generate the ramp by counting 2^n times. The second is an extension of the ramp counter; where, instead of a unidirectional one, a bidirectional counter is allowed to track the input (up and down). The

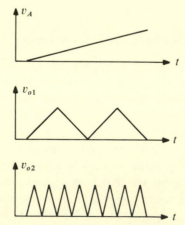

Fig. 6.15. Idealized waveforms for Fig. 6.14. Adapted from [11].

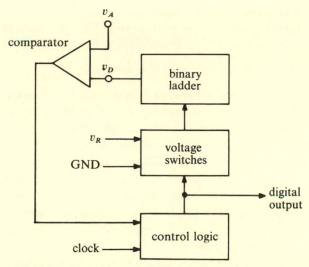

Fig. 6.16. Basic feedback type A/D converter. Adapted from [1].

latter method can be operated asynchronously and is especially suited for
a non-multiplexed single-channel input mode of operation.

(*a*) *Binary ramp*
Fig. 6.17 illustrates the basic block diagram for the *binary ramp* counter
converter. The simplest way to operate the system is to start the counter

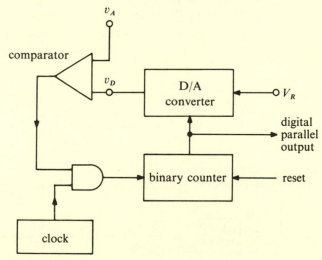

Fig. 6.17. Basic binary ramp counter A/D converter. Adapted from [1].

initially at a zero count and allow it to count until the output of the D/A converter equals or exceeds the analog input. One point worth mentioning is the restriction on the clock period which must satisfy

$$T_{\text{clock}} > \Delta T + t_{rLN} + t_{rVC}, \tag{6.6}$$

where ΔT represents the worst-case propagation delay time of the n-bit binary counter, t_{rLN} the worst-case response time of the ladder network, and t_{rVC} the worst-case response time of the voltage comparator.

If the clock period agrees with (6.6) then the output of the comparator will have time to arrive at the AND gate before the next clock pulse arrives. The counter will continue to accumulate until the output of the D/A converter exceeds the input voltage (v_A), at which time the voltage comparator will change states and inhibit the counter from further action.

(b) Continuous balance method

A slight modification of the ramp counter method is to replace the simple unidirectional counter with a bidirectional up–down one. Now, once the proper digital representation has been found, the converter can continuously follow the analog voltage. This method is particularly useful when only a single channel of information is being monitored, but is less effective for multiple inputs or when the input changes faster than the converter is able to follow. Fig. 6.18 illustrates a basic block diagram of the continuous counter encoder.

One trick sometimes used to reduce the initial conversion time with multiple inputs is to reset the MSB to 0 and all other bits to 1. In other words, start in the middle and go both ways.

This type of encoder has reasonable resolution up to about ten bits and is fairly easy to implement. The 'count up' output yields directly the delta

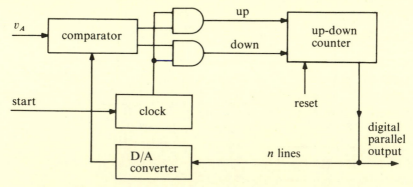

Fig. 6.18. Basic continuous balance encoder. Adapted from [1].

modulation code of the analog input, which is an attractive feature of this simple encoder.

6.3.2 Successive approximation

The feedback type successive approximation converter differs from the open-loop version (Section 6.2.5) insofar as the successive bits are generated sequentially in a synchronized operation. Fig. 6.16 illustrates a valid block diagram. The successive approximation process is controlled by the control logic. The unknown analog input is first compared with one half of the FS voltage by setting the most significant voltage switch to 1. If the analog input is greater than this first approximation, the second most significant voltage switch is also set to 1. This causes the input to be compared with three fourths of the FS voltage. On the other hand, if the analog input is less than the first approximation, the first voltage switch is reset to 0 while the second is set to 1, causing the input to be compared with one fourth of the FS voltage. In like manner, the analog input is compared to successively better approximations until the LSB has been determined.

This class of A/D converters represents an excellent compromise between circuit complexity, speed, and ability to produce high-accuracy codes. This technique is widely used in applications requiring generation of six to fourteen digits at output bit rates from about 100 kHz to 10 MHz.

For this type of conversion the digital output corresponds to some previous value of the analog input during the conversion. Thus, the aperture time depends on the value of the analog input and its change as a function of time during conversion. It can be reduced by connecting a sample-hold circuit to the input of the converter.

6.3.3 Multiple comparison subranging method

A closed-loop (feedback) version of the multiple comparison subranging method is shown in Fig. 6.19. Here a number of comparators C are referenced at equally spaced intervals in the range between the output voltages of two D/A converters. The system starts with the lower D/A converter at zero and the upper one at maximum voltage. The output of the comparators indicates which range contains the input, say between the reference applied at C_K and the reference applied at C_{K+1}. Then the code producing the reference voltage at C_K is applied to the lower D/A converter, and that producing the reference voltage at C_{K+1} is applied to

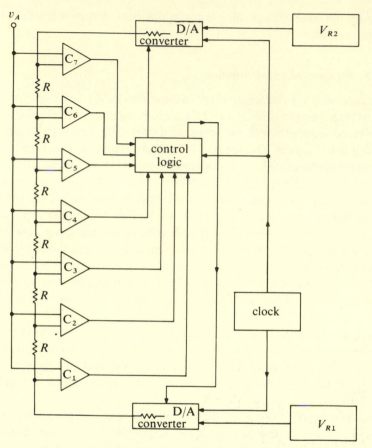

Fig. 6.19. Subranging feedback converter. C: comparator. Adapted from [1].

the upper converter. A new smaller set of ranges is produced. The process is then repeated. A discussion on subranging converters is found in reference [12].

6.4 Errors[13]

Since it is easier to determine the location of a transition than it is to determine a midrange value, errors and settings of A/D converters are defined and measured in terms of the analog values at which transitions occur, in relation to the ideal transition values.

Like their D/A counterparts, A/D converters have *offset error* (the first transition may not occur at exactly $+\frac{1}{2}$ LSB in the case of level splitting

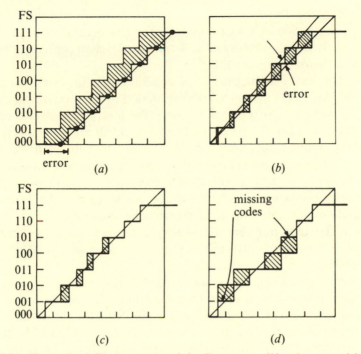

Fig. 6.20. Errors in A/D converters: (*a*) offset error; (*b*) gain error; (*c*) non-linearity; (*d*) excessive non-linearity. Adapted from [2].

encoding), scale factor or *gain error* (the difference between the values at which the first transition and the last transition occur may not equal FS minus 2 LSB), and *linearity error* (the differences between transition values may not all be equal). Figs. 6.20(a), (*b*) and (*c*)[2] depict the corresponding relationships in the presence of offset, gain error and non-linearity. The hatched areas relate to the error bands, i.e. the differences between the ideal and actual relationships.

Integral and *differential linearity* of a converter have been defined in Section 5.3. Whereas integrating converters exhibit excellent linearity, this is difficult to achieve in feedback types using a switched resistor D/A converter. An elegant method of improving the differential linearity of a successive approximation converter through a statistical averaging method has been developed by Gatti and co-workers[14,15] and is widely employed in nuclear electronics.

Whereas some degree of non-linearity occurs in all A/D converters, the transfer relationship of those employing a D/A converter in the feedback path deserves special attention if the D/A converter is nonmonotonic. Depending on the kind of converter and the particular code combination at which non-monotonicity occurs, there may be a certain number of *missing codes*. A transfer function containing missing codes is shown in Fig. 6.20(*d*).[2]

Though there may be gaps in the code at the major carry points (e.g. code 011111111111 going to 100000000000 etc.), the converter can still be monotonic (see Section 5.3.1). But if the specified analog accuracy is not within $\pm \frac{1}{2}$ LSB, then one should not expect all codes to be present at the carry points. Thus, a twelve-bit A/D converter with relative accuracy of 0.05% may miss some of the major carries and thus will not necessarily present all 4096 of the possible code combinations. To ensure that all 4096 combinations are available the unit would need a relative accuracy better than $\pm 0.012\%$ with negligible deadband and noise.

At high speeds, transient effects eclipse the static errors: a statistically monotonic converter may miss some codes beyond a certain rate of conversion.

Absolute accuracy is a measure of the difference between a converter's output in response to a given input and the value of the actual input voltage as it would be measured by the NBS (National Bureau of Standards) if there were no restraints on the time allowed for the measurement. This means that the converter's reference voltage ought to be traceable to NBS. Most manufacturers, therefore, use the term 'absolute accuracy' only with respect to the reference voltage, while output code relationships are specified in terms of 'relative accuracy'. Rarely will a packaged A/D converter have a nominal absolute accuracy better than 0.005% to 0.01%, and even this is difficult to achieve.

Relative accuracy can be defined as the A/D converter's ability to yield accurate output codes for all possible readings relative to FS. Therefore a converter could have high absolute accuracy for its reference, and low relative accuracy. Or the reverse could be true.

If a unit is said to have a relative accuracy of, say, 0.02%, then it should not be possible to put in a known voltage and yield an output code relatively differing from the known input by more than 0.02% of the nominal FS voltage. But relative accuracy, among other factors, depends heavily on the quantization and code splitting techniques used in the circuit. For example, relative accuracy is often specified as an analog accuracy $\pm \frac{1}{2}$ LSB. For this type of specification to be true, the code splitting must be such that the transitions occur exactly in the middle of a theoretical quantization level (*level splitting encoding*, Fig. 6.22(*b*) below). Sometimes, however, it is desirable to have the transitions fall exactly at the quantization level (*non-splitting encoding*, Fig. 6.22(*a*) below). In this case, relative accuracy should be specified as a percentage plus or minus a full bit.

Accuracy must include all sources of error (quantization, non-linearity, noise, and short-term drift). Relative accuracy is often defined as the deviation from a straight line passing through zero and the nominal FS value (very similar to integral linearity). A typical accuracy specification might be 0.05% $\pm\frac{1}{2}$ LSB at +25 °C.

Deadband is defined as a region of uncertainty in which the output is not uniquely related to the input. This is mainly due to noise, drift, thermal effects and, in certain types of converters, comparator hysteresis.

Resolution is the ability of a converter to distinguish between adjacent values of the quantity being measured. Normally the resolution would be considered to be limited only by the number of bits carried. In practice, however, the ultimate resolution of a given design is limited in addition by the noise in the various analog and switching circuits and by the linearity and monotonicity of the converter. Specifications for the resolution of a converter should be compatible with the number of bits and vice versa, otherwise the specification would imply that the readings convey a higher degree of resolution than could actually exist.

Settling time is defined as the interval required for a converter to recover from its transient response to such a degree that its output approaches its steady state value within plus or minus one LSB. Total settling time may include sample-and-hold time, multiplexing time, converter settling time, plus the actual conversion time. Conversion time specifications must be read carefully as some may include all, and others only some of the above mentioned.

Speed is sometimes specified as the time needed to perform a conversion and sometimes as the number of independent conversions that can be performed per second. These look like different ways of saying the same thing but they are not: succeeding conversions do not necessarily have the same accuracy as the first. There may be input transients which must settle out before the specified accuracy is achieved. Also, the unit may suffer from hysteresis at high operating speeds. A poor circuit layout can cause high-speed errors which may not be obvious from the schematic. A converter might, for example, perform a single conversion at the specified speed; but if commanded to repeat the measurement immediately, it may give a second output that differs slightly from the first. This type of error can be caused by interference between analog and digital portions of the circuit. Ground-loop currents and fields set up by one logic state may influence subsequent performance of the analog circuitry.

6.5 Noise

Quantization error

The *quantization error* is defined as the difference between the input voltage and the decoded value of the digital number produced by the A/D converter. Unlike the errors representing a deviation from ideal characteristics, the quantization error is inherent in the operation of every A/D converter.

The quantization error may be looked upon as noise introduced into the system as a result of the digitization. The absolute maximum peak-to-peak value of the quantization error is one LSB, but its rms value is much less. For a uniform probability distribution of the quantization error, its standard deviation can be shown to equal $(q/12)^{1/2}$, where q is the value of a single quantum. This yields, for a sine wave of FS peak-to-peak amplitude and n bits, an SNR of $(6n + 1.8)$ dB. The additive constant term in this relationship is a function of signal definition. It becomes insignificant for high values of n, in which case the SNR approaches 6 dB per bit – a rule of thumb mentioned by Stockham.[16]

Bruce,[17] in a simulation of a multistage converter, obtained the results shown in Fig. 6.21. The straight line shows the relationship between the SNR and the number of bits, due to quantization error. Deviation from linearity, in terms of multiplier errors in the binary weighting factors, may also be looked upon as noise, and reduces the SNR as shown by the curves deviating from the straight line.

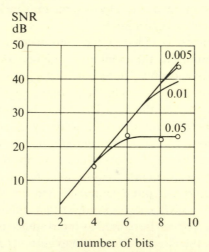

Fig. 6.21. Deterioration of SNR due to quantization noise and multiplier errors in binary weighting factors. Parameter: multiplier errors. Adapted from [19].

Aperture error

The aperture error of an A/D converter is defined as the difference between the actual encoded value and the value that would have been obtained if the input signal during conversion had retained the value it had at the start. Hence, it is a function of the signal slope and can be computed if the statistical distribution of the latter is known. If the converter is preceded by an *ideal* sample-hold, the signal is constant during conversion and the aperture error is zero.

Dunbridge[18] has presented an analytical evaluation of the average absolute aperture error for successive approximation type converters, based on a statistical approach. His results are restricted to the case in which the signal slope is constant during conversion. More general results could be obtained by computer simulation of the converter.

Small-signal loss in the presence of noise[19]

Signal processing can extract very small signals from a noisy background. However, as Goldstone and Hertz have pointed out, if the LSB is stable, this background is missing and small-signal drop-out can occur. Depending on the character of the expected signal, two cases are of interest, each of which tends toward a different form of threshold placement.

No dc signal. In this case, a sensitive zero cross detector or comparator is desired. The noise level of the comparator should be low and its sensitivity should be sufficient to permit switching by input noise. The transfer function of such a converter is shown in Fig. 6.22(a) (non-splitting encoding). It will display random variations in the value of its LSB as reflected by changes in sign in the absence of a signal, and a solid reading when tested with voltages remote from a threshold. This converter

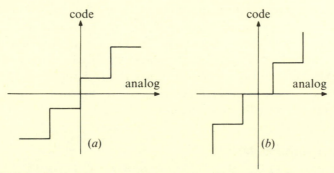

Fig. 6.22. (*a*) Non-splitting encoding. (*b*) Level splitting encoding. Adapted from [19].

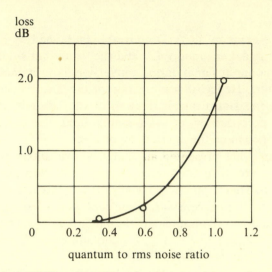

Fig. 6.23. Loss in SNR as function of quantum-to-rms noise ratio. Adapted from [19].

operates as a one-bit converter on signals of peak amplitude insufficient to switch any threshold beyond the one at zero. Such converters display a 2 dB loss in SNR for small signals.

This small-signal loss can be reduced by arranging to switch additional thresholds beyond that at 0 V for the no-signal input condition. Fig. 6.23 is a plot of the small-signal SNR loss as a function of the ratio of quantum size to rms noise.

dc input present. The second case occurs when dc is present in addition to the small signal and noise. In this case, small-signal drop-out can occur if the input noise is insufficient to switch the nearest threshold. In addition, the ability to estimate the value of the small signal is also lost. To remedy this, quantum size should be reduced until the rms noise with no signal can cause the necessary switching. The number of bits added to the system once again depends on the allowable loss in small-signal SNR.

In this instance, the placement of a threshold at 0 V loses its importance, and a transfer function such as shown in Fig. 6.22(*b*) is satisfactory (level splitting encoding).

6.6 Converter codes[20]

The language of A/D and D/A converters is essentially one of codes, the most widely used being the natural binary, which easily accommodates the two states of digitization: on (1) or off (0). The *bits*, or basic units of information, are combined to form words, which may appear in parallel or in series. Apart from the binary code, other codes are available depending on the specific requirements: signal range, polarity, conversion technique, special characteristics, and the origin or destination of the digital data. These codes include the BCD, the 2-4-2-1 BCD, the Gray code, complementary codes, and bipolar codes, the most significant of which are the sign magnitude, offset binary, two's complement, and one's complement.[20]

The binary coded decimal

The BCD is a code in which each decimal digit is represented by a group of four binary coded digits. The LSB of the most significant group, or *quad*, has a weight of 0.1, the LSB of the next quad has a weight of 0.01, the LSB of the next has a weight of 0.001, and so on. Each quad has ten permissible levels with weights 0 to 9. Quad values of more than 9 are not permitted.

A/D converters with the BCD code are used primarily in digital voltmeters and panel meters, since each quad's output may be decoded to drive a numeric display using the familiar decimal numbers.

Overranging. Many BCD A/D converters have an additional bit, with a weight equal to FS, in a position 'more significant' than the MSB. This additional bit provides a maximum of 100% 'overrange' capability. The overrange bit is most commonly used in digital voltmeters and panel meters to indicate that nominal FS has been exceeded and that the visual reading may be erroneous.

Overrange bits need not be restricted to BCD. They are useful as flags in any conversion process for which an overrange input would give an ambiguous reading, or where an overrange input indicates anomalous analog system behavior. The overrange bit, of course, must be of suitable accuracy, since it is, in effect, the MSB.

The 2–4–2–1 binary coded decimal

A code still in common use, the 2–4–2–1 BCD is one in which the bit in the MSB position in each quad has a weight of 2 instead of the usual 8. It is found, for example, at the digital output of some Hewlett–Packard digital voltmeters. This code, which was more economical to implement in the days before IC logic became common, still has the advantage that all 1s indicate FS minus LSB and requires a smaller range of resistances than the more common 1–2–4–8 BCD.

Fig. 6.24. Gray versus binary shaft encoder. Adapted from [2].

The Gray code

The *Gray code* is a binary code in which the bit position does not signify a numerical weighting; however, in converters using it, each code still corresponds to a unique portion of the analog range. As the number value changes, the transitions from one code to the next involve only one bit at a time, making the gray code useful for shaft encoders (angle-to-digital converters) because false intermediate codes that could occur in natural binary conversion are eliminated. For comparison, the results of Gray code and binary optical shaft encoders for four-bit resolution are illustrated in Fig. 6.24.

With the Gray code converter, if the edge of a shaded area is slightly out of line, the coding will be in error only by a single LSB, since there is only a one-bit change at each transition. In the binary converter, however, all four bits change at once at the 180° and 360° transitions. If bit 2's shaded area were to end a little to the left of the 180° transition, the code, in a small region, would be 0011, indicating the $67\frac{1}{2}°$ range, or a fictitious progression from $157\frac{1}{2}°$ to $67\frac{1}{2}°$ instead of 180°. We leave the catastrophic implications of this to the reader's imagination.

The shaft encoder is a simultaneous converter: all bits appear at once and can be read in parallel at any time. There is an electrical equivalent form of simultaneous A/D converter having a Gray code output, the flash converter (Section 6.2.4).

A variation of this scheme, the cyclic converter (see Section 6.2.5), which also has a Gray code output, uses fewer comparators but requires more time to perform the conversion; it continuously tracks the analog input.

The use of the Gray code in fast asynchronous converters that provide continuous conversions has the same rationale as that of the shaft encoder. Any Gray code output value that is latched into a register will always be within ±1 LSB of the correct value, even if the latching occurs just as a bit is switching. With binary codes, however, where many bits can switch at a single transition, it is possible to latch in midflight and, because of the 'skew' between turn-on and turn-off speeds, an utterly false code can be locked in.

Complementary codes

Certain converters, such as D/A converters using monolithic current switching, may provide *negative true coding*, in which the logic 1 is the more negative of the two voltage levels. Such a converter requires a code in which all bits are represented by their complements, i.e. a complementary code.

Analog polarity

So far, the conversion relationships mentioned have been unipolar: the codes have represented numbers, which in turn represent the normalized magnitudes of unipolar analog variables,* without regard to sign. A unipolar A/D converter will respond to analog signals of only one polarity, and a unipolar D/A converter will produce analog signals of only one polarity.

The analog signal polarity is determined either by using a converter whose reference and switches (or specifications) are compatible with the desired analog polarity or (if, for reasons of economy or availability, a converter is obtained having predetermined polarity) by operating on the analog signal before A/D conversion – or after D/A conversion – to invert its polarity, and by performing any necessary scale changes if range must also be adapted.

Bipolar codes

For conversion of bipolar analog signals into a digital code or vice versa, one extra bit – the sign bit – is necessary to retain the sign information. The sign bit is added to the digital word to the left of the MSB. Figs. 6.25 and 6.26 show the mutual relationship between the analog quantity and the corresponding three-bit digital word for three commonly used bipolar codes, with the dots relating to the values of the analog quantities at which transitions occur.

Fig. 6.25 shows the offset binary and two's complement codes, which mostly employ level splitting encoding, and thereby place zero at half

---

* The Gray code is an exception. Since it is not quantitatively weighted, it can represent any arbitrary range of magnitudes of any polarity.

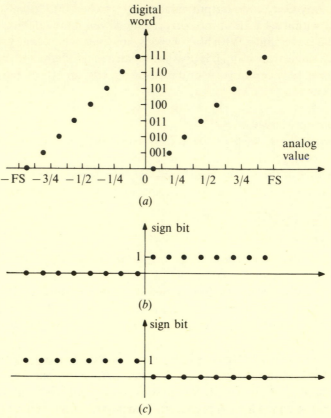

Fig. 6.25. (a) and (b) Offset binary code; (a) and (c) two's complement. Both shown for level splitting encoding.

scale. They have the same magnitude code but complementary sign bit code.

The *offset binary* shown in Figs. 6.25(a) and (b) is the easiest one to embody in converter circuitry, since it is obtained directly from a unipolar A/D converter which has been turned into a bipolar one by applying to the input an offset equal to one half of FS. The computationally more useful two's complement shown in Figs. 6.25(a) and (c) is easily obtained from the offset binary by complementing the MSB (sign bit).

Both codes have a single unambiguous code for zero. Their principal drawback is that a major bit transition occurs at zero (all bits change), which can lead to glitch problems in D/A converters. Both codes are also suitable for non-splitting encoding for which the dots in Fig. 6.25 must be shifted by half an LSB to the left.

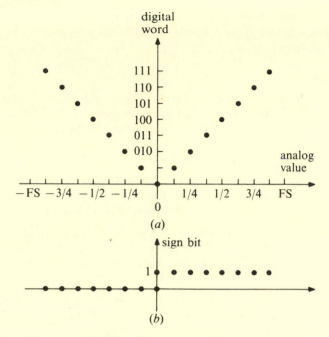

Fig. 6.26. Sign magnitude code.

The sign magnitude code shown in Fig. 6.26 appears to be the most straightforward way of expressing signed analog quantities digitally: simply determine the code appropriate for the magnitude and add a polarity bit. This can be directly obtained from a unipolar A/D converter preceded by an absolute value circuit (analog full wave rectifier), which also provides the sign bit. The sign magnitude code has no major transition at zero and can therefore be used advantageously when the application calls for smooth and linear transitions from positive small to negative small voltages.

Its disadvantage is that it has two codes for zero. This is indeed not surprising, since zero can be interpreted as the starting point for both the positive and the negative ranges. Because of this ambiguity it is harder to interface digitally, but in applications such as digital voltmeters its simplicity is unsurpassed.

Serial coding
Certain A/D converters such as the closed-loop successive approximation and V/F converters deliver a serial output (coded and uncoded, respectively) which can be directly transmitted through a single optical

isolator. This may be a particular advantage in the presence of high-level interferences, which make a digital transmission path and interruption of the ground-loop currents mandatory.

Problems

6.1. Plot the response of a four-bit binary ramp for a slowly rising analog ramp input, if the error of the weighting factors for the second and third bit is plus 3% and plus 4% of FS, respectively. The weighting factor of the two remaining bits is accurate.

Repeat for an error of plus 3% for both bits.

6.2. In the dual slope encoder shown in Fig. 6.6, both the integrator and the comparator exhibit an offset voltage ΔV_{BE} and an offset current ΔI. Find the effect of each of the above terms on the output N of the converter.

6.3. In the dual slope converter shown in Fig. 6.6, it is proposed to connect the input FET switches directly to the virtual ground, and to use a matched pair of resistors connecting each switch to the corresponding source v_A and V_R respectively. Enumerate the pros and cons for this modification.

6.4. Draw a complete block diagram for the V/F converter shown in Figs. 6.7 and 6.8. Pay particular attention to the connection between the logic waveforms a_1 and a_2, the state of the switches S_1 and S_2, and the gating voltage for the diodes.

6.5. Find a simple modification, which converts the continuous balance A/D converter into a peak-detector.

7

Converters, selected topics

7.1 Voltage-to-frequency and frequency-to-voltage converters [1]

Although belonging to the general classes of A/D and D/A converters respectively, V/F (*voltage-to-frequency*) and F/V (*frequency-to-voltage*) converters deserve a separate section. Available as integrated modular units, these versatile building blocks can serve as interfaces between analog and digital circuits and are capable of performing a large variety of functions.

V/F converters can be implemented in various forms. Two kinds of integrating V/F converters (ramp and dual slope) were discussed in Section 6.2. Here we shall describe the *charge balancing technique*, mostly employed with commercially available V/F converters.

The heart of the charge balancing technique is a current pulse genera-tor driven by a PTRC (*precision timing reference circuit*), delivering pulses of constant charge at a frequency determined by external synchro-nization. This building block, in conjunction with an integrator, is basic-ally a F/V converter as shown in Fig. 7.1. Using the same building block in a feedback arrangement yields the V/F or I/F (*current-to-frequency*)

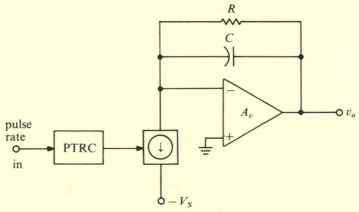

Fig. 7.1. Charge balancing F/V converter. Adapted from [1].

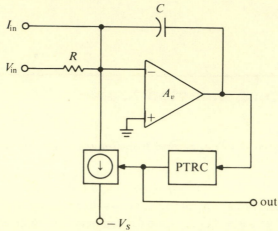

Fig. 7.2. Charge balancing V/F or I/F converter. Adapted from [1].

converter shown in Fig. 7.2. Each time the output voltage of the integrator reaches a certain reference level, the PTRC emits a current pulse of constant charge Q, which resets the integrator. The amplitude of the resulting sawtooth ramp, due to repeated resets, equals Q/C, and its frequency equals V_{in}/QR.

Unlike the V/F converter shown in Fig. 6.5, no time is lost during resetting the ramp to zero, since the capacitor C is continuously charged by the signal. Furthermore, the shape of the current pulses delivered by the PTRC is irrelevant as long as the charge remains constant. Hence, accuracy is inherently high.

In practice, commercial modules can be programmed for V/F or F/V conversion by making the appropriate pin connections.

The charge balancing technique, whilst being comparatively slow, exhibits two important properties inherent in its operating principle: it is monotonic, and linearity does not fall off near zero as with some other schemes such as A/D converters employing the linear ramp technique. Commercially available modules have an integral linearity of typically 0.002% to 0.05% over the input–output operation range, and excellent temperature stability – typically 10 to 100 ppm over the operating temperature range. On the input side, F/V converters take analog inputs in the −10 V to +10 V range. Typical output pulse rates are between 10 kHz and 5 MHz.

V/F conversion
While the V/F module is a relatively slow A/D converter, its cost is low and accuracy can be high. The digital output of the converter is in discrete

serial form, and must be counted over some period to give a final conversion value in parallel form.

To get a complete digital measuring instrument, it is only necessary to precede a V/F converter with a signal conditioning circuit, such as a high input impedance amplifier, and follow it with a digital counter and display. Then, if the time base for the counter is set to one second, the actual output pulse rate of the V/F converter will be displayed. If a 10 kHz converter is used, an FS value of 10 000 would be displayed with a one second time base; with a ten second time base an FS value of 100 000 would be displayed, although the counting time would be too long for many applications.

It is useful to discuss the characteristics of V/F converters in terms of A/D converter specifications. For a V/F converter, conversion time is determined by the time base, one second being a convenient time base for many applications. For faster conversion time, a 0.1 second time base could be used, giving an FS count of 1000 for a 10 kHz converter.

For an A/D converter, resolution is expressed in bits and is determined by the number of parts into which the FS range is divided. By comparison, a 10 kHz V/F converter has a resolution of 1 part in 10 000, assuming a one second conversion time. This is equivalent to a resolution of greater than thirteen bits (1 part in 8192). A 100 kHz converter with a one second time base gives greater than sixteen-bit resolution (1 part in 65 536).

Linearity is another important A/D converter specification. A good A/D converter has an integral linearity of $\pm\frac{1}{2}$LSB over its FS input range. For a 10 kHz V/F converter with a typical linearity figure of 0.002%, the linearity is equivalent to that of a fourteen-bit A/D converter. Therefore, a 10 kHz V/F converter, as described, performs at least as well as a thirteen-bit A/D converter in both resolution and integral linearity.

Differential linearity, if compared with that of a successive approximation D/A converter, is incomparably better, since the problem of critical code transitions is absent.

Converters employing pulse rates of 100 kHz or more, while offering better resolution, generally have worse linearity than 10 kHz converters. This is because circuit parasitic time constants vary with the pulse duty cycle. At high output pulse rates the small variations in pulsewidth with duty cycle will be proportionately more significant, thus increasing the amount of non-linearity. Therefore, the best resolution and linearity are achieved with slower pulse rates, namely the 10 kHz converters with a ten second time base. These achieve better than sixteen-bit resolution and better than fourteen bits of integral linearity.

Another useful way of looking at V/F converters is in terms of dynamic range. This specification is critically dependent on linearity. A 10 kHz V/F converter that holds its linearity down to zero can measure voltages as low as 1 mV, which corresponds to a dynamic range of 4 decades, or 80 dB. Similarly a 100 kHz converter has a dynamic range of 5 decades, or 100 dB, if its linearity is maintained through zero. The practical lower limit of 1 mV is chosen because of drift in the zero adjust potentiometer, long-term drift of the circuit, and noise at the input to the integrator.

V/F converters have two other significant features when utilized for analog conversion. First is their monotonicity. A V/F converter is naturally monotonic because its output pulse rate must increase with increasing input voltage. Second is the excellent noise rejection inherent in using a reasonably long time base like one second. Random and periodic noise are effectively integrated over the conversion period. Periodic noise, such as a 50 or 60 Hz power pick-up, is effectively integrated when the conversion period is long compared to the period of the power frequency. For a 60 Hz noise integrated over an unsynchronized one second measurement period, the noise rejection is approximately 46 dB; for a 0.1 second period, the rejection is 26 dB.

Remote monitoring

Remote data monitoring is one application well suited to the V/F technique. Remote monitoring can be difficult, especially when analog signals pass through an environment with high levels of electrical noise, as in a manufacturing facility where there is heavy equipment. If a high degree of accuracy must be maintained, analog signal transmission becomes prohibitive.

One solution is to use a V/F converter to transmit the data directly in serial form. This is a simple and effective way to achieve an accurate system of ten to thirteen bits resolution (0.1% to 0.01%), if the data rate is slow. At the monitoring end, the pulse train can be simply counted for a one second period and then displayed to show the analog value. This can be done with a low-cost four-digit counter if a 10 kHz V/F converter is used.

Some instrumentation problems involve parameters that must be derived from high-voltage measurements. In these circumstances, transmission of the desired information back to normal ground-potential circuits requires some form of isolation. One answer to this is to use an isolation amplifier powered from a non-isolated supply. If the data are desired in digital form, the output from the amplifier would then go to an A/D converter.

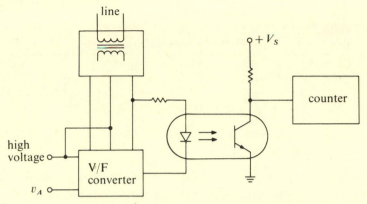

Fig. 7.3. Measurement of signal at high voltage using floating V/F converter and optical isolator. Adapted from [1].

An effective alternative is shown in Fig. 7.3. It employs a V/F converter with a floating power supply while optically coupling the digital data back to ground-level circuitry. The V/F converter output is a serial pulse train and therefore requires only one low-cost optical isolator. The isolated power supply cost must also be factored in. This can be relatively low if the voltage is not too high (up to 1500 V peak).

Ratiometric measurements
Ratiometric measurements are important for applications in which a transducer output might be affected by variations in the exciting power supply voltage, as, for example, in a resistor bridge. This can be overcome by the measurement system shown in Fig. 7.4, which determines the ratio

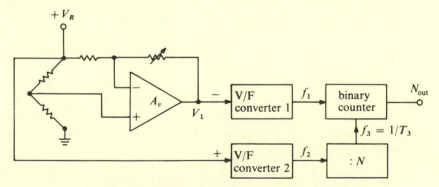

Fig. 7.4. Ratiometric measurement system.

Table 7.1

V_1(V)	V_R(V)	f_1(Hz)	f_2(Hz)	$T_3 = N/f_2$(s)	$N_{out} = f_1 T_3$	Ratio
10^{-1}	10	10^2	10^4	10^{-1}	10	10^{-2}
1	10	10^3	10^4	10^{-1}	10^2	10^{-1}
10	10	10^4	10^4	10^{-1}	10^3	1
10	1	10^4	10^3	1	10^4	10
10	10^{-1}	10^4	10^2	10	10^5	10^2

of transducer output to excitation voltage, thereby normalizing the measured result with respect to the excitation voltage and thus cancelling its drift and temperature coefficient. The resistance bridge is excited by a reference voltage V_R which also goes to the input of V/F converter 2. The output of the bridge is amplified and goes to V/F converter 1. The resulting pulse rate is fed to a binary counter. The output pulse rate of V/F converter 2 representing V_R is fed to a divide-by-N circuit, and the resulting pulse train is used as the time base for the counter. The parallel output of the counter may drive a numerical display. If the counting time equals the output period of the divide-by-N circuit, the output count is proportional to NV_1/V_R.

The dynamic range of the circuit far exceeds the requirements of the particular application shown in Fig. 7.4. Table 7.1 illustrates the operation of this ratio circuit over a dynamic input range of $1:10^4$, employing V/F converters with a pulse rate of 10^4 Hz and a divider of $N = 10^3$.

A/D integration
Ideal integration over a wide dynamic signal range is difficult to perform purely analogically, especially over an extended period like several minutes: drift error in the operational integrator causes problems. Even if a very expensive, low input current amplifier with low drift is used with an expensive, stable capacitor that has low leakage and low dielectric absorption, the operational integrator cannot work well when the integration period exceeds ten minutes. A simple alternative is a hybrid A/D integrator using a V/F converter. The analog signal is applied to the input of a V/F converter, and the output goes to a counter operated in the totalizing mode to give a total count equal to the time integral of the signal:

$$(1/RC) \int_0^T v_A(t)\, dt = (k/C) \int_0^T [dN(t)/dt]\, dt = kN_T/C$$

where N_T is the total count during the period T of integration, and k is the charge/count conversion factor. This is a simple and accurate realization of a pole at the origin of the complex frequency plane.

Because of the superior linearity, the integration is accurate for a signal dynamic range of 10 000 to 1. Since the output is an accumulated pulse count, there is no integrator drift as there would be with an operational integrator. Also, the counter can be stopped at any time for an indefinite period without affecting the integrated value. The limitation on the total integral is the total count capacity of the counter. Therefore, counter capacity must be based on the signal values and period of integration.

The actual integration time can be days if the counter has sufficient capacity. Assume, for example, a signal with an average value around 2 V but with occasional high peaks up to 10 V (FS input of the V/F converter). The output frequency of a 10 kHz converter is then 2 kHz, on average. If an eight-decade counter is used (99 999 999 FS count), the integration period can be as long as 50 000 seconds, or 13.88 hours. The counter itself can be made from low-cost ICs and be operated manually or by an external logic signal.

Digital logarithmic conversion of V/F output
An implementation of a digital-to-digital converter changing a pulse train into the parallel coded equivalent of its logarithmic rate has been described by D. Arnold.[2] In conjunction with a V/F converter this becomes a simple and accurate logarithmic averaging A/D converter. An example for such a requirement is a power meter where signal power is measured by a thermopile over a given time interval, and the result is displayed digitally in decibels.

The key to implementing such a processor is the logarithmic counter. On the curve $y = \log x$ in Fig. 7.5, x represents the number of pulses fed into the counter from the V/F converter and y is the corresponding counter output. Since the counter output is digital, the output is quantized into equal increments Δy. The corresponding increments Δx, however, vary according to the instantaneous value x.

Consider the point $y' = \log x'$. To increment y', the following relationship must hold:

$$y' + \Delta y = \log (x' + \Delta x') = \log x' + \log (1 + \Delta x'/x').$$

Therefore

$$\Delta y = \log (1 + \Delta x'/x') \simeq 1 + \Delta x'/x'.$$

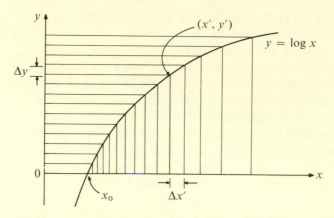

Fig. 7.5. Incrementing the logarithmic counter. Adapted from [2].

Since Δy is constant, $\Delta x / x$ is constant for all values of x. Thus, $\Delta x = Kx$, where K is an arbitrary constant. Since K will be less than unity, Δx may be found for every value of x by division.

A method for implementing the logarithmic counter is shown in Fig. 7.6. Here, the output of the V/F converter is fed to both the x and Δx counters. The x counter is a straightforward up counter whereas Δx is a presettable down counter.

Whenever the count in Δx goes to zero, the y counter is incremented. A number equivalent to kx is then loaded into the Δx counter and counting continues until Δx again goes to zero. This process repeats until the selected time interval elapses, terminating the count. The contents of the y counter are then equal to the logarithm of the count in x.

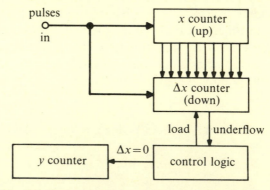

Fig. 7.6. Logarithmic counter. Adapted from [2].

By making K a negative power of 2, Δx is easily derived by shifting the contents of a register containing the binary equivalent of x or, as shown here, by simply transferring the MSBs in the x counter to the LSBs in the Δx counter.

During operation, counting proceeds for the selected time interval, at the end of which the contents of the y counter are outputted and all counters are reset for the next counting cycle. This raises a problem, however, because $\log 0 = -\infty$, hardly a practical number to set into the y counter. The difficulty is overcome by selecting the point x_0 for which $y = 0$ (Fig. 7.5), and leaving the y counter clear until the x count reaches this point. This places a constraint on the minimum number for which the conversion is valid but causes no problem because the upper end of the scale can be extended as far as necessary to give the range desired.

Sources of errors. Δy can be made arbitrarily small to reduce quantizing errors to negligible proportions. The only other source of errors comes from the loading of Kx into the Δx counter. Since division is accomplished by truncating the LSBs, this does not represent an exact division. The errors can be minimized by making the values of Δx as large as possible. As a result, the number in the y counter is several orders of magnitude smaller than the number in the x counter. The errors can be further reduced by making x_0 slightly smaller, since analysis has shown that this circuit always give a result slightly larger than the true logarithm.

This system is used in the Model 3745A selective level measuring set manufactured by Hewlett-Packard. It uses a VCO that operates over a frequency range of 0.1 to 1 MHz. The x counter has twenty binary stages with the ten MSBs transferred to the Δx counter, which makes K equal to $1/1024$. To get an answer with three-digit resolution, 10^5 to 10^6 pulses are counted, and this takes about 800 ms in the long-averaging mode. In the short-averaging mode, the division ratio is changed and the counting period is 100 ms.

It is worth pointing out that in systems where a logarithmic amplifier is first in the chain, a bias error must be compensated for because the average of a logarithm is not identical to the logarithm of the average. No such error occurs in the system described here.

F/V conversion

The F/V converter is basically an analog pulse counter since the output voltage is linearly proportional to input rate. Once the pulse rate is in analog form at the F/V converter output, other analog operations can be performed. Subtracting the outputs of two F/V converters gives an analog frequency difference, a quantity difficult to obtain by other means.

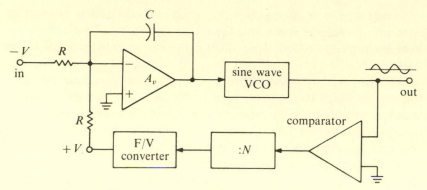

Fig. 7.7. Linearization of sinusoidal VCO. Adapted from [1].

Applications employing the F/V converter include frequency measurements in flowmeters and tachometer read-outs in motor speed controls. Output pulse rates from these devices are used to develop an analog voltage proportional to speed or flow. The voltage, in turn, is usually fed back to regulate the process or system.

Another application of the F/V converter is in the stabilization and linearization of a sinusoidal VCO. VCOs with a high degree of linearity and low temperature coefficients are quite expensive, especially if a wide variation of output frequency is needed. Very high quality VCOs use an oven controlled inductance–capacitance element (LC) to stabilize the frequency. On the other hand, low-cost VCOs have only moderate linearity and temperature stability.

A low-cost VCO can be combined with a low-cost F/V converter to achieve a linearity of better than 0.005% and a temperature coefficient of 20 ppm/°C maximum. As shown in Fig. 7.7, the F/V converter is used in a feedback loop to control the VCO frequency. The controlled frequency is higher by a factor N than the operating frequency of the F/V converter. Of course, if a pulse output is satisfactory for a system, a V/F converter could be used directly. A large proportion of VCOs, however, are used with sinusoidal outputs and, in addition, at frequencies higher than those available in V/F converters.

7.2   Hybrid computational building blocks

D/A and A/D converters can be employed singly or combined with each other and/or various analog and digital modules, to create a multitude of functional relationships.[3] Here we shall mention only a few selected examples.

A widely employed property of D/A converters is their multiplying feature according to (5.1), the reference voltage V_R serving as one multiplicand and the digital input as the other. If V_R is derived from another D/A converter, the analog product of two digital numbers is obtained.

Applying the same idea to an A/D converter employing a D/A module in the feedback path, we expect an inverse functional relationship (i.e. division). Accordingly, the digital output number of the converter depends on the ratio of the analog input to the reference voltage of the D/A. Special-purpose multiplying D/A converters are commercially available and may be used in a variety of applications.[4]

A logarithmic transcoder with a logarithmic-to-linear characteristic has been described.[5] It yields an exponential relationship if employed directly, and a logarithmic one if used as feedback element in an A/D converter.

Practically any functional relationship or waveform generator may be obtained by cascading an A/D converter with a ROM (*read-only memory*) or with a PROM (*programmable ROM*), with possible reconversion to analog by a cascaded D/A converter. Digital shift registers in conjunction with input A/D and output D/A conversion provide a non-volatile and variable analog delay.

A particularly attractive implementation of an analog delay employs a continuous balance A/D converter, whose output is in the form of the delta modulation code and can be delayed by a digital shift register. Reconversion of the delta modulation code is effected by substituting 'space' by negative pulses whose area equals that of the positive 'mark' pulses, and integrating the resulting pulse train. The delay of the recovered analog signal can be varied by changing the length of the digital shift register.

As a final example, we consider the D/A converter which is part of a feedback type A/D converter employed in the Hewlett-Packard Model 1722A oscilloscope. The D/A converter[6] is built around a rate multiplier, a digital device that outputs pulses in proportion to the BCD number at its input. For example, if the input number is 6, the rate multiplier outputs 6 pulses for every 10 input clock pulses. Four rate multipliers are connected to a clock of 140 kHz, and their combined output pulses are interleaved in such a way that their total number per 1/14 s corresponds to the digital setting of the four-decade rate multiplier. Finally, the output pulses are integrated by an operational integrator, yielding the desired analog equivalent of the digital input.

Resolution is $1 : 10^4$, with inherent monotonicity. Apart from the obvious speed limitations, the main design problem was to maintain the area under each pulse constant, as temperature changes. It was found that an increase in ambient temperature caused the pulses to get narrower and slightly higher. The clock repetition rate was chosen so that these effects compensate each other. The resulting temperature coefficient of 0.005%/°C bears witness to the ingenuity of the designer of this converter.

8

Data acquisition systems

Signal transmission from remote sources to a central location for processing and ensuing evaluation, display, decision making and storing are the main tasks of a data acquisition system. As a typical example we shall consider the system shown in Fig. 8.1. Delivering the information obtained from the various sources to the digital data processor unit at the desired accuracy involves several aspects.

The signal is transmitted from the various sources $(1, 2, \ldots, n)$ to the individual amplifiers with due care for keeping noise and interference pick-up at the necessary low level. The signal is then amplified to a level sufficiently high to prevent the imperfections of the following system from reducing the SNR below specifications, filtered to reduce aliasing of signal and noise due to the following sampling process, and multiplexed in order to feed the signals sequentially to a centralized signal processing system. Then follows a sample-hold module (optional), an A/D converter, a digital demultiplexer for feeding the digitized signals from the individual sources into separate channels, and a digital data processor in each channel, which may include a recovery filter to smooth the sampled data. Finally, the data may be stored, displayed or fed into another unit for automatic evaluation and possible control of the process being monitored.

In choosing the various components, the foremost considerations are the number of sources, the signal amplitude and its spectrum. The latter determines the *sampling* or *Nyquist rate*, which is theoretically slightly more than twice the highest frequency of interest. In practice it will be significantly higher, the actual value depending on the design of the low-pass filters and the desired suppression of spurious responses due to aliasing.

Assuming that the signal spectrum of all signal sources is the same, all channels will be sampled at the same *frame rate*. Multiplication of the frame rate by the number of channels yields the *data rate* or *throughput rate* of the system, which determines the required speed of the A/D converter.

266

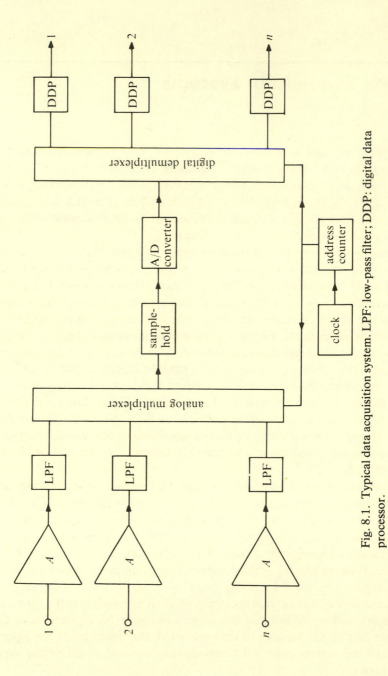

Fig. 8.1. Typical data acquisition system. LPF: low-pass filter; DDP: digital data processor.

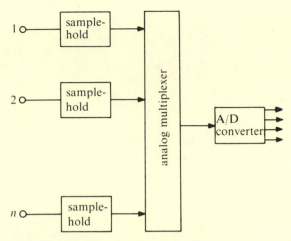

Fig. 8.2. Simultaneous sampling system.

Simultaneous sampling system

A possible variation of the system is shown in Fig. 8.2, which is termed a simultaneous sampling system. A typical application of such a system is in monitoring several parameters of a process initiated by some excitation. Another application is the monitoring of randomly occurring data. In this case, the sample-hold module holds the data until interrogated by the analog multiplexer. Subsequently, the sample-hold module is returned to the sample mode to accept the next data.

Decentralized data acqusition[1]

Availability of low-cost IC A/D converters makes it economically feasible to design decentralized acquisition systems. As shown in Fig. 8.3, A/D conversion is performed separately in every channel, and further signal transmission and multiplexing is entirely digital. This approach has several interesting aspects.

Signal transmission from the digital multiplexer to a central digital data processor and storage facilities is entirely noise-free, a special advantage where the multiplexed data are to be transmitted over long distances (highway) in a noisy environment.

Signal processing may be adapted to the requirements of each channel and can be applied individually at the analog or digital level.

Adaptive bandwidth allocation to the different channels can be implemented by employing buffers which are sampled proportionally to the rate at which they accumulate data – the 'squeaky wheel' philosophy. This not only may result in significant reduction of the overall throughput

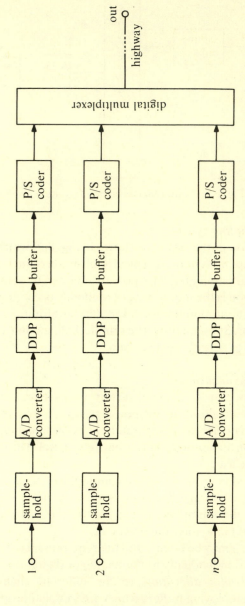

Fig. 8.3. Data acquisition for noisy environment: decentralized approach. DDP: digital data processor; P/S: parallel-to-series. Adapted from [1].

rate, but also has a derandomizing effect on each channel, an important feature in special cases. Care should be taken, in this case, that the time at which the measurements are taken accompanies each sample through the channel buffer in addition to an identifier for each individual channel. The latter is essential in adaptive channel sampling, where the channel cannot be identified from its location in the frame.

Comparison between centralized and decentralized systems

Comparing the cost of the two schemes, we note that the decentralized system employs several basic ICs, which repeat themselves in each channel. Their individual specifications are much less stringent than the specifications of units used in a *centralized system*, which need to have higher speed, higher accuracy and be less susceptible to noise. Note that analog multiplexers suffer from offset voltage, switching noise, non-linearity and crosstalk, all of which are practically non-existent in a digital multiplexer. The repetitiveness of a large number of relatively simple units opens the possibility of realizing *decentralized systems* using large- or medium-scale ICs, making them highly competitive with centralized schemes. The multiplicity of identical units also helps with the maintenance of the system. Last but not least, a catastrophic failure is less likely, which may be of decisive importance in spacecraft and military applications.

9

Estimation of statistics of waveforms and events

In Appendix A.1 we introduce the amplitude distribution function and its derivative, and in Appendix A.2 the correlation function and its equivalent in the frequency domain, the spectral power density function.

In the present chapter we deal with the case in which the statistical properties of a signal are unknown *a priori*, and need to be determined empirically.

Statistical distributions can be estimated by an analog or a digital computer which can compute the average of two or more sequential arrays, element by element. By averaging the computed arrays over a sufficient time period, statistical fluctuations may be filtered out to any desired level, provided that the process under analysis is stationary and the measuring apparatus does not contribute any error.[1]

With non-stationary processes, the averaging time may be adjusted and plots of ensemble-averaged arrays versus time may be used. Alternatively, averaging may be performed in the exponential mode. In an analog computer, this is accomplished by connecting a resistor in parallel with the analog storage capacitors, thereby obtaining an exponentially weighted average of the stored analog samples. In a digital computer, averaging is accomplished by continuously updating the current average by an algorithm which reduces expoentially the contribution of the 'old' samples as time advances (see Appendix F).

9.1 Probability density and distribution functions [2]

9.1.1 Estimation

The PDF (*probability density function*) of voltage waveforms varying with time can be measured using a multichannel analyser. The analyser samples the voltage waveform at constant time intervals Δt compatible with its bandwidth, converts the amplitude of each sample to a digital number in its A/D converter, adds one count to the memory location corresponding to the digital number, and then re-samples when the next sampling command occurs.

If the sampling interval is not coherent with the sampled waveform, every time interval of the waveform is equally likely to be a sampling interval; thus the analyser is, in effect, recording the relative frequency of occurrence of voltage levels of the incoming waveform. If, for example, a ten-bit A/D converter is used and the highest quantized voltage level is 10.23 V, the channel width equals 10 mV. Thus the frequency ratio of the signal occurring within any of the 10 mV intervals during one measuring cycle approaches the probability that the value of the signal lies within that particular interval as the duration of the measuring cycle and hence the number of samples is increased.

It should of course be kept in mind that whether or not a good estimate can be obtained depends strictly upon just how concentrated the probability distribution of an individual sample is about its mean value. In order to evaluate the variance of this measurement, the particular law of distribution must be known.

If the channels (representing voltage levels) are displayed in the horizontal axis and counts in the channels are displayed vertically, we obtain a plot of frequency of occurrence versus voltage level. This plot (when normalized) approaches a plot of the PDF of the voltage input as the number of samples increases. Hence, the probability of any one

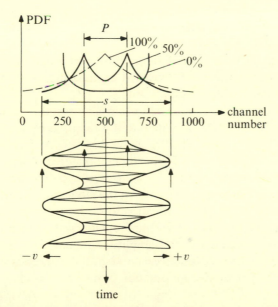

Fig. 9.1. PDF of amplitude modulated sine wave for various modulation depths. Adapted from [2].

voltage level is the count read on the vertical axis in the channel corresponding to the voltage level, divided by the total number of samples taken.

Another kind of PDF analyser employs an A/D converter which follows the input signal continuously (continuous balance). This yields directly the time intervals during which the signal has dwelt between the various quantized voltage levels.

For the particular channel $v_1 < v \leqslant v_2$, yielding m time intervals $\Delta\tau_j$, we obtain

$$\int_{v_1}^{v_2} p_v(v)\,\mathrm{d}v = \frac{1}{T}\sum^m \Delta\tau_j \qquad (9.1)$$

where $\mathrm{d}v = v_2 - v_1$ and T is the total measuring time.

After a probability function has been stored in the multichannel analyser memory, the probability distribution function or exceedance distribution function can be obtained by summing the numbers stored in the corresponding memory locations, and then dividing this by the total number of counts stored.

9.1.2 Applications

Amplitude modulation
Much of the needed information about an amplitude-modulated waveform is available in its PDF. Both qualitative information about the form of the modulating signal and quantitative information (for example, percentage modulation) are provided by measurements with a multichannel analyser.

Fig. 9.1 demonstrates the relationship between the peaks moving in and the skirts of the PDF moving out, if a carrier is modulated between 0% and 100%. The top of Fig. 9.1 shows three PDFs of a carrier which is unmodulated, modulated by 50%, and modulated by 100%. The carrier shown below is modulated by 50%. Referring to the three PDFs, the amount or (percentage) that the peak of the modulated waveforms moves from its original unmodulated location toward the center is a measure of the percentage modulation. If the unmodulated waveform is not available for comparison to compute the amplitude modulation, the modulated carrier waveform density function contains all the necessary information to compute the percentage modulation of the carrier, which is

$$100\frac{(S-P)}{(S+P)}$$

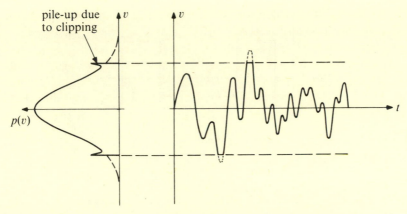

Fig. 9.2. PDF of clipped Gaussian noise. Adapted from [3].

where P is the separation of the peaks and S the separation of the maximum excursion of the skirts of the PDF. When this information is digitized in an analyser memory, simple channel locations of the skirts and the peaks enable the measurements to be made quickly.

Non-linearity
The PDF can also yield important information about non-linearities in a sytem. If, for example, a Gaussian type signal is applied to an amplifier having an insufficient dynamic range, the distorted output has a PDF with 'ears' indicating that clipping has occurred. This is shown in Fig. 9.2.[3]

Non-linearity can also be measured employing a periodic pulse train as a test signal.[4] Fig. 9.3 shows the ideal waveform and its PDF, and Fig. 9.4 shows the PDF of a distorted pulse train. The heights k_1 and k_2 represent the fraction of time the signal exists at these levels. Areas A, B, and C represent non-linearity of three different types. Areas A and C represent an overshoot of the negative and positive pulses respectively. Area B results from the finite rise and fall times of the waveform. Sensitivity depends on the resolution of the A/D converter used to quantize the voltage levels (i.e. the number of channels of the multichannel analyser).

Structural design problem
The variety of applications for probabilistic information could be continued for many more pages. To indicate the range of applicability, we shall conclude this section with an application of the probability distribution function to a structural design problem.[4]

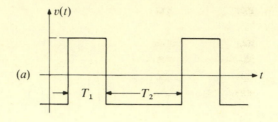

(a)

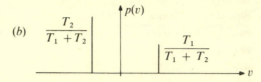

(b)

Fig. 9.3. (a) Pulse train. (b) PDF of (a).

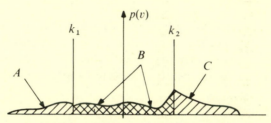

Fig. 9.4. PDF of distorted pulse train. Adapted from [4].

A structural member to be tested is subjected to random loading conditions in a controlled environment. The loading levels are varied in a programmed manner and the test continues until fracture occurs. The probability distribution function of load conditions is drawn with the abscissa in physical units of load and one minus the ordinate corresponding to the probability of exceeding a given loading condition or, if multiplied by the number of times a load was applied, to the number of times the structure was subjected to loads exceeding a given value. By taking the data both during the test and at fracture, valuable design information concerning safe and maximum load levels is obtained.

9.2 Probability functions of events

9.2.1 Estimation[2]

If n events occur in t seconds, their distribution with respect to time can be measured with n or t operating as the random variable, and t or n fixed as a parameter. For example, one can measure either the probability that n pulses from a nuclear source will occur in 1 ms, or the probability that t seconds will elapse for every five pulses. In the first case the PDF is discrete since the random variable n is discrete, and in the second it is continuous since the variable t is continuous. The PDF of n events occurring in t seconds is written $p_n(n, t_0)$ when n is the random variable and $p_t(n_0, t)$ when t is the variable.

The probability and exceedance distribution functions of n events occurring in t seconds with either n or t as the random variable are defined similarly:

$$P_n(n_j, t_0) = \sum_{n=0}^{n_j} p_n(n, t_0) \quad \text{and} \quad 1 - P_n(n_j, t_0) = \sum_{n=n_j+1}^{\infty} p_n(n, t_0), \qquad (9.2)$$

or

$$P_t(n_0, t_j) = \int_{t=0}^{t_j} p_t(n_0, t)\, dt \quad \text{and} \quad 1 - P_t(n_0, t_j) = \int_{t=t_j}^{\infty} p_t(n_0, t)\, dt. \qquad (9.3)$$

Probability functions of events can be estimated using a multi-channel analyser and a scaler timer operating in the preset count or preset time mode. A suitable block diagram is shown in Fig. 9.5. The scaler counts voltage pulses arriving at its count input terminal during a time interval corresponding to a fixed number of pulses arriving at its clock input (hence preset count). The number of voltage pulses counted (i.e. the random variable) is fed into the address counter of the multi-channel analyser and thereby determines the address location for each measurement. The number stored at this address is increased by one, the

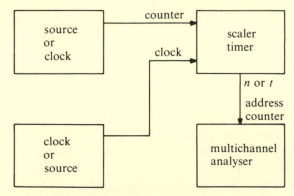

Fig. 9.5. Measurement of $p_n(n, t_0)$ or $p_t(n_0, t)$.

timer and address counter are reset and another measurement is initiated. After a sufficient number of measurements has been made, the normalized contents of the memory correspond to the probability that n pulses arrive at the count terminal during the time determined by the preset number of clock input pulses.

Hence, if a random source of events is connected to the count input and a clock to the clock input of the scaler, we measure the probability $p_n(n, t_0)$ that n pulses from the random source occur during a time $t_0 = m/f$ determined by the clock pulse frequency f and the preset count m (if the scaler has an internal clock, it can be operated in the preset time mode and no external clock is required). If, however, the connections from the random source and the clock are reversed, we measure the probability $p_t(n_0, t)$ that t seconds elapse while a preset number $m = n_0 + 1$ of pulses is emitted from the random source.

The number of pulses (clock or random) falling within the pulse gating the count input events is a possible source of confusion. Assuming the gating pulse turns on immediately after an external input pulse and turns off immediately after the preset mth external input pulse that occurred while the gating pulse was turned on, why does $t_0 = m/f$ if $p_n(n, t_0)$ is measured, but $n_0 = m - 1$ if $p_t(n_0, t)$ is measured?

In the first case, the pulses are coherent (not random) and m is simply the number of clock pulses that occurred during the time of the gating pulse. In the second case, the pulses are random (Poisson distributed, for example) and $n_0 = m - 1$. The beginning and end of the gating pulse, which determine t, are random with respect to events within the gate, excluding the last event (this triggered the gate to turn off and therefore is obviously coherent with the gate). Therefore, the number of events from a random source that occurs in a random time period t is $m - 1$, where m is the preset count setting of the scaler. The last (mth) event serves only as a random signal to trigger the end of the time period t and cannot be counted as one of the n_0 events occurring within the time period of t seconds.

The mean (or expected value) and variance of the measured values is important. If the pulses obtained from the source are Poisson distributed, then the mean and variance are the same. Thus, in measuring $p_n(n, t_0)$, the mean counting rate and its variance approach n_T/t_{0T}, where n_T is the total number of pulses counted and t_{0T} the total time of counting. Conversely, in measuring $p_t(n_0, t)$, the mean time interval between pulses and its variance approach t_T/n_{0T}.

A related problem is the measurement of constant time intervals, a deterministic signal. This is discussed in Section 10.2.

9.2.2 Applications

In addition to the most obvious application, namely measurement of the time statistics of a random source of events, this measurement technique can be put to many other uses.

Error probability density functions
If a voltage pulse can be triggered when an error occurs in a system, a multichannel analyser can be used to measure the PDF of the number of errors (or successes) with the system shown in Fig. 9.5. Moreover, even though the trials (in which either successes or failures can occur) are not linearly related to time, the PDF of the number of failures or successes n in a fixed number t_0 of trials or of the number of trials t it takes for a fixed number n_0 of failures or successes to occur can be measured. In this case the clock pulses in Fig. 9.5 would become pulses indicating trials. Thus, for example, if $p_t(n_0, t)$ is desired, the input to the address counter of the multichannel analyser would be pulses representing the trials, while the input to the clock terminal of the scaler timer would be the pulses representing the failures or successes, and the preset count setting would be n_0. The percentage distribution of errors in a digital communication system, for example, could be measured with this system.

Nuclear spectroscopy
An important application of probability measurement of events occurs in nuclear spectroscopy:[5] namely, measuring the statistical distribution of the energy levels of nuclear emissions from a radioactive source.

If a nuclear event is detected, the resulting signal is converted into a pulse whose amplitude is proportional to the energy measured. These pulses can be fed into a comparator, with those whose amplitude exceeds the threshold voltage V_R producing a standard output. For a linear signal processing system, the threshold voltage can be calibrated in terms of energy. Hence, a plot of the normalized number of pulses $N(E)/N_T$ counted at the output of the comparator versus energy E yields the exceedance distribution function $1 - P(E)$ for those events whose energy is greater than E, with $E = KV_R$ and K, the factor of proportionality, to be determined in each particular case:

$$1 - P(E) = N(E)/N_T, \tag{9.4}$$

where N_T is the total number of pulses measured during the experiment. In practice, experimental data are not normalized, and the resulting relationship is given by (9.5) which is commonly defined as the *integral*

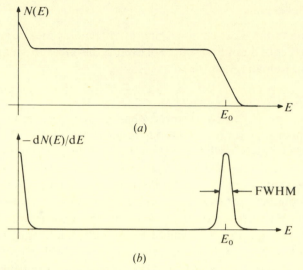

Fig. 9.6. Energy spectrum: (*a*) integral counting rate; (*b*) differential counting rate.

counting rate:

$$N_T[1 - P(E)] = N(E). \qquad (9.5)$$

Differentiation of (9.5) with respect to E yields the *differential counting rate*:

$$N_T p(E) = -dN(E)/dE. \qquad (9.6)$$

Equations (9.5) and (9.6) are plotted in Figs. 9.6(*a*) and (*b*), from which the physical significance of (9.6) is recognized as a plot of the spectral energies. Typical for such a spectrum is the broadening of the spectral line, due to statistical fluctuations in the system. It is customary to measure this effect as FWHM (*full-width half-maximum*), the width of the spectral line at half its amplitude. For Gaussian noise, FWHM = 2.35σ.

The energy spectrum of Fig. 9.6(*b*) shows a single energy peak whose expected value is E_0, and an increase at low amplitudes in the counting rate due to noise. Since only pulses of positive polarity are counted, the negative half of the noise amplitude density function is missing.

In practice, plotting the integral counting rate and subsequent differentiation is a rather archaic way of measuring an energy spectrum. A statistically superior method is to employ two comparators which measure the differential counting rate within an energy window $\Delta E =$

$E_2 - E_1$ and sliding the window over the energy range of interest. Still, this is a tedious procedure, and the method commonly employed is to use a multichannel analyser operating in the pulse height analysis mode. Such an instrument quantizes the incoming pulse amplitudes, assigns a channel to each quantized level and stores the number of events occurring at each of the respective levels. After sufficient statistical information has been collected, the number of counts recorded in each channel, displayed versus channel number, yields the desired energy spectrum. Initially built as hard-wired special-purpose instruments these analysers are being gradually replaced by sophisticated systems performing a variety of functions such as peak stripping of spectra, smoothing of data, computing peak centroid location and FWHM, and featuring different kinds of two- or three-dimensional displays.

Frequency distribution
Since the frequency is a measure of the number n of events occurring in one second, a measure of its PDF can be made with either of the two procedures discussed above. Such frequency distribution measurements might be useful in determining the instabilities of a microwave signal source. If the long-term aging effects are essentially negligible for the period of time that data are being accumulated, the resultant plot will show the distribution of the instabilities around the center frequency of the oscillator. Great care should be taken that the jitter introduced by measuring n (or n_0) and t_0 (or t) is negligible compared with the standard deviation of the measured distribution. This not only concerns the frequency standard employed, but also the associated timing circuits.

Period and phase distribution
The period distribution of a waveform is simply the PDF of the time elapsed during the occurrence of $n_0 = 1$ event (or period). This can be applied to measure the jitter of a pulse train. Similarly, the phase distribution between two waveforms or between two points of the same waveform could be measured. Further examples are: monitoring of production rates; measurement of telephone call distribution for optimizing telephone line routing; distance distribution in radar measurement; distribution of zero crossings and switching time of a random binary source.

As a final example, the jitter in the elapsed time between two consecutive events occurring in processes such as the emission of particles due to a decaying radioactive element can be measured. This kind of measurement makes it unnecessary to compare the random time variable with a timing standard, since the successive time intervals can be compared with each other by converting them into voltage pulses of amplitude proportional to the time intervals (employing a TAC) and measuring the distribution of the voltage pulses in a multichannel analyser. This technique is widely used in nuclear spectroscopy, where probability density functions with standard deviations of the order of 10^{-11} s are routinely measured.

9.3 Correlation functions and power density spectra

9.3.1 Estimation[4]

A basic configuration for estimating correlation functions is shown in Fig. 9.7. With the switch set to the auto position, the output is

$$R_x(\tau) = \frac{1}{T} \int_0^T x(t)x(t-\tau)\, dt \qquad (9.7)$$

and, with the switch on cross,

$$R_{xy}(\tau) = \frac{1}{T} \int_0^T x(t)y(t-\tau)\, dt. \qquad (9.8)$$

There is of course a great deal of flexibility built into this basic configuration. The primary processing block consists of a delay or storage element, a multiplier and an integrator or averaging circuit. As shown, this configuration is processing continuous signals. Its realization employing analog building blocks is straightforward.

The advancement in digital processing techniques has brought about a digital or sampled data equivalent of this processing system which is instrumented by a digital correlation analyser and conforms to the processing equations given below:

$$R_x(\pm\tau) = \frac{1}{m} \sum_{i=1}^m x(i\,\Delta t)x(i\,\Delta t - j\,\Delta t), \qquad (9.9)$$

$$R_{xy}(\tau) = \frac{1}{m} \sum_{i=1}^m x(i\,\Delta t)y(i\,\Delta t - j\,\Delta t), \qquad (9.10)$$

where $\tau = j\,\Delta t$.

The mechanization of this equation is shown pictorially in Fig. 9.8(a), and Fig. 9.8(b) shows its practical implementation. Both signals $x(t)$ and $y(t)$ (just $x(t)$ or $y(t)$ for autocorrelation) are sampled and digitized at increments Δt of the desired delay variable τ. The sampling rate $1/\Delta t$ is chosen to be consistent with the signal bandwidth. In Fig. 9.8, $x(i\,\Delta t)$, the ith sample of $x(t)$, is being multiplied with samples $(i-j)\,\Delta t$ of $y(t)$.

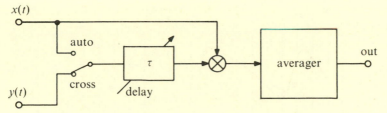

Fig. 9.7. Analog correlator. Adapted from [4].

Assume that we wish to compute 100 points of the correlation function – a reasonable number for most practical purposes. In this case $j = 0, 1, \ldots, 99$, and we need to store repeatedly the value of 100 samples of $y(t)$, taken at time intervals Δt. This is the function of the delay store implemented in most practical cases by a closed-loop shift register, in which the circulating words become externally available as they pass serially through the output.

For each updating, the correlator goes through two cycles of operation: the *acquisition cycle* in which a pair of samples is being taken and digitized, and the *process cycle* in which the sampled value $x(i\,\Delta t)$ of $x(t)$ is consecutively multiplied by all 100 values $y[(i-j)\,\Delta t]$. These digital multiplications are performed between the samples being taken.

In the next acquisition cycle, the new digitized sample of $x(t)$ replaces the previous common multiplicand, and that of $y(t)$ shifts all the numbers in the delay store by one position to the right. Each previous sample bit $y(t)$ now becomes $y(t-\Delta t)$, and the old $y(t-99\Delta t)$ value is discarded.

The products resulting from the 100 multiplications performed during the process cycles are each stored and averaged in the corresponding location of the main store. In the straightforward averaging mode the instrument goes through m complete cycles of operation, the products are summed up and the final result is divided by m.

Each value of m corresponds to a different integration or averaging time. If the processing is accomplished in the exponential averaging mode (see Appendix F), each different value of m corresponds to a different RC time constant. Moreover, the exponential decay can be shown to result in a distortion of the correlation function or loss in processing gain. (In some applications, however, this distortion is accepted in order to achieve a sliding window type of integration.)

Note that only half of the crosscorrelation function is computed by the arrangement shown in Fig. 9.8(b). The other half is computed by exchanging the inputs, since $R_{xy}(-\tau) = R_{yx}(\tau)$.

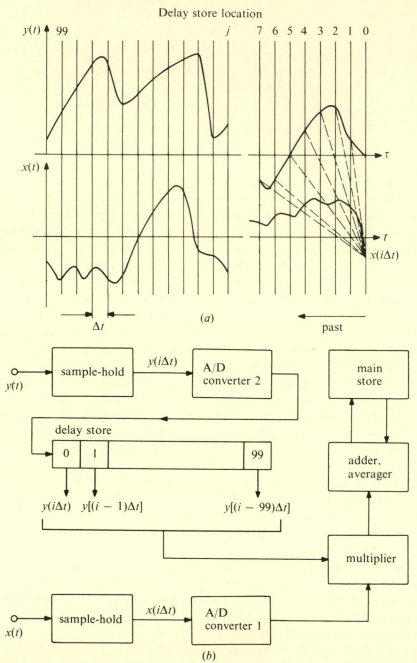

Fig. 9.8. (*a*) Single sampling cycle of sampled data correlator. Adapted from [4]. (*b*) Digital crosscorrelator.

Once a correlation function is known, it can be converted into the power density spectrum, since the two are Fourier transform pairs. This is a standard procedure if the correlation function is available in digital form.

The autocorrelation function of noise can be observed on an oscilloscope, as has been demonstrated by Radeka.[6] This technique is useful in making a rough estimate of the effect of a parameter variation on the noise spectrum under consideration.

9.3.2 Applications[7]

Not only are the autocorrelation and crosscorrelation processes powerful techniques for extracting signals buried in noise, but the inherent properties of the crosscorrelation function enable the user to study the transmission or propagation delay characteristics of various systems, whether they be electrical, mechanical, or biological in nature. This information is complemented by the power spectrum which identifies the signal frequency content and the relative strength of each frequency. This is important in the classification of complex aperiodic wave shapes.[8] Examples of such wave shapes are: monopulse radar returns, which must be separated into target and non-target types; microphone responses to spoken words, which must be classified as one of several commands; EKG time histories, which must be interpreted as normal or abnormal heart functioning; sonar echoes, which must be classified into fish or plant life types. In addition, the crosspower spectrum provides valuable information concerning the existence and relative phaseshift of frequencies common to two signals.

Characterisation of linear systems
Y. W. Lee[9] showed that it is possible to obtain the unit impulse response of a linear system by driving it with broadband (white) noise and crosscorrelating this input with the system output. A basic block diagram is shown in Fig. 9.9.

This technique has important practical implications. Consider, for example, the effect of spurious noise when trying to determine an impulse response or when the system is in constant use and is being driven by external signals. Even under these circumstances it is possible to obtain the crosscorrelation function between the system output and a test white noise signal. Since there is no correlation between the test signal and the control signal or extraneous noise, the control signal and noise will not affect the crosscorrelation function obtained – which is the impulse

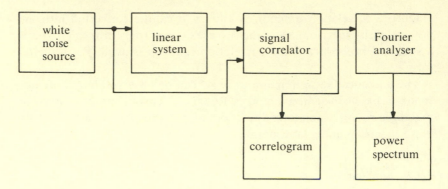

Fig. 9.9. Measurement of linear system impulse response.

response undisturbed by the extraneous signals. This immunity to internal system noise also allows the response to be obtained with very small exciting noise signals which do not interfere with the signals that the system normally handles. Thus the impulse response function can be determined while the system is in operation.

With correlators operating in real-time it is possible to keep the response of critical systems under constant surveillance and to make optimizing adjustments. Self-optimizing systems can be constructed by introducing feedback from a subsystem that evaluates the impulse response. And, because the correlation function is comprised only of frequencies common to the white noise test signal and the system output signal, a Fourier analyser can be used to process the correlation function to provide the system transfer characteristics as a function of frequency.

Vibration analysis
Vibration studies employ random noise as a realistic simulation of actual conditions (Fig. 9.10). A strain gauge attached to the specimen under study at critical points provides an output which, when correlated with the random noise input, displays the period of the resonant frequency of the sample under study. It is usually difficult to discern between random and coherent components in a vibratory system. Using autocorrelation techniques one can determine the degree of periodicity or randomness by evaluating the correlogram. In addition, the Fourier analyser can be used to extract spectral density information; studying the spectral response of a vibratory system often provides additional insight into the physical phenomenon producing the vibration.

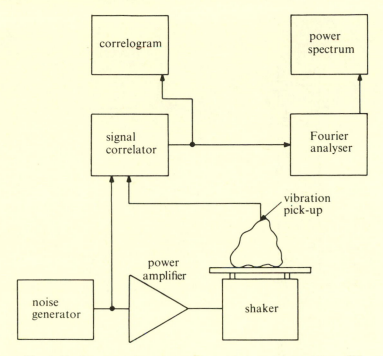

Fig. 9.10. Resonant frequency determination. Adapted from [7].

Turbulent fluid flow

One of the most useful applications of crosscorrelation is in the study of hydrodynamic turbulence (Fig. 9.11(*a*)). Turbulence studies are usually made by repeatedly inserting a velocity-responding probe into the turbulent stream of fluid at different points. Analysis of data from a single probe is, however, only of limited value. By contrast, the crosscorrelation of signals from probes of variable separation gives more meaningful information about turbulence and diffusion. If two velocity responding probes are inserted in a stream of fluid undergoing turbulent flow and the ac components of the signals produced by the probes are crosscorrelated, one obtains results that are qualitatively indicated in Fig. 9.11(*b*).

We note that as the probe *B* is moved downstream relative to the fixed probe *A*, two things happen. First, the time delay for maximum correlation increases with separation, as would be expected from the general flow of the liquid in the pipe. Second, the magnitude of the correlation diminishes and the width of the peak increases, giving information on the amplitude and coherence time of the turbulating eddies.

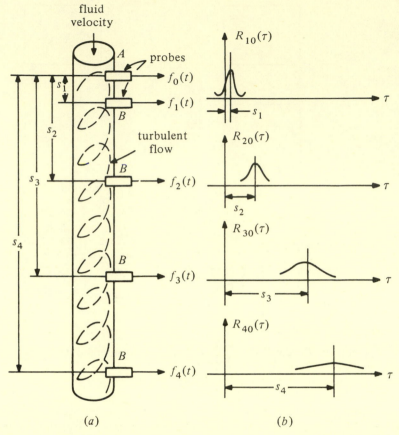

Fig. 9.11. Turbulence measurement by velocity probes. Adapted from [7].

Another method of obtaining information about turbulence is through the use of a correlator and Fourier analyser in conjunction with a hot wire anemometer system as shown in Fig. 9.12. The hot wire anemometer produces an electrical current sufficient to maintain the probes at a constant temperature (see Chapter 2). If the flow is laminar, the currents in probes A and B will be constant. However, if the fluid flow is turbulent, the small eddies cause the electrical current in probe A to be a random nature. If probe B is sufficiently close to probe A, the same eddies flowing by probe B cause a similar electrical current to be generated in line B. Crosscorrelating the signals in A and B produces the crosscor-relating functions $R_{ba}(\tau)$, which have a peak at a specific value of time delay dependent upon the average velocity of the fluid flow and

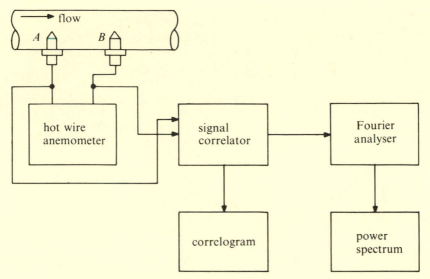

Fig. 9.12. Turbulence measurement by hot wire anemometer.[7]

the distance between the probes. The Fourier analyser will then provide information concerning the frequencies of the turbulence. By comparing the power density spectra from each probe, one can determine whether new frequencies were created in the passage of fluid between A and B. In addition, the crosspower spectrum shows all those frequencies detected by probe A which remained in the fluid as it passed by probe B.

This two-probe crosscorrelation technique for investigating turbulence and diffusion has applications in hydrodynamics, rocket exhaust studies, plasma physics, etc. It is probably the most important single tool available to investigators in these fields.

Identification of crosstalk components

To measure the amount of crosstalk between pairs of cables, amplifier and cable, two amplifiers, or to test the shielding effectiveness of coaxial cable, both a signal correlator and a Fourier analyser are used.

In a typical application (Fig. 9.13), white noise is fed into an amplifier A_1. The noisy output of amplifier A_2 is crosscorrelated with the white noise source and the crosscorrelation function fed into a Fourier analyser which provides the crosspower spectrum of the crosstalk between the amplifiers. In this manner the frequency components of A_1 which have permeated A_2 are easily identified and measured.

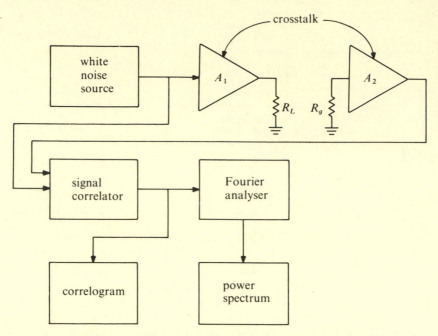

Fig. 9.13. Crosstalk measurement. Adapted from [7].

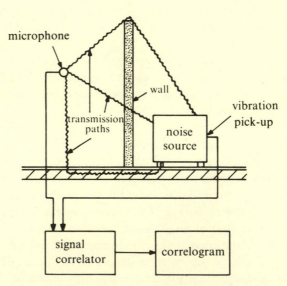

Fig. 9.14. Noise transmission path investigation. Adapted from [7].

Noise transmission studies

Operation of factory machinery will often induce undesirable noise and vibration in neighboring areas, to which energy may be transferred in several ways through the building structure or through the air (Fig. 9.14). Before the noise and vibration can be controlled, it is necessary to establish the transmission paths. Such problems can often be solved by placing a microphone in the noisy environment and crosscorrelating the input and output for the system under study. Since each transmission path through the system will generally be associated with a different delay time, a separate peak will occur in the crosscorrelogram for each path which contributes significantly to the output. If the expected time delays associated with the various possible paths can be calculated, these delays can then be compared to the measured time dispacements of peaks in the crosscorrelogram to identify the paths contributing significantly to the output.

Other applications

The above applications for correlation techniques form only a partial list of possible uses. Other applications include plasma physics studies, radar signal retrieval, sonar studies, geophysical and oceanographic investigations, biomedical studies, noise source direction finding, torsion measurement in rotating shafts, contactless velocity measurement, speed research and others.

10

Signal enhancement

A very significant role in signal processing is played by waveform averaging computers. This technique, termed signal averaging or enhancement, utilizes the additional information conveyed by the repetitive nature of certain signals to improve their SNR. Similarly, time interval averagers improve the accuracy of time measurement.

10.1 Signal averagers[1,2]

Signal averaging or *signal enhancement* provides a means of estimating the shape of a periodic or repetitive response buried in non-coherent interference. By using the redundant information inherent in repetition, the SNR can be improved. Multiple exposure photography and persistent CRT displays have been used for years to superimpose repetitive signals in order to enhance their SNR.

Signal averaging can be instrumented by repeatedly sampling the noisy signal at fixed time intervals, and storing the time-correlated samples in a memory. Fig. 10.1 shows a repetitive signal which is sampled n times per cycle. Each sampling cycle is initiated by a trigger derived from the signal if it is repetitive, or from the stimuli if the signal is due to an evoked response. The n time-correlated sets of samples are stored in n separate

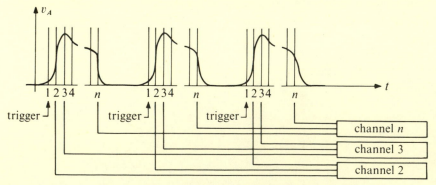

Fig. 10.1. Waveform averaging. Adapted from [3].

memory locations, indicated by the n channels in the figure. If the sampled noise values are uncorrelated, then the improvement attained in the presence of stationary noise is the square root of the number of repetitions. If, however, the sampled noise values are correlated, the improvement could be either smaller or greater, depending on whether the correlation factor between the samples is positive or negative (i.e. anticorrelation, in the latter case).

If the sampling cycle is repeated a sufficient number of times, the average value of the samples in each memory location approaches the expected value of the signal plus that of the noise, the latter being zero. Hence, theoretically the SNR may be enhanced arbitrarily if the sampling process is repeated a sufficient number of times. In practice, however, this improvement is limited by instrument noise, i.e. by noise introduced at various stages, in the signal averager.

The improvement of the SNR, due to signal averaging, can be shown formally as follows.

Let the input be $f(t)$, composed of a repetitive signal portion with a period T, $S(t)$ and a noise portion $N(t)$:

$$f(t) = S(t) + N(t). \tag{10.1}$$

We shall sample m repetitions of the signal with the ith repetition beginning at t_i, $i = 1, 2, \ldots, m$ and $t_1 = 0$. Assume the samples are taken every Δt seconds. The sampled values are therefore

$$f(t_i + j\Delta t) = S(t_i + j\,\Delta t) + N(t_i + j\,\Delta t)$$
$$= S(j\,\Delta t) + N(t_i + j\,\Delta t) \tag{10.2}$$

with $j = 0, 1, 2, \ldots, (n-1)$. For a given j, the set of samples $N(t_i + j\,\Delta t)$ is a random variable. It is reasonable to assume in a real situation – where the noise is thermal, shot or low-frequency – that the mean value of this variable is zero and that the samples are taken at time intervals T sufficiently long to make the sampled noise values uncorrelated.

If the rms value of the noise is σ, then the SNR of any single sample equals

$$\text{SNR} = S(j\,\Delta t)/\sigma. \tag{10.3}$$

After m repetitions, the value stored at the jth memory location is

$$\sum_{i=1}^{m} f(t_i + j\,\Delta t) = \sum_{i=1}^{m} [S(j\,\Delta t) + N(t_i + j\,\Delta t)] \tag{10.4}$$

where

$$\sum_{i=1}^{m} N(t_i + j\,\Delta t) = m^{1/2}\sigma.$$

Hence the SNR after m summations is

$$(\text{SNR})_m = mS(j\,\Delta t)/m^{1/2}\sigma = m^{1/2}\,\text{SNR}. \tag{10.5}$$

Averaging in the frequency domain
It is instructive to consider a pulse train of m pulses, taking the jth sample of a repetitive waveform,

$$h_j(t) = \sum_{i=1}^{m} \delta(t - iT - j\,\Delta t) \tag{10.6}$$

where T is the time interval between the sampling cycles.

Transforming $h_j(t)$ into the frequency domain, we obtain the transfer function of the averager:

$$|H_j(j\omega)| = \sin(\omega m T/2)/\sin(\omega T/2). \tag{10.7}$$

Equation (10.7) describes a *comb filter*, which exhibits peaks of amplitude m at frequencies $\omega = \pm 2r\pi/T$, $r = 0, 1, 2, \ldots, \infty$.

For large values of m, the 3 dB width of one tooth of the comb filter is

$$\Delta f \approx 0.89/mT. \tag{10.8}$$

It is the averager's comb filter behavior that makes this kind of signal processing so eminently suitable when the desired signal and the noise are in the same frequency range. The noise power remaining in the averaged waveform is the sum of the powers contained in each comb tooth. Since the tooth width is inversely proportional to the number m of repetitions, the ENB is also inversely proportional to m. Here, then, is another demonstration that the rms noise reduces by a factor $1/m^{1/2}$ with respect to the signal if the noise spectrum is white.

Three kinds of averaging
The choice among the three possible averaging algorithms depends on the nature of the signal.

For stationary signals, the equally weighted averaging mode should be applied. The total number of summations m taken will depend on the desired improvement in the SNR, which is limited only by noise introduced by the averager.

If the characteristics of the signal are not strictly uniform, however, the signal will change somewhat during a long averaging period. For this case, the longer averaging periods are not very informative, since the change in signal characteristics over such a period reduces the theoretical effectiveness of the averaging process. Thus, the number m of repetitions should be chosen judiciously, and the summing process should be repeated if a slowly changing signal is to be observed.

Finally, some signals such as certain evoked biological responses exhibit significant transient changes and hence are not even quasi-stationary, but still persist for a sufficient number of repetitions to make signal enhancement possible. Here the exponential averaging mode should be employed.

Choosing the instrument
The three categories of signal averaging computers are the boxcar integrator, the multichannel signal averager, and special optical signal averagers. To specify the system needed in a given application, one requires information about:

– the duration of the transient to be averaged;
– the number of segments into which the transient must be divided to achieve the desired time resolution (bandwidth);
– the number of repetitions of the transient that will be available for averaging (does it change with time or does it self-destruct?);
– the magnitude and nature of the noise that masks the effect being studied;
– the SNR (amplitude resolution) required in the final result.

The effective SNR one achieves in waveform averaging is very much dependent on the above considerations as well as on instrumental factors. If, as an example, slowly changing bioelectric signals are averaged in the exponential mode, the presence of very low frequency noise may cause significant correlation between noise samples. This will reduce the factor $m^{1/2}$ of improvement and thereby significantly reduce the ultimate SNR obtainable in the exponential averaging mode. A suitably dimen-sioned baseline restorer will practically eliminate correlation between noise samples and thereby optimize the performance in such a situation. If the bandwidth of the noise is wider than that of the signal, then an antialiasing filter connected between the signal source and the signal averager affects the SNR in two ways: (a) it improves the SNR before signal averaging; (b) it affects the value of the autocorrelation between successively taken noise samples and therefore modifies the factor of SNR improvement due to averaging. In most practical cases, however, correlation between individual samples will be insignificant, leaving the improvement in the SNR due to prefiltering as the dominant factor. The moral of this is that the chosen bandwidth of the antialiasing filter should not be greater than necessary to pass the signal spectrum.

Of course the above considerations assume a perfect instrument which is, in practice, never obtainable. The three main instrumental qualities that determine the idealness of a signal averager are *time efficiency* (by which we mean the ratio of number of repetitions an ideal averager would require to achieve a given SNR to the actual number required), *dynamic range* (a measure of the limiting SNR ratio improvement that can be achieved), and *holding time* (a measure of the volatility of the memory in which the average is being computed).

Time efficiency is a measure of the use made of the available information. In the boxcar averager, the time efficiency is very poor since only a small part of each waveform is averaged at each repetition (see below). Often it is worthwhile to make this compromise to achieve ultra-short gate widths (windows). Also, some multichannel digital averagers only add a few counts per channel at a pass, resulting in poor time efficiency. The analog multichannel averagers have an inherently good time efficiency.

Dynamic range is characterized by the ratio of maximum channel capacity to baseline noise. Digital instruments should have upwards of twenty-four bits per channel of memory to achieve a dynamic range of better than 1:1000. In analog instruments, the dynamic range is usually determined by instrumental factors.

Holding time is a measure of an instrument's ability to retain information for a long period of time. For digital instruments this is infinite for all practical purposes. Analog systems, while having finite memories generally in the form of charge on a capacitor, are adequate for most applications where the repetition rate is not too low.

Boxcar integrators
As stated earlier, boxcar integrators are designed to average only a tiny segment of the repetitive waveform. One way to look at boxcar integration is to compare it with synchronized strobe lighting. At precisely the proper moment in time the strobe lamp fires, illuminating a rotating object for a split second and making it appear to freeze in its motion. The boxcar integrator operates in essentially the same manner. As shown in Fig. 10.2, a repetitive waveform and a synchronous trigger is presented to the boxcar. At precisely the selected moment, the sample-hold gate opens for a very short selected period (aperture time), and the portion of the waveform existing during this time is retained and successively

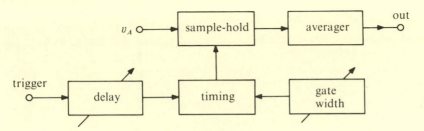

Fig. 10.2. Boxcar integrator.

averaged. The rest of the waveform is ignored. Since many repetitions of
the signal are sampled, the output of the boxcar will be proportional to
the average level of the input signal minus any random noise which
accompanied the signal.

The actual point in time when the gate is opened is determined by an
external trigger signal, usually the same one employed to drive the
experiment, and a time delay circuit built into the boxcar. By increasing
the time delay, segments occurring further into the waveform can be
averaged.

The length of time during which the gate remains open depends on the
setting of a gate width or aperture control. Selection of the optimum
gatewidth for a given application is determined by a trade-off between the
SNR improvement required and the time inefficiency that can be
tolerated.

Some boxcar integrators can be set to scan automatically across the
entire input waveform during a period of time in which many repetitions
of the waveform occur. Gate width is selected by internal circuitry so that
each point along the waveform is sampled at least ten times to provide a
reasonably close average of its amplitude. This mode of operation is used
in conjunction with chart recorders to provide a hard-copy of the actual
waveshape in its entirety.

Multichannel averagers
Where the entire waveform must be averaged in the shortest possible
time a multichannel instrument must be used. The input signal plus the
noise or interference associated with it is amplified and fed into a number
of sequentially switched averaging circuits. As these switched channels
sequentially sample the waveform on each repetitive signal, the noise –
since it is random – averages to a minimum level while the contents of
each channel tend to the average level of the signal during the corresponding interval.

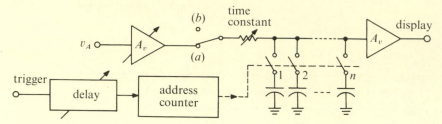

Fig. 10.3. Analog multichannel averager: (a) record; (b) display. Adapted from [2].

There are three types of multichannel averager available. The analog type shown in Fig. 10.3 is the fastest and most time efficient of any multichannel analyser. Fifty and one-hundred channel versions are available which, because they are analog, can provide continuously variable window widths – not the fixed widths inherent in digital instruments.

The volatility due to leakage of the capacitor type storage used in the analog multichannel averager limits its use to relatively short period averages of less than fifteen minutes duration. The low cost and high speed, however, make the analog averager ideal in many applications. The exponential averaging mode is realized simply by connecting a resistor in parallel with each storage capacitor.

The typical digital multichannel averager shown in Fig. 10.4 consists of an optional sample-hold input circuit (not shown here), an A/D converter, control logic and memory banks.

In many cases these functions are divided between a main frame and plug-in modules. These modules provide the flexibility of performing one, two or four channel signal averaging (sharing the memory), multiparameter analysis, probability density analysis, time interval distribution and time trend analysis histograms, auto- and crosscorrelation analysis, and exponential or normalized signal averaging.

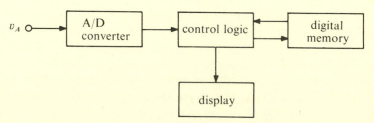

Fig. 10.4. Digital multichannel averager.

They usually have 1000 to 8000 eighteen to twenty-four bit words and a minimum channel width of 1 μs (usually with reduced A/D conversion resolution). These units can best be described as digital computers with hard-wired programs. As such, they are very compatible with computer interfacing so that additional processing of average data can be accomplished.

Because of the limited size of the memories in these systems, little more than signal averaging can be accomplished on-line. Additional functions can be realized, however, by adding more memory and software. The most flexible of all multichannel signal averagers – but not the most convenient – is the general-purpose programmable digital computer used on-line with the necessary peripheral equipment and software. The best compromise would seem to be the use of a dedicated signal averager for the actual raw signal processing and, if needed, a minicomputer to perform any additional data manipulation, such as Fourier transformation and normalizing.

Optical signal averager
In the field of optical spectroscopy, a special-purpose signal averager – the optical multichannel analyser – combines the parallel data recording capability of a traditional photographic emulsion with the speed and signal processing facilities of modern electronic instrumentation.

The optical multichannel analyser differs from conventional signal averagers in that the signal of interest is not inherently a time-varying voltage, but must be made into such. Since a conventional spectrophotometer uses dispersive optics to project spectral information as 'lines' spread out along a dimension perpendicular to the lines themselves, it is possible to use any of several types of television camera tubes to record the spectra and transform them into time-varying voltages suitable for further processing.

Applications of signal averagers
Fluorescence decay studies provide insights into the composition, structure, and energy levels of atomic particles. In most instances, a sample excited by a light stimulus as shown in Fig. 10.5 can be induced to fluorescence. A photoelectric transducer output is usually applied to an oscilloscope to obtain a trace of waveform decay. Unfortunately, noise in the photodetector electronics, photon noise, and limited resolution of the oscilloscope time scale produce errors in obtaining a faithful decay curve characteristic of the material under test. By signal averaging repeated decay curves, the SNR is improved.

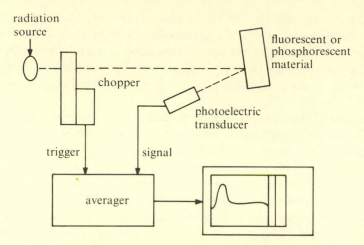

Fig. 10.5. Fluorescence and phosphorescence studies. Adapted from [4].

An important application is the detection of mechanical equipment malfunctions.[4] For example, consider the output of an accelerator, vibration pick-up, an engine or any machine with rotating components displayed as a function of shaft angle or time from some periodic reference event. The vibrations or noise emitted from gears, bearings, pistons, valves or other internal structures will have a particular sound characteristic for their condition and position. This is called a mechanical signature. Plots or displays of amplitude versus time or shaft angle are used to compare normal and defective equipment and monitor operating equipment for the purpose of detecting incipient malfunctions due to mechanical wear, and scheduling equipment shutdown for maintenance.

A popular application area for signal averagers is in biomedical research experiments in which a visual, audible, or tactile stimulus is applied to a subject. As an example, the evoked response that occurs in relation to the stimulus is buried among other EEG (*electro-encephalogram*) signals, which are picked up via scalp electrodes (Fig. 10.6). After a sufficient number of stimuli have been applied, the averaged evoked response is extracted from the extraneous non-repetitive and non-synchronous EEG signals.

Noise resulting from investigations involving lasers has plagued researchers since the phenomenon was first discovered. Signal averagers have been applied effectively to recover output wave shapes.

Pulsed ultrasonics is a technique used to characterize materials by studying the attenuation and transit time characteristics of acoustic waves

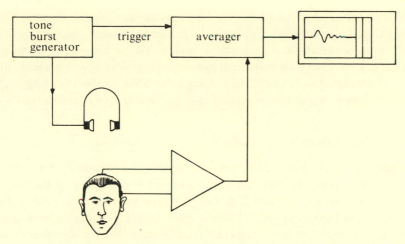

Fig. 10.6. EEG evoked response studies. Adapted from [4].

passing through a sample, be it for use in textiles or more exotic applications as in space re-entry vehicles.

In the typical situation shown in Fig. 10.7, the sample under investigation is interfaced with a buffer material whose acoustic properties are well defined, for example, aluminum. A transmitter and receiver piezoelectric crystal is mounted at each end of the sandwich. A train of pulses is sent

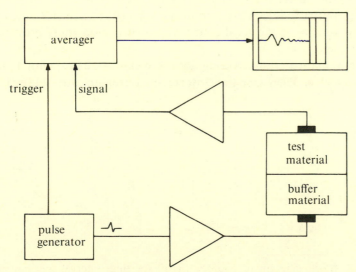

Fig. 10.7. Pulsed ultrasonic studies. Adapted from [4].

through the sandwich and detected by the receiver crystal. By studying the velocity characteristics, one can determine the elastic constants which relate stress and strain relationships, such as Young's modulus. In many instances, the materials under test provide a great deal of attenuation and signal averaging is required to enhance the SNR of the resulting very weak signal to make the necessary characterisations of the material.

10.2 Time interval averagers [5]

Time interval averaging provides a powerful yet economical method of greatly increasing the accuracy and resolution of time interval measurements on repetitive signals. The basis of time interval averaging is the statistical reduction of the ± 1 count error inherent in digital measurements. As more and more intervals are averaged, the measurement will tend toward the true value of the unknown time interval. Fig. 10.8 shows the basic circuit arrangement of a typical universal counter when making a time interval measurement. The time base generates a pulse train with a very accurate and stable period. To measure the unknown time interval, the AND gate must open when the start signal is received and close when the stop signal is received. While the gate is open, the time base (clock) signal passes through the gate and is counted. The total count provides the measure of the time interval.

To average N time interval measurements, the counter accumulates the counts for the individual measurements until N intervals have passed. Since N is usually selectable in decade steps ($N = 1, 10, 100, \ldots$) the displayed reading is the total count with a correspondingly positioned decimal point.

Assume that we are measuring a pulse width of 225 ns, with a clock period of 100 ns. Then each time interval measurement will yield either 2

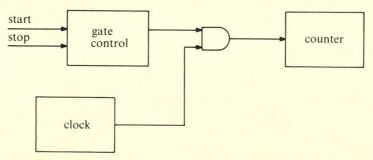

Fig. 10.8. Block diagram of counter circuit for time interval measurement. Adapted from [5].

or 3 counts, that is, a reading of 200 ns or 300 ns. But if we average 100 measurements, we may obtain the following results:

Counts during interval	Number of intervals	Total counts
2	75	150
3	25	75
Total:	100	225

The total count is 225 for 100 intervals, an average of 2.25 counts per measurement. Thus the counter will display 225 ns. The statistical aspect of this measurement will be discussed in the sequel.

The ±1 count error

The fundamental error for all digital measurements is the ±1 count error. Fig. 10.9 illustrates its source. In Fig. 10.9(*a*), a time interval equal to 4.5 clock periods opens the gate to allow 4 counts. In Fig. 10.9(*b*) the same interval allows 5 counts through the gate. The only difference in the two

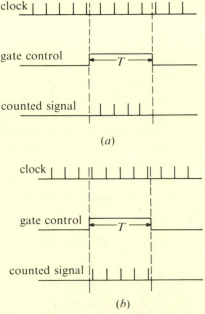

Fig. 10.9. The ±1 count ambiguity of time interval measurement: (*a*) 4 counts; (*b*) 5 counts. Adapted from [5].

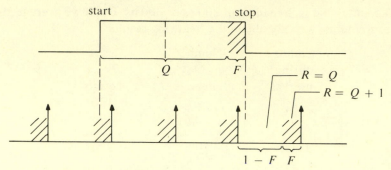

Fig. 10.10. Measurement zones. Adapted from [5].

situations is the phase of the clock relative to the start of the time interval. In this example, the ±1 count error is a significant percentage of the measurement.

In discussing time interval measurements, we express both the measured time interval T and the resultant counter reading R in units of the counter's clock period. Thus the clock period equals 1 count. A time interval equal to 4.5 clock periods (4.5 counts) will produce a counter reading of 4 counts or 5 counts.

As indicated in Fig. 10.10, T can be broken into integral and fractional parts:

$$T = Q + F, \qquad Q = 0, 1, 2, \ldots; \quad 0 \leqslant F < 1. \tag{10.9}$$

After making a time interval measurement, the counter will display either $R = Q$ or $R = Q + 1$, depending on when the start signal occurs in relation to the clock pulse train. If it coincides with the shaded portion of the pulse train, $R = Q + 1$. Otherwise, $R = Q$. In either case, however, the counter will register at least Q counts on every measurement.

Assuming the start of each time interval occurs with equal probability over all phases of the clock (asynchronous repetition rate), the expected counter reading will be equal to the measured time interval. Looking at Fig. 10.10, we can see that

$$\text{Prob } (R = Q) = 1 - F, \tag{10.10}$$

$$\text{Prob } (R = Q + 1) = F. \tag{10.11}$$

If $\hat{R}$ is the expected counter reading, then

$$\hat{R} = Q + F = \bar{T}. \tag{10.12}$$

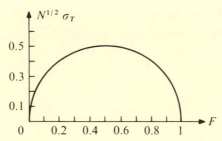

Fig. 10.11. Standard deviation of counter reading as function of fractional part of time interval. Adapted from [5].

Equation (10.12) assumes that the average time interval $\bar{T}$ is known. In practice, the problem is just the reverse: once a counter reading is obtained, the problem is to determine as much as possible about the time interval. It can be shown[5] that the best estimate of the time interval is simply the average counter reading:

$$\hat{T} = \bar{R}. \tag{10.13}$$

The standard deviation of $\hat{T}$ for $N \geqslant 10^4$ is

$$\sigma_T = [F(1-F)]^{1/2}/N^{1/2}. \tag{10.14}[5]$$

Equation (10.14) is displayed in Fig. 10.11.

For a smaller number N of intervals averaged, resolution deteriorates towards extreme values of F (0 and 1, respectively) and slightly improves in the middle ($F = 0.5$).

Other sources of error are as follows.

Internal trigger error. Internal noise in a counter's amplifier trigger circuits can cause the counter randomly to start or stop a time interval measurement slightly early or late. Internal trigger error is generally much less than ±1 count error. Furthermore, time interval averaging reduces it. Hence, this error can be virtually ignored for most measurements.

Time base error. The accuracy of the counter's reference time base, or clock, can limit the accuracy of a measurement for long time intervals. In averaging situations, time base error is rarely a significant factor.

Systematic error. Systematic error includes differences in the propagation times of the start and stop sensors, differential delays in the start and stop channel amplifiers of the counter, and errors in trigger level settings of the start and stop channels of the counter. The systematic error is fixed for a given measurement set-up and constant waveforms. Therefore it affects accuracy but has no effect on resolution. Most systematic errors can be virtually eliminated by calibrating the measurement set-up.

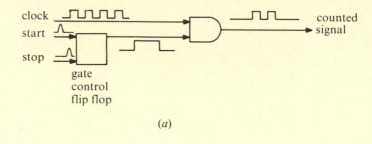

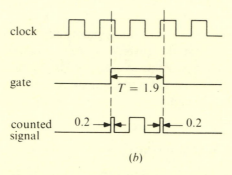

Fig. 10.12. Error due to direct gating: (*a*) block diagram; (*b*) waveforms. Adapted from [5].

Direct gating error. Direct gating can cause an unacceptable bias in time interval measurement by truncating clock pulses. Fig. 10.12 shows what can happen using direct gating.

The clock signal is actually a pulse train. When the gate opens it may truncate some fraction of a clock pulse. When closing, the gate may again truncate a clock pulse. The counter does not know which of the truncated pulses should be counted. In Fig. 10.12, if the minimum countable pulse width of the counter is less than 0.2, then the counter will display $R = 3$ which produces an error of greater than 1 count. Such errors can produce a significant bias in the expected counter reading.

In summary, direct gating has the following disadvantages for time interval averaging measurements.

(1) Truncation of clock pulses can produce more than 1 count error.
(2) Time interval measurements will be biased.
(3) The time interval can be too short to be measured; the counter will never count intervals shorter than the minimum countable pulse width.

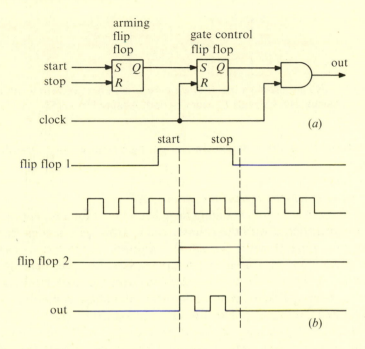

Fig. 10.13. Synchronized gating: (a) block diagram; (b) waveforms. Adapted from [5].

Synchronized gating

Synchronized gating solves the problem of bias in time interval measurements, and produces reliable and accurate measurements. Fig. 10.13 shows a representative synchronized gating circuit and resultant gate timing. In practice several variations of the circuit may be used.[6] The resulting gate is 'synchronized' to the clock. The start and stop signals properly arm the gate either to open or to close; an edge of a clock pulse actually switches the gate control flip flop. Thus only integral clock pulses can pass the gate. No clock pulses are truncated. Since synchronized gating operates only on an edge of a clock pulse, the clock becomes effectively a train of zero width pulses, as shown in Fig. 10.14. Thus synchronized gating provides the following advantages for time interval measurements.

(1) The expected measurement is unbiased.
(2) The synchronized gate can be designed to measure time intervals even shorter than the minimum countable pulse width of the counter.

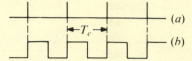

Fig. 10.14. Synchronized gating of clock pulses: (*a*) effective zero width clock pulses; (*b*) actual clock signal. T_c: clock period. Adapted from [5].

These advantages make synchronized gating essential for time interval averaging.

The case of synchronous repetition rates

So far all the results for averaging have been based on a repetition rate that is asynchronous with the counter clock. However, in some cases a synchronous repetition rate can actually improve the resolution of a time interval averaging measurement; in other cases, however, it limits the resolution. A repetition rate is synchronous with the clock if the start of each time interval always occurs at one particular phase of the clock, or at some limited number of points during the phase of the clock.

Time interval averaging is based on the start of N time intervals tending to occur uniformly throughout the clock period. Fig. 10.15 compares an asynchronous with a synchronous repetition rate. There are an infinite number of repetition rates that produce perfectly valid averaging; there are only a finite number of cases where synchronous repetition rates can cause difficulties.

Assume, as an example, that the counter has a 100 ns clock. A repetitive sequence of time intervals starts every 250 ns. If the first start

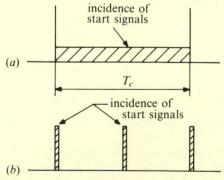

Fig. 10.15. Comparison of (*a*) asynchronous and (*b*) synchronous repetition rates. Adapted from [5].

signal occurs simultaneously with a clock pulse, then the second start signal will occur exactly midway between two clock pulses. Each successive start will occur either at a clock pulse or midway between clock pulses, but at no other time. Thus the repetition rate is synchronous with the clock.

This leads to a definition of the *class number M*. In our example, the repetition rate has class number $M = 2$. That is, the repetitive start signals occur only at two points in the clock period. In general, if the repetitive start signals occur only at M points during the clock period, then the repetition rate has class number M.

The theoretical significance of M is a key issue. No matter how many time intervals are averaged, the resolution of the time interval measurement can never be increased by more than a factor of $1/M$. In theory, if $M = 1$, then averaging does not actually improve the measurement error, but rather the error equals zero just because $M = 1$.

In practice, the occurrence of synchronous repetition rates is most improbable. Furthermore, should they occur, short-term fluctuations and phase jitter from the driving source and the counter clock will break the synchronous relationship unless the two are phase-locked together. If enough periods are averaged, instabilities in the repetition rate will usually produce valid averaging.

Problems

10.1 A repetitive signal is contaminated with white noise. It is connected to a signal averager through a filter whose bandwidth has been chosen in accordance with the bandwidth of the signal. The signal is averaged over m repetitions, and the time intervals T between corresponding samples are chosen to be sufficiently long to make the sampled noise values uncorrelated. For a given improvement in the SNR, show that the penalty for choosing a too wide bandwidth is a proportionate increase in m.

10.2. A signal is contaminated with noise whose autocorrelation coefficient is $\rho_n(\tau)$. Show that, for a given number m of repetitions, the improvement in the SNR due to averaging equals

$$(m)^{1/2}\left[1+\frac{2}{m}\sum_{i=1}^{m-1}(m-i)\rho_n(iT)\right]^{-1/2}.$$

11

Computer aided design

11.1 Introduction [1]

CAD (*computer aided design*) immensely enhances the engineer's capability to deal with complex design problems.

The computer's particular attributes include speed, memory, and reliability. It computes and makes logical decisions, and performs these functions in an extremely reliable fashion.

The man, on the other hand, is not good at computation. He is slow; he makes mistakes; and he has a poor memory. He does, however, have some attributes that are indispensable to the design process. The design engineer brings to the design problem his experience, imagination, and a tolerance for ambiguity. He has a massive store of information about design and relationships. He can be creative, innovative, and inventive. The built-in randomness of his thinking process and his associative memory allows the designer to sift through a large number of possible combinations in his search for a solution. The final choice of the solution best suiting the requirements is not unlike the principle of natural selection by survival of the fittest, with the natural selection being replaced by intelligent judgement and intuition.

But the design process is characterized by only a few brief moments of creativity and comparatively lengthy periods of mechanical computation and documentation. Whereas the conceptualization of a solution requires the engineer to draw on all the experience and knowledge at his disposal, computations and tabulation of results constitute a mechanical process with correspondingly less opportunity for creativity. It is during this phase that a final solution will in certain cases be obtainable only by employing CAD.

During the last few years, CAD has matured into a tool capable of taking over practically all chores in the design of sophisticated systems. Graphic displays,[2] in conjunction with touch-sensitive devices and light pens, serve as versatile communication links between man and computer. Such systems are capable of eliminating time consuming procedures such as breadboarding of electronic circuits, and provide computerized instructions for the automatic manufacture of mechanical parts or masks for the production of integrated circuits.[3]

However, the effort which goes into the computer programming of a completely automated design system as referenced in [3] is so considerable that the cost of designing it is prohibitive in many cases. Such systems have been designed in the past mainly where a suitable market ensured the sale of a sufficiently large number to amortize the initial investment.

Design problems in the field of instrumentation occur on a much smaller scale. If the problem on hand is too involved to be solved by iterating between simplified computations and measurements on the complete system or an analog model, then simulation by a computer will be justified. The following example is a case in point.

11.2   Application of CAD to signal processing for position-sensitive radiation detectors

The problem to be discussed deals with signal processing for dissipative unidimensional position-sensitive radiation detectors. The reader wishing to familiarize himself with the physical background is referred to the relevant literature.[4,5,6]

Here we shall restrict ourselves to a description of the model for the detector and the instrumentation employed, as shown in Fig. 11.1. A current $i(t)$ models the signal produced by radiation interacting with the detector. It enters the distributed RC line modelling the detector at a distance x from terminal 1, with x constituting the desired position information.

In the rise time method (the object of the investigation to be described), position information is derived by comparing the rise-time of the signals obtained from the two detector terminals. As shown in Fig. 11.1, this comparison is effected by twice differentiating the signals obtained from the terminals. The time difference between the two resulting zero crossings is

$$\Delta t_z = t_{z1} - t_{z2} = f_1(x),$$

i.e. Δt_z is a (not necessarily linear) function of x and contains the required position information. The signal from terminal 1 is delayed by a time interval t_D in order to prevent the time difference becoming zero when $x = L/2$, and changing polarity for $x < L/2$.

Finally, a TAC converts the difference in arrival times of the stop and start pulses into a pulse whose amplitude V is proportional to that

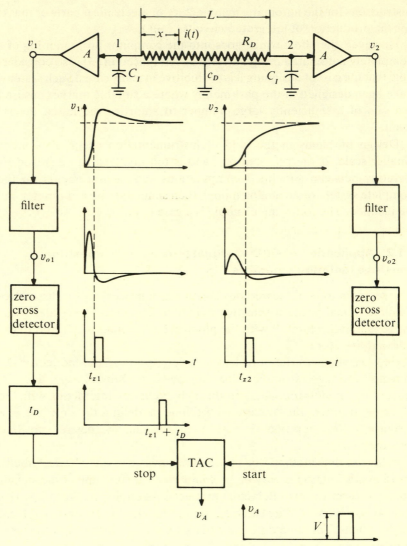

Fig. 11.1. Block diagram of signal processing system for position-sensitive detector.

difference and thereby conveys the position information to the following instrumentation (not shown).

Definition of problem
System performance is defined by its position resolution and position linearity, the first being a function of the standard deviation of the

amplitude measurement V, and the second depending on the relationship between V and x. Thus, a complete evaluation of the system requires the following:

(1) The time response of the signal processing channel must be calculated as a function of x, for differentialy shaped signals. The signal shape is determined by the kind of detector used. The channel itself consists of a distributed RC line and a filter. The solutions required are the time at which the signal crosses zero, and its derivative at that instant.

(2) The standard deviation of the measurement of x as a function of V must also be found. This is due to detector noise.

An analytical treatment of the time domain computations is impractical, whereas a computerized numerical solution is readily available. The noise computations would, if unaided, involve numerous algebraic computations and extensive use of integral tables, a tedious process unjustifiable in a computer age.

Before going into a detailed description of the computations involved, we shall briefly review the general design considerations for the system to determine the parameters to be optimized.

Design considerations

We want the system to have good position linearity and high position resolution. Both are governed by the interplay between the detector time constant $R_D C_D$ and the characteristics of the signal processing filter; the latter is characterized by the second time moment T_2 (see Appendix D), which is related to the half-power frequency ω_h.

We will show later that good position linearity is obtained for those values of T_2 which ensure the monotonicity of the filtered detector pulses for all values of x. Position resolution, in contrast, can be improved by reducing T_2 (i.e. by increasing the half-power frequency) until the attendant increase in noise and reduction in position sensitivity outweigh the improvement in rise time. For certain values of x, response will then become non-monotonic resulting in two zero crossings of the twice differentiated signal.

Sacrificing linearity to improve resolution is particularly attractive because linearity may be restored at a later stage by digital techniques. Thus, the purpose of this investigation is to compute the position resolution and linearity as a function of T_2 for various kinds of detectors and filters. From the computed results, we can compare the relative merits of the filters, and assess the possible improvement in position resolution if position linearity is disregarded.

Since we are concerned only with basic properties of the system, we shall disregard bandwidth limitations due to the amplifier and zero cross detector, and assume that the signal processing filter alone is responsible for signal shaping. Noise due to the various building blocks will be lumped into a single equivalent series noise source referred to the input of the preamplifier. Parallel noise due to the leakage current of the detector and preamplifier, however, is assumed to be negligible at practical time constants.

Choice of signal processing filters

The unfiltered response of a distributed RC line to a current impulse at the position $x = L/2$ is always monotonic. For $x < L/2$, the resulting response becomes non-monotonic at terminal 1 as shown in Fig. 11.1.

As already mentioned, for certain values of T_2 of the signal processing filter short enough to preserve the non-monotonicity of the detector time response, position resolution is improved. We may conclude intuitively that the position information contained in the non-monotonic detector response will be obliterated if the response of the signal processing filter alone is non-monotonic. Such a response could yield a non-monotonic relationship between Δt_z and x.

From the various monotonic filters, we choose integration of varying order n for simplicity. No qualitative modification of results is expected for more sophisticated filters. A maximum value of 7 is used for n; this yields a reasonably close approach to Gaussian response.

The integrating time constant is normalized with respect to T_2. Hence,

$$H_I(j\omega) = (1 + j\omega T_2/n^{1/2})^{-n}. \tag{11.1}$$

Two kinds of double differentiation can be used: either two successive time derivatives – RC shaping – or two cascaded delay line differentiations – DDL *(double delay line)* shaping. The latter performs the operation

$$
\begin{aligned}
f_{D2}(t) &= f(t) - f(t - \tau_D) - [f(t - \tau_D) - f(t - 2\tau_D)] \\
&= f(t) - 2f(t - \tau_D) + f(t - 2\tau_D).
\end{aligned}
\tag{11.2}
$$

In the frequency domain, the transfer function of the filter performing the two successive time derivatives is

$$H_{D1}(j\omega) = (j\omega T_2)^2, \tag{11.3}$$

and that of the DDL differentiations is

$$H_{D2}(j\omega) = 1 - 2\exp(-j\omega\tau_D) + \exp(-2j\omega\tau_D). \tag{11.4}$$

The minimum number of integrations applicable to RC shaping is $n = 3$, since noise for $n = 2$ becomes infinite because of the double differentiation. With DDL shaping, noise is finite for $n = 1$.

Definition of position linearity
The normalized coordinate of interaction x/L between radiation and detector is obtained as a function of $\Delta t_z = t_{z1} - t_{z2}$. An ideal plot of $x/L = f(\Delta t_z)$ should be a straight line. A practical plot can be approximated by the expected value of x/L, defined as the straight line passing through $\Delta t_z = 0$, $x = L/2$, for which the average error between the measured values of x/L and those lying on the straight line becomes zero. Deviation from linearity can then be defined as σ_L, the normalized standard deviation of the error in the measured values of x/L.

Qualitative considerations relating to position linearity
To compute the relationship between Δt_z and x/L, we need to find the time response of the distributed RC line in cascade with the particular filter under consideration. The general case is evaluated by a computer program discussed in a later section. Some intuitive insight may, however, be gained from an analytical method of computing Δt_z.

If t_z equals the delay time of the detector response (defined by Elmore as the first time moment T_1 of the singly differentiated detector signal (see Appendix D)), it can be shown[7] that

$$\Delta t_z = R_D(C_D + 2C_I)(x/L - \tfrac{1}{2}) \qquad (11.5)$$

i.e. Δt_z is a linear function of x. Strict validity of this relationship depends, however, on two conditions.
(*a*) Elmore's relationship is valid only if the time response is monotonic.
(*b*) Since the zero crossing of the second derivative of a step response occurs at the time when the impulse response $h(t)$ reaches maximum, t_z will equal T_1 only if $h(t)$ is symmetrical with respect to the point of maximum response, in which case the first moment of $h(t)$ with respect to $t = 0$ is equal to the time at which $h(t)$ is maximum. This condition is satisfied in the case of a Gaussian response, which is approached to a varying degree by filters commonly used in nuclear electronic instrumentation. Hence, position linearity will improve for higher order filters (i.e. with increasing n in (11.1)), which yield a near Gaussian response.
DDL shaping behaves similarly, since it does not affect the asymmetry of the detector response (the integrated impulse response) with respect to

the inflection point of its slope. However, an additional factor, not present with RC shaping, is that the delay must be sufficiently long for $f(t)$ in (11.2) to approach its steady state value when $f_{D2}(t)$ crosses zero. Practically, this means $\tau_D > \tau_R$. If this condition is not satisfied, the decaying portion of $f(t)$ interferes with the position of the zero-crossing point because of the term $-2f(t - \tau_D)$ in (11.2).

Any deviation from these two conditions introduces additional terms in (11.5) which interfere with the linearity of the relationship between Δt_z and x/L. This conclusion is confirmed both by practical experience and by computational results, which show that linearity improves for higher order filters if the time constant is sufficiently long to ensure monotonicity, but deteriorates rapidly for short filter time constants yielding non-monotonic response.

Position resolution
The position resolution of the system is defined as the standard deviation of the position measurement, normalized with respect to the length of the detector L:

$$\eta = \sigma(x)/L, \tag{11.6}$$

with

$$\sigma(x) = \sigma(\Delta t_z)/S(\Delta t_z) \tag{11.7}$$

and

$$S(\Delta t_z) = \partial(\Delta t_z)/\partial x, \tag{11.8}$$

where $\sigma(\Delta t_z)$ is the standard deviation of the time difference measurement Δt_z.

We define $S(\Delta t_z)$ to be the position sensitivity, i.e. the change in Δt_z relative to x.

Quantitative noise considerations
Important conclusions for a practical design may be drawn from a computation of the noise figure F of the system. To this end we shall compare the variance of the time difference measurement $\sigma^2(\Delta t_z)$ due to preamplifier noise with that due to detector noise, both being measured at the output. Their ratio yields the noise figure $F - 1$ directly since both preamplifier noise and detector noise are processed by the same channel.

The value of $\sigma^2(\Delta t_z)$ is obtained as follows. The variance of a zero crossing time measurement is derived in Section 4.2 as

$$\sigma^2(t_z) = \overline{v_{no}^2}/[\dot{v}(t_z)]^2, \tag{11.9}$$

where $\dot{v}(t_z)$ is the signal derivative at the zero crossing time t_z.

The variance due to detector noise of the time difference measurement between two zero crossings is from $\Delta t_z = t_{z1} - t_{z2}$,

$$\sigma_D^2(\Delta t_z) = \lim_{T \to \infty} \frac{1}{2T} \int_{-T}^{T} \left[\frac{v_{no1}(t)}{\dot{v}_1(t_{z1})} - \frac{v_{no2}(t - \Delta t_z)}{\dot{v}_2(t_{z2})} \right]^2 dt = Y_D \overline{v_{noD}^2},$$

$$(11.10)$$

with

$$Y_D = \left\{ \frac{1}{[\dot{v}_1(t_{z1})]^2} + \frac{1}{[\dot{v}_2(t_{z2})]^2} - \frac{2\rho(\Delta t_z)}{\dot{v}_1(t_{z1})\dot{v}_2(t_{z2})} \right\},$$

where $v_{no1}(t)$ and $v_{no2}(t)$ are the noise voltages at the two output terminals, $\overline{v_{noD}^2}$ is the mean squared noise voltage due to detector noise at either output and $\rho(\Delta t_z)$ the crosscorrelation coefficient between the two.

The variance of Δt_z due to preamplifier noise is

$$\sigma_A^2(\Delta t_z) = Y_A \overline{v_{noA}^2},$$

$$(11.11)$$

with

$$Y_A = \left\{ \frac{1}{[\dot{v}_1(t_{z1})]^2} + \frac{1}{[\dot{v}_2(t_{z2})]^2} \right\},$$

and $\overline{v_{noA}^2}$ the mean squared noise voltage due to preamplifier noise at either output. The term containing $\rho(\Delta t_z)$ is absent in (11.11) since the noise of the two preamplifiers is uncorrelated and noise transmitted from each preamplifier input to the opposite end of the detector is negligible in most practical cases.

Denoting the equivalent series noise resistances of the detector and preamplifier by R_{eqD} and R_{eqA},

$$\overline{v_{noA}^2} / \overline{v_{noD}^2} = R_{eqA} / R_{eqD}.$$

Finally,

$$F = 1 + \sigma_A^2(\Delta t_z) / \sigma_D^2(\Delta t_z) = 1 + R_{eqA} / R'_{eqD}(\Delta t_z) \qquad (11.12)$$

where

$$R'_{eqD}(\Delta t_z) = R_{eqD}[1 + f(\Delta t_z)] \qquad (11.13)$$

and

$$f(\Delta t_z) = -2\rho(\Delta t_z)/(a + 1/a), \qquad a = \dot{v}(t_{z1})/\dot{v}(t_{z2}) \quad \text{and} \quad \Delta t_z = f(x).$$

The term $f(\Delta t_z)$ signifies the effect of $\rho(\Delta t_z)$ on F. If F is to satisfy a certain design target, we need to calculate the minimum value of $R'_{eq}(\Delta t_z)$, the weighted detector noise resistance.

An examination of $f(\Delta t_z)$ reveals that $\rho(\Delta t_z)<0$, since the noise voltages at the two detector terminals are partly anticorrelated. This is easily visualized if we assume that $C_D = 0$, in which case the noise voltage due to R_D across the two terminating capacitors C_I is entirely anticorrelated, i.e. $\rho(0) = -1$. Hence, with $\rho(\Delta t_z)<0$, the numerator of $f(\Delta t_z)$ in (11.13) becomes maximum for $x = L/2$, corresponding to $\Delta t_z = 0$. Similarly, the factor $(a + 1/a)$ becomes minimum for the same condition, since $a = 1$ for $\Delta t_z = 0$. Consequently, $f(\Delta t_z)$ attains its maximum value for $x = L/2$ or $\Delta t_z = 0$. Furthermore, it can be shown that, with increasing Δt_z, $f(\Delta t_z)$ decreases monotonically. Hence $R'_{eq}(\Delta t_z)$ becomes minimum, or F becomes maximum, for $x = 0$ or $x = L$. It follows that the relationship to be satisfied for negligible preamplifier noise contribution is

$$R_{eqA}/R'_{eqD}(\Delta t_z) \ll 1, \tag{11.14}$$

with $\Delta t_z = f(0)$ or $\Delta t_z = f(L)$ (i.e. Δt_z is the same for $x = 0$ and $x = L$).

In this context it is interesting to note that an increase in R_D, if accompanied by a proportional scaling of the time constants of the signal processing filter, results merely in a corresponding extension of the time scale and thus increases the signal duration, whilst preserving the basic properties of linearity and position resolution.

This can be shown from the following relationships.

(a) $R_{eqD} \propto R_D$; $\Delta f \propto 1/T_2 \propto 1/R_D$ (due to scaling of time constants). Hence,

$$\overline{v_{noD}^2} = 4k\mathcal{T}R_{eqD}A^2(0)\,\Delta f \neq f(R_D), \tag{11.15}$$

where $A(0)$ is the gain of the preamplifier, which is assumed to be independent of frequency, and Δf the ENB of the filter.

(b) $\dot{v}(t_z) \propto 1/R_D$; hence $\sigma(\Delta t_z) \propto R_D$.

(c) $S(\Delta t_z) \propto R_D$, since Δt_z is proportional to the time scale. Finally,

$$\eta = \sigma(\Delta t_z)/S(\Delta t_z)L \neq f(R_D) \tag{11.16}$$

QED. Thus, although position sensitivity is intimately connected with the presence of R_D and C_D, the value of R_D has no effect on position resolution. The latter is only a function of C_D, C_I, and the shaping network.

In cases where signal duration is only of secondary importance, this opens an interesting aspect. If R_D can be increased sufficiently to make the noise due to R'_{eqD} dominant, then preamplifier noise contribution will become negligible. The only detrimental effect will be a deterioration of the time resolution, accompanied by a corresponding reduction in the maximum counting rate of the detected radiation.

R'_{eqD} is obtained from (11.13) and (11.15) as

$$R'_{eqD}(\Delta t_z) = \overline{v_{no}^2}[1 + f(\Delta t_z)]/4 k \mathcal{T} A^2(0) \, \Delta f, \tag{11.17}$$

where $\Delta t_z = f(0) = f(L)$.

A plot of $R'_{eqD}(\Delta t_z)$ as defined by (11.17), versus T_2 for various kinds of filters enables the designer to evaluate the relative noise contribution of a preamplifier whose R_{eqA} is known. Clearly, if two filters yield the same position resolution, the one for which the value of $R'_{eqD}(\Delta t_z)$ is greater will be the preferable one. This consideration may affect the choice of signal filter in a practical design.

Time domain computations

Evaluation of linearity and a in (11.13) requires the filtered time response to be calculated. The transfer functions involved are those of a distributed RC line in cascade with the two kinds of filters employed.

The CSMP (*continuous system modelling program*)[8] was chosen to compute the time response. This is an IBM System/370 problem-oriented program designed to aid the simulation of continuous processes. The program allows these problems to be prepared directly from either a block diagram or a set of ordinary differential equations. It accepts Fortran statements, which facilitates a definition of the input driving function for the system. A fixed format is provided for printing tables and plotting graphs at selected increments of the independent variable.

Since all computations are performed directly in the time domain, DDL differentiation performing $f(t) - 2f(t - \tau_D) + f(t - 2\tau_D)$ is conveniently modelled in this language.

The exact solution for the time response of the distributed RC line involves partial differential equations. Since this computation cannot be modelled by the CSMP, the distributed line was replaced by a lumped constant line consisting of n identical sections, as shown in Fig. 11.2. The equations are arranged as follows:

$$\left.\begin{aligned}
\dot{v}_a &= 1/(C_I + C/2)(i_a + 0 && - v_a G + v_b G) \\
\dot{v}_b &= 1/C && (i_b + v_a G && -2v_b G + v_c G) \\
\dot{v}_c &= 1/C && (i_c + v_b G && -2v_c G + v_d G) \\
&\;\vdots \\
\dot{v}_n &= 1/(C_I + C/2)(i_n + v_{n-1} G - && v_n G + 0)
\end{aligned}\right\}. \tag{11.18}$$

Sample evaluations of the response of a model with sixteen sections showed only insignificant changes in the signal shape in comparison with computations based on eight sections, at the expense of considerably

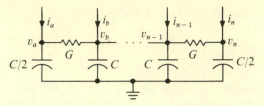

Fig. 11.2. Lumped model for distributed RC line.

increased computer time. Hence the calculation involving eight sections yielding nine positions for the input signal was chosen as the point of diminishing returns, beyond which the increase in computer time was not considered justifiable by the greater accuracy obtained.

The possibility of using Fortran instructions in conjunction with CSMP, facilitates introduction of arbitrary signal current shapes, which is important in the simulation of certain kinds of detectors.

The values computed by the CSMP are t_{z1}, t_{z2}, $\Delta t_z = t_{z1} - t_{z2}$, $\dot{v}(t_{z1})$ and $\dot{v}(t_{z2})$ as functions of five of the nine input positions. The remaining four yield symmetrical results with respect to the middle of the detector.

Pitfalls in computing higher order derivatives
In the RC shaping mode, three differentiations must be performed in calculating $\dot{v}(t_z)$. If these derivatives are taken successively, excessive high-frequency noise is introduced by the computation. This noise may completely mask the signal unless each differentiation is followed by an integration to smooth the accumulated high-frequency noise. Though this effect is due to digital computation, its remedy is similar to what could be termed 'good engineering practice' in analog signal processing, where each differentiator would be cascaded with one integrating section of the overall filter function, in order to reduce the noise contribution of the active differentiator.

Since the RC shaping mode includes a minimum of three integrations, the required sequence of operations is obtained simply by alternating differentiation and integration. No such problem is observed in the DDL differentiating mode, which involves computation of a single time derivative, namely that at zero cross.

Noise computations
Noise computations were made in the frequency domain. The PLI program by IBM[9] was employed, which was developed to serve both the scientific and data processing areas. Its modularity facilitates its

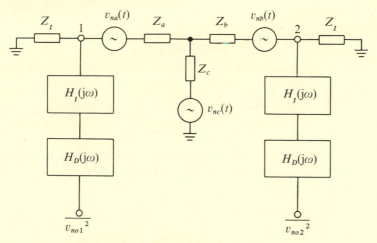

Fig. 11.3. Model for distributed *RC* line used in noise computations.

application to a variety of problems. Among its many features the program performs arithmetic functions and comparisons. Its built-in functions include hyperbolic ones, which allow accurate noise computations for the distributed *RC* line to be made in the frequency domain.

The results obtained from the CSMP were brought into suitable form to serve directly as input data for the PLI.

The quantities to be computed are $\overline{v_{no}^2}$, the filtered noise output voltage, and the crosscorrelation coefficient $\rho(\Delta t_z)$ between $\overline{v_{no1}^2}$ and $\overline{v_{no2}^2}$.

Fig. 11.3 shows the equivalent *T*-model of the distributed *RC* line cascaded with the shaping filters $H_I(j\omega)$ and $H_D(j\omega)$ performing integration and differentiation, respectively. In practice, the order of these operations is not as shown since differentiation and integration alternate for reasons explained in the preceding section. The relationships involved in Fig. 11.3 are:

$$Z_a = Z_b = Z_0[\cosh(u) - 1]/\sinh(u); \qquad Z_c = Z_0/\sinh(u);$$

$$u = (j\omega R_D C_D)^{1/2}; \qquad Z_0 = (R_D/j\omega C_D)^{1/2}; \tag{11.19}$$

$$Z_I = 1/j\omega C_I; \qquad \overline{v_{na}^2} = \text{Re}\{Z_a(j\omega)\}, \quad \text{etc.}$$

R_D and C_D represent the total resistance and capacitance of the distributed line. $H_I(j\omega)$ and $H_D(j\omega)$ are defined in (11.1), (11.3) and (11.4). Instead of going through the rather tedious detailed computations, we shall merely point out the various problems encountered.

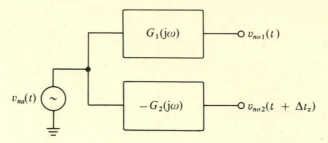

Fig. 11.4. Transmission of noise through different filters. Serves as basis for computation of crosscorrelation between $v_{no1}(t)$ and $v_{no2}(t + \Delta t_z)$.

Although the noise voltages $v_{na}(t)$, $v_{nb}(t)$ and $v_{nc}(t)$ are uncorrelated, polarity should be assigned to each of them in the computation of the crosscorrelation coefficient between $\overline{v_{no1}^2}$ and $\overline{v_{no2}^2}$. The two output noise voltages due to $v_{nc}(t)$, for instance, will be fully correlated, which is quite obvious if we consider the symmetry of the two transfer functions.

On the other hand, the output voltages due to $v_{na}(t)$ or $v_{nb}(t)$ are anticorrelated, as indicated in Fig. 11.4 for $v_{na}(t)$.

Precautions. Fractions need to be inspected carefully at $\omega = 0$ for poles in the denominator due to a term $1/j\omega C$ and for identical zeros in the numerator and denominator. The poles in the first case should be transformed into zeros of the numerator by multiplying both numerator and denominator by $j\omega C$, and the zeros in the second case should be cancelled out. A case in point is the transmission of the noise v_{na} to output terminal 2. This involves the function

$$G_2 = [(Z_a + Z_c + Z_I)/(Z_a + 2Z_c + Z_I)(1 + Z_a Y_I)]H_D(j\omega)H_I(j\omega),$$

which needs to be manipulated as indicated above. To show that this is so is left to the reader as a problem.

Verification of computations
An essential step in 'running-in' a computer program involves verification of the computed results. Frequently, a simplified model makes possible a comparison of the computed results with an algebraic evaluation 'by hand'. As an example the two noise output voltages, evaluated with $C_D = 0$ and a simple filter by hand, were compared with computed values for $C_D = 10^{-3}$ pF, which corresponds practically to zero detector capacitance. This kind of verification is particularly attractive,

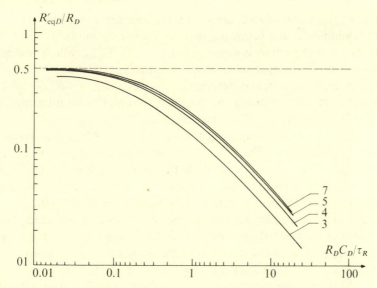

Fig. 11.5. Weighted noise resistance for $x = L$. RC differentiation; parameter n; $R_D = 9.6$ kΩ; $C_D = 100$ pF; $C_I = 7.5$ pF.[10]

since it involves two different algorithms (sinh in the case of the computer, and the Laplace transform of differential equations by hand).

In many cases, an interpretation of plotted results leads to a better understanding of the underlying physical relationships. The plots of R'_{eq} are a case in point, a typical sample being shown in Fig. 11.5. The asymptotic behavior of R'_{eq} as a function of the relationship between the corner frequency f_{c1} of the detector noise voltage spectral density and Δf_2, the ENB of the filter, can be explained by the following.

Let $S_{v1}(f)$ be the noise spectral density at the detector output and $H_2(j\omega)$ the response of the filter; then, over the frequency range $0 \leqslant f \ll f_{c1}$, S_{v1} is not a function of frequency. It follows that for $\Delta f_2 \lll f_{c1}$ the noise voltage at either detector terminal $\overline{v_{noD}^2} = 4k\mathcal{T}R_{eqD}\,\Delta f_2 \propto \Delta f_2$, and R_{eqD} is constant (see problem 11.1).

Within the region of practical filter time constants, however, $\Delta f_2 \ggg f_{c1}$, in which case $\overline{v_{noD}^2} \neq f(\Delta f_2)$ (see problem 11.2). Hence, in accordance with (11.15), $R_{eqD} \propto 1/\Delta f_2$, i.e. R_{eqD} reduces with increasing Δf_2. These two regions of asymptotic behavior can be clearly discerned in Fig. 11.5.

Furthermore, various quantitative predictions can be made if $\Delta f_2 \ll f_{c1}$, which serve well as an additional check on the noise computations. The first two relate to the fact that $\overline{v_{no1}^2}$ and $\overline{v_{no2}^2}$ are partly crosscorrelated. Consequently, R_{eqD} consists in accordance with (11.13) of two terms, one modelling the uncorrelated noise voltage and the other the crosscorrelated one. Hence the following predictions can be made.

(a) For low-capacitance detectors $(C_D \ll C_I)$, R_{eqD}, which is due to uncorrelated noise, approaches $R_D/4$ (see problem 11.3).

(b) For high-capacitance detectors $(C_D \gg C_I)$, $R_D/4 < R_{eqD} < R_D/2$, its accurate value depending on the ratio C_I/C_D. For an unterminated detector $(C_I = 0)$, $R_{eqD} = R_D/2$ (see problem 11.4).

(c) The value of R'_{eqD} approaches $R_D/2$ if $\tau_R \gg R_D C_D$, which can be clearly recognized from Fig. 11.5. This can be verified by reasoning that for long filter time constants, at which the shunting effect due to C_D and C_I becomes negligible, the total noise power (due to uncorrelated *and* correlated noise) measured at both terminals is supplied by R_D, or the noise power measured at one terminal is half its value as indicated above. The latter is a particularly useful criterion, since, in accordance with (11.13), R'_{eqD} is a function of two differently computed terms and a computational error in either term will show up immediately.

Time domain computations by the CSMP were verified using a dummy detector of eight sections in conjunction with a complete signal processing system. Finally, position resolution was computed for the same system and verified by practical measurements for a limited range of shaping filters.

Final results were printed in tabulated form, each table including:
  (a) input data, i.e. input current shape, detector and filter parameters;
  (b) intermediate results, such as Δt_z, $\rho(\Delta t_z)$, ENB etc., for checking;
  (c) output data for plotting the graphs.

Plotting of results
Position resolution and linearity were plotted against the normalized second time moment of the signal processing filter $\tau_R/R_D C_D$, henceforth to be called the NTM (*normalized time moment*), with $\tau_R = (2\pi)^{1/2} T_2$. The normalized equivalent detector noise resistance was plotted against $(\tau_R/R_D C_D)^{-1}$, in order to emphasize its relationship to a Bode plot. The plots were drawn for high- and low-capacitance detectors, for different signal shapes and for the RC and DDL differentiating modes.[10]

High-capacitance detectors, such as surface barrier silicon detectors, have a total capacitance C_D of an order of 100 pF and a resistance R_D of an order of 10kΩ. Typically, $C_D \gg C_I$. Signal duration is relatively short and can be approximated by an impulse. Low-capacitance detectors, such as proportional counters, are characterized by $C_D \ll C_I$. Signal duration is not negligible compared with signal processing time and should be taken into account.

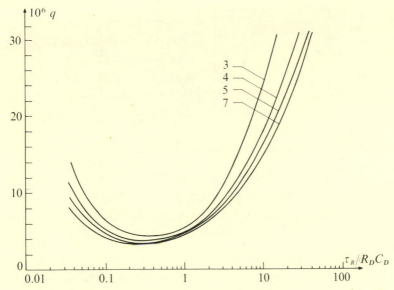

Fig. 11.6. Resolution for $x = L/2$. RC differentiation; parameter n; $R_D =$ 9.6 kΩ; $C_D = 100$ pF; $C_I = 7.5$ pF.[10]

A detailed presentation of the computed plots is beyond the scope of this treatment. Here we shall show only a few typical plots, taken for a high-capacitance detector and the RC shaping mode.

A typical plot of position resolution versus NTM is shown in Fig 11.6 for the RC shaping mode. Position resolution is expressed in terms of the equivalent number of ion pairs q required to obtain a normalized standard deviation $\eta = 10^{-3}$, with η being given by (11.6) and q designating the charge of a single electron. The parameter is n, the number of integrations. The plot is drawn for $x = L/2$. Position resolution versus x is plotted in Fig. 11.7, with the NTM serving as parameter.

A linearity plot for the same kind of shaping and n as parameter is shown in Fig. 11.8. The ordinate is calibrated in units of $100\ \sigma_L$, the percentage standard deviation from a perfect linearity plot over the total length of the detector.

Discussion of results
The trend of improving resolution in the RC shaping mode with increasing n is clearly recognizable from Fig. 11.6. The same tendency can be observed in Fig. 11.8, where increasing n is seen to improve linearity. The latter has indeed been predicted from qualitative observations discussed

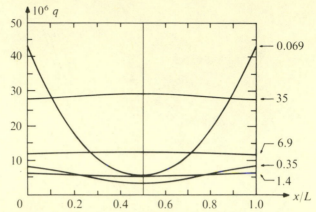

Fig. 11.7. Resolution versus x, with NTM as parameter. RC differentiation; $R_D = 9.6$ kΩ; $C_D = 100$ pF; $C_I = 7.5$ pF.[10]

before. Excellent linearity is obtained at NTMs above 5. Also predicted has been the improvement in resolution for values of NTM shorter than those yielding good position linearity.

Fig. 11.7 provides a good estimate for the interplay between the NTM and resolution as a function of x. Noting from Fig. 11.8 that excellent linearity is obtained for $NTM > 5$, a significant improvement in resolu-

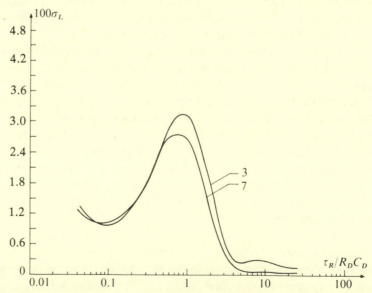

Fig. 11.8 Linearity plot. RC differentiation; parameter n; $R_D = 9.6$ kΩ; $C_D = 100$ pF; $C_I = 7.5$ pF.[10]

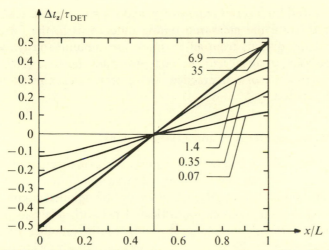

Fig. 11.9 Normalized time difference (computed) versus x/L, with NTM as parameter: $\tau_{DET} = R_D(C_D + 2C_I)$.[10]

tion is seen to be available at NTM $= 1.4$ if linearity is sacrificed. Further reduction of NTM improves resolution in the middle, but reduces it towards the ends of the detector.

Fig. 11.9 is a plot of Δt_z versus x/L. Δt_z has been normalized in accordance with (11.5). The plot shows clearly that (11.5) holds for NTM > 5 and thus corroborates the qualitative considerations made there. Three important observations can be made regarding resolution.

(*a*) At NTMs yielding monotonic response and hence good position linearity, resolution is reasonably uniform over the total length of the detector. This is because for this condition the response time of the filter is dominant, so that the filtered detector pulses do not change their shape significantly as a function of x. Hence, position resolution, which is a function of the signal derivatives at cross-over, changes only slightly.

(*b*) At shorter NTMs the rise time of the detector pulses comes into play, but improvement of position resolution is limited due to gradually increasing noise and decreasing position sensitivity. Improvement is greatest for $x = L/2$, where the signal rise time at the two detector terminals is equal. It is much less noticeable towards $x = L$, where the signal rise time at the far end is greatly reduced and limits the resolution.

(*c*) The NTM for optimal resolution is a function of x. It is minimum for $x = L/2$ and maximum for $x = 0$ and $x = L$. Because of this fact, a compromise must be struck. As an alternative, one could envisage a system employing several signal processing channels, each consisting of a filter optimized for a certain range of x, and a TAC. Since a TAC uses a capacitor to store the measured timing differences in the form of an amplitude, the outputs from all channels are simultaneously available after processing a signal and the one providing the best SNR can be chosen as a function of x. In practice, such a degree of sophistication may rarely be justified.

11.3 Macromodelling of amplifiers for CAD

We shall conclude this chapter with a short review on macromodelling, an important technique in the simulation of large systems.

Computer simulation of systems, based on the detailed circuits, yields accurate results but requires sophisticated network analysis programs and therefore rapidly becomes unwieldy as the number of circuits involved increases. In the design of systems incorporating a large number of linear ICs and LSI digital circuits, classic CAD techniques are therefore being replaced by macromodelling which simulates the performance of complete circuits employing a greatly simplified model: each self-contained circuit is represented by switches, diodes, passive components and dependent sources which model CMRR, noise, distortion and slewing of linear circuits. Noise sensitivities and transient response of logic circuits and interconnection networks between them may be simulated as well. Thus, analysis of the complete performance of LSIs is possible by greatly simplified computational means.

A simple macromodel for an operational amplifier simulating input and output impedances, voltage gain and phase shift as functions of frequency, offset voltage and currents, CMRR and slew rate has been developed by Trelaven[11] for use with the ECAP circuit analysis program. The model is readily adaptable to other programs. Macromodelling has been treated in further detail by Boyle *et al.*[12] and reviewed by Rabbat *et al.*[13] Ranfft and coworkers[14] have proposed an analog computer for simulation of high-speed ICs; an IC comparator has been macromodelled by Getreu *et al.*[15]

Problems

11.1. This problem relates to the asymptotic behavior of R'_{eq} as discussed in this chapter.

Show that

$$\overline{v_{noD}^2} = \int_0^\infty S_{v1}(f)|H_2(j\omega)|^2\, df \simeq \int_0^{\Delta f_2} S_{v1}(f)\, df \simeq S_{v1}(0)\, \Delta f_2,$$

if Δf_2 is the ENB of $H_2(j\omega)$, $\Delta f_2 \ll f_{c1}$, and f_{c1} is the noise corner frequency of S_{v1}.

This can be shown: (a) algebraically substituting

$$S_{v1}(f) = 4k\mathcal{T}R'_{eq}/(1 + \omega^2\tau_1^2) \quad \text{and} \quad H_2(j\omega) = 1/(1 + j\omega\tau_2);$$

(b) graphically, on a log–log scale.

11.2. Show that, for $\Delta f_2 \gg f_{c1}$,

$$\overline{v_{noD}^2} = \int_0^\infty S_{v1}(f)|H_2(j\omega)|^2 \, df \simeq [H_2(0)]^2 \int_0^\infty S_{v1}(f) \, df \neq f(\Delta f_2).$$

11.3. Consider the system shown in Fig. 11.1, with $C_D = 0$. Find the equivalent noise resistance $R_{eq}(0)$ of the detector as defined in Appendix A, referred to the input of the preamplifier A.

11.4. (a) Show that the input impedance of an unterminated distributed RC line approaches

$$\lim_{\omega \to 0} [Z_a(j\omega) + Z_c(j\omega)] = R_D/2 + 1/j\omega C_D. \tag{11.20}$$

Z_a and Z_c are defined by (11.19) for Fig. 11.3.

Appendix A

Random data analysis

Probability theory is an essential tool in analysing the properties of random data. Here we shall deal with relationships which are widely encountered in signal processing. Their basic properties and physical significance will be emphasized. A mathematical derivation is given in references [1] and [2], and a simpler approach is presented in reference [3].

A.1   Single random variable

A *discrete random variable* $x(k)$ is a function of the outcomes k of an experiment. Each possible outcome has a specified probability of occurring, and the sum of the probabilities of all possible outcomes is equal to unity if the outcomes are disjoint (i.e. mutually exclusive).

Discrete random variables can only assume a discrete range of values. But random variables are not necessarily discrete. For example, the random variable which represents the value of a thermal noise voltage at a specified instant of time is a *continuous-parameter random variable*, since it may assume any value between two bounds.

Random processes

The outcomes of an experiment, or the measured *sample values* of a continuous parameter random variable, define the *sample function* represented by that experiment. The collection of all possible sample functions (or ensemble) which the random phenomenon might have produced forms a *random* (or stochastic) *process*.

Random processes may be stationary or non-stationary.[1] Here we shall deal exclusively with stationary processes. (The statistical properties of a stationary process are invariant under a shift of the time origin. See also Section A.2, equation (A.25).)

Probability distribution function

Let $x(k)$ denote a certain random variable. Then, for any fixed number x_i, the random events $x(k) \leq x_i$ are defined as the set of all possible outcomes k such that $x(k) \leq x_i$. We define a (first order) *cumulative probability distribution function* $P(x_i)$ as the probability assigned to the set of all possible outcomes satisfying the inequality $x(k) \leq x_i$, i.e.

$$P(x_i) = \text{Prob}\,[x(k) \leq x_i], \tag{A.1}$$

with

$$P(-\infty) = 0 \quad \text{and} \quad P(\infty) = 1. \tag{A.2}$$

Probability density function

If the random variable assumes a continuous range of values, then the (first order) PDF (*probability density function*) is defined by

$$p(x)\,dx = \text{Prob}\,[x < x(k) \leq x + dx], \tag{A.3}$$

where $p(x) \geq 0$,

$$P(x_i) = \int_{-\infty}^{x_i} p(x)\,dx \tag{A.4}$$

and

$$P(\infty) = \int_{-\infty}^{\infty} p(x)\,dx = 1. \tag{A.5}$$

In the case of a discrete random variable, $p(x)$ is a set of discrete values, and

$$P(x_i) = \sum_{\text{over } x \leq x_i} p(x). \tag{A.6}$$

Exceedance probability distribution

The exceedance probability distribution function is defined as one minus the probability distribution function:

$$P_{\text{exc}}(x_i) = 1 - P(x_i) = \int_{x_i}^{\infty} p_x(x)\,dx = \sum_{\text{over } x > x_i} p(x). \tag{A.7}$$

Expected value and variance of a single random variable

The expected value of a random variable or its mean value μ is defined as the first moment of its *PDF* around zero:

$$\mu = E[x] = \int_{-\infty}^{\infty} xp(x)\, \mathrm{d}x$$

or

$$\mu = \sum_{\substack{\text{over} \\ \text{all } x}} xp(x) \quad \text{if } x \text{ is discrete.} \tag{A.8}$$

The variance σ^2 of a random variable is defined as the second moment of its PDF around the mean value:

$$\sigma^2 = \mathrm{Var}\,[x] = \int_{-\infty}^{\infty} (x-\mu)^2 p(x)\, \mathrm{d}x$$

or

$$\sigma^2 = \mathrm{Var}\,[x] = \sum_{\substack{\text{over} \\ \text{all } x}} (x-\mu)^2 p(x) \quad \text{if } x \text{ is discrete.} \tag{A.9}$$

Gaussian density function

The most commonly encountered PDF, that models exceptionally well naturally occurring random phenomena, is the Gaussian (or normal) distribution:

$$p_x(x) = \frac{1}{(2\pi)^{1/2}\sigma} \exp\left[-(x-\mu)^2/2\sigma^2\right]. \tag{A.10}$$

Referring to Fig. A.1, let $v(t)$ be a noise voltage whose PDF is Gaussian. To establish this property one may observe the frequency of occurrence of $v(t)$ as a function of amplitude by dividing its voltage range into n intervals or windows (assuming finite bounds). Let T be the total time of observation, which is measured by a train of clock pulses. Then, if we count the number of clock pulses occurring during the time the noise voltage has dwelt within a particular window j, and divide this by the total number of clock pulses occurring during the time T, we obtain the proportion of time which the noise voltage has spent in that particular window. For T approaching infinity, the resulting number yields $(\sum^m \tau_i)/T$, where τ_i represents the duration of each of the m observed time intervals.

The relative number of observations (counted clock pulses, in our case) within a window $v_j < v \leqslant v_j + \Delta v$ is called the observed frequency ratio f_j,

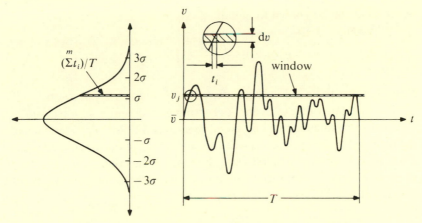

Fig. A.1. PDF of Gaussian noise. Adapted from [4].

and a plot of f_j versus j forms a frequency histogram for the observed data.

If certain conditions are satisfied,[1] the frequency histogram for the observed data fits the Gaussian density function $p(v)$ as indicated in Fig. A.1.

This method of measurement is referred to in Chapter 9 as estimation of the (one-dimensional) statistics of a waveform.

The mean power W dissipated by the ac component of v in $R = 1/G$ across which it is being measured, is the second moment of v around $\bar{v}$ (i.e. its variance):

$$W = G\int_{-\infty}^{\infty} (v - \bar{v})^2 p(v)\, dv = G\sigma^2. \qquad (A.11)$$

The standard deviation σ of $p(v)$ corresponds to the effective noise voltage v_{rms} measured across a resistor of one ohm. The probability of a Gaussian noise voltage lying between values of $\pm v_{\text{rms}}$ is

$$\int_{-\sigma}^{\sigma} p(v)\, dv \approx 0.68 = 68\%. \qquad (A.12)$$

Poisson distribution

Many random occurrences in time, such as emission of electrons by a heated cathode or radioactive decay, are Poisson-distributed.

For a stationary Poisson process, the probability that n events will occur within a time interval of t_0 seconds at an average rate of μ events per second can be shown to equal

$$p_n(n, t_0) = \frac{(\mu t_0)^n}{n!} \exp(-\mu t_0). \qquad (A.13)$$

Note that n is the random variable, while μ and t_0 are parameters.

The expected value of events occurring per second is $E[n]/t_0 = \mu$.

The reverse question can be asked regarding a Poisson process: what is the probability that t seconds will elapse during the occurrence of a fixed number (n_0) of events? In this case, the random variable is the continuous variable of time t rather than the discrete variable of number of events. The resulting PDF can be shown to equal

$$p_t(n_0, t) = \mu \frac{(\mu t)^{n_0}}{n_0!} \exp(-\mu t) \quad \text{for } 0 \leqslant t \leqslant \infty. \qquad (A.14)$$

The expected value of the time interval between two events is given by $E[t]/n_0 = 1/\mu$.

Practical applications for the two functions defined by (A.13) and (A.14) are given in Chapter 9.

A.2 Two random variables

Two-dimensional probability theory

The preceding section deals with probability functions of a single variable. These functions can provide statistical information about the relative frequency of occurrence of magnitudes, but none about the statistics of the random variable in the time domain.

Many engineering problems, however, require a comparison between data taken at different instants of time. The data can be pressure, temperature, amount of rainfall, bioelectric signals, etc. A set of data can be represented by a sample record obtained from observing the physical process over a finite time interval. Time is usually the independent variable, but not always.

The successive data values in each of these records could be related in some way to the other values in the record, and perhaps to values in some other records. Two-dimensional probability theory analyses these relationships so that the physical process can be better understood. In what follows we first define the statistical relationships between two random variables, and then show how these relationships enable us to introduce the additional dimension of time.

As in the one-dimensional case, we define the joint (second order) cumulative probability distribution function $P(x_i, y_i)$ as the probability associated with the subset of sample values satisfying both $x(k) \leqslant x_i$ and $y(k) \leqslant y_i$. Formally,

$$P(x_i, y_i) = \text{Prob} [x(k) \leqslant x_i \quad \text{and} \quad y(k) \leqslant y_i]. \tag{A.15}$$

Similarly, we define $p(x, y)$, the joint (second order) PDF associated with $x(k)$ and $y(k)$, respectively, to be that function which, when multiplied by the infinitesimal product $dx\, dy$, gives the probability that the value of the first coordinate is in the range x to $x + dx$ and the value of the second coordinate is simultaneously within the range y to $y + dy$:

$$p(x, y)\, dx\, dy = \text{Prob} [x < x(k) \leqslant x + dx \quad \text{and} \quad y < y(k) \leqslant y + dy]. \tag{A.16}$$

$$P(-\infty, y) = 0 = P(x, -\infty),$$

$$P(\infty, \infty) = \int\!\!\int_{-\infty}^{\infty} p(x, y)\, dx\, dy = 1. \tag{A.17}$$

Two random variables are said to be statistically independent if

$$P(x_i, y_i) = P(x_i)P(y_i) \quad \text{and} \quad p(x, y) = p(x)p(y). \tag{A.18}$$

The *cross-covariance* C_{xy} between $x(k)$ and $y(k)$ is defined as

$$C_{xy} = E[(x(k) - \mu_x)(y(k) - \mu_y)] = \int\!\!\int_{-\infty}^{\infty} (x - \mu_x)(y - \mu_y)p(x, y)\, dx\, dy,$$
$$\tag{A.19}$$

where μ_x and μ_y are the mean values of $x(k)$ and $y(k)$, respectively.

The expressions for the covariance C_{yx}, C_x and C_y are defined in the same manner. Comparison with (A.9) shows that the autocovariance C_x corresponds to σ^2, the variance of $x(k)$, where for briefness we are using C_x instead of C_{xx}.

The *normalized cross-covariance*, also known as the *correlation coefficient*, is defined as

$$\rho_{xy} = C_{xy}/\sigma_x \sigma_y. \tag{A.20}$$

From $|C_{xy}| \leqslant \sigma_x \sigma_y$ it follows that

$$-1 \leqslant \rho_{xy} \leqslant +1. \tag{A.21}$$

Random variables whose correlation coefficient is zero are said to be uncorrelated. If two Gaussian distributed variables are uncorrelated, they are also statistically independent, but this is not necessarily true in the general (non-Gaussian) case.

Time-dependent random variables

In measuring the statistical properties of a random process involving time as independent variable we may apply two methods. The first is based on a series of measurements throughout a very long time (ideally infinitely long), whereas the second calls for a simultaneous set of measurements on a large number (ideally an infinite number) of similar systems. Computation by the second method is often called *ensemble averaging*. Both methods yield the same result if the process is ergodic.* In actual practice, random data representing stationary physical phenomena are generally ergodic. Hence, their properties can be measured, in most cases, from a single observed time history record.

Let us consider an ensemble of k continuous-parameter random variables $\{x_k(t)\}$. If the random process thus defined is stationary, then their mean value $\mu_x = E[x_k(t)]$ is independent of t.

If the process is also ergodic, then its mean value may be computed from a single sample function:

$$\mu_x = \lim_{T \to \infty} \frac{1}{T} \int_{-T/2}^{T/2} x_k(t)\, dt. \tag{A.22}$$

Similarly, the mean square value, or the average of the squared values of the time history, is given by

$$\psi_x^2 = \lim_{T \to \infty} \frac{1}{T} \int_{-T/2}^{T/2} x_k^2(t)\, dt. \tag{A.23}$$

The variance, or the mean square about the mean, is given by

$$\sigma_x^2 = \lim_{T \to \infty} \frac{1}{T} \int_{-T/2}^{T/2} [x_k(t) - \mu_x]^2\, dt = \psi_x^2 - \mu_x^2. \tag{A.24}$$

Now consider the case in which two variables $x_k(t_1)$ and $y_k(t_1 - \tau)$ represent measurements made on two waveforms at two specified instants of time. Again assuming stationarity, the statistics of x and of y do not depend on t, but only on the time difference. Hence, in this case we

---

* An ergodic process is always stationary, but a stationary process can be non-ergodic. Ergodicity always applies to stationary noise originating from a physical source.

may substitute t for t_1 in x and y, and their cross-covariance function is given by

$$C_{xy}(\tau) = E[x_k(t) - \mu_x)(y_k(t-\tau) - \mu_y)]$$

$$= \lim_{T \to \infty} \frac{1}{T} \int_{-T/2}^{T/2} [x_k(t) - \mu_x][y_k(t-\tau) - \mu_y] \, dt. \quad (A.25)$$

The expressions for $C_{yx}(\tau)$, $C_x(\tau)$ and $C_y(\tau)$ are defined in the same manner.

Correlation functions

In order to compare the statistical properties of $x_k(t)$ with those of $y_k(t)$ as a function of the delay between them, it is customary to define the *crosscorrelation function*

$$R_{xy}(\tau) = E[x_k(t)y_k(t-\tau)] = \lim_{T \to \infty} \frac{1}{T} \int \int_{-T/2}^{T/2} x_k(t)y_k(t-\tau) \, dt,$$

$$(A.26)$$

and the *autocorrelation function*

$$R_x(\tau) = E[x_k(t)x_k(t-\tau)] = \lim_{T \to \infty} \frac{1}{T} \int_{-T/2}^{T/2} x_k(t)x_k(t-\tau) \, dt. \quad (A.27)$$

The covariance functions are related to the correlation functions by the equations

$$C_{xy}(\tau) = R_{xy}(\tau) - \mu_x \mu_y, \quad (A.28)$$

and

$$C_x(\tau) = R_x(\tau) - \mu_x^2$$

or

$$C_y(\tau) = R_y(\tau) - \mu_y^2. \quad (A.29)$$

Hence, for zero mean values, $C(\tau)$ and $R(\tau)$ are identical.

Excluding special cases such as sine waves, it can be shown that

$$\mu_x = [R_x(\infty)]^{1/2}. \quad (A.30)$$

It can also be shown that the autocorrelation function is an even function of τ, i.e.

$$R_x(-\tau) = R_x(\tau), \quad (A.31)$$

and that the crosscorrelation function satisfies the relationship

$$R_{xy}(-\tau) = R_{yx}(\tau). \quad (A.32)$$

An upper bound for the crosscorrelation function is given by the inequality

$$|R_{xy}(\tau)|^2 \leqslant R_x(0)R_y(0). \tag{A.33}$$

The correlation function coefficient (or normalized crosscovariance function) is given by

$$\rho_{xy}(\tau) = C_{xy}(\tau)/[C_x(0)C_y(0)]^{1/2}, \tag{A.34}$$

which satisfies

$$-1 \leqslant \rho_{xy}(\tau) \leqslant +1. \tag{A.35}$$

If the mean values μ_x and μ_y are zero, then $\rho_{xy}(\tau)$ becomes

$$\rho_{xy}(\tau) = R_{xy}(\tau)/[R_x(0)R_y(0)]^{1/2}. \tag{A.36}$$

From a comparison of (A.36) with (A.32) it follows that

$$\rho_{xy}(-\tau) = \rho_{yx}(\tau). \tag{A.37}$$

So far we have considered the statistical properties of stationary random processes. These results are directly applicable to stationary noise voltages and currents encountered in electronic circuits. Any reference to noise will henceforth imply that it is represented by a single time history record, which has been observed over a time interval sufficiently long to estimate its statistical properties.

Now let us assume that the two random variables represent the value of noise voltages as a function of time: $x(t) = v_x(t)$ and $y(t) = v_y(t)$. Computation of the crosscorrelation function

$$R_{xy}(\tau) = \lim_{T \to \infty} \frac{1}{T} \int_{T/2}^{T/2} v_x(t)v_y(t-\tau) \, dt \tag{A.38}$$

amounts to displacing one waveform with respect to the other by a time interval τ, measuring their instantaneous values as a function of time, multiplying the corresponding measured values with each other, and finally summing all products. This number, when normalised by the number of independent measurements, estimates the similarity in the statistics of the two waveforms, or their crosscorrelation, as a function of their displacement τ with respect to each other.

Similarly, the autocorrelation function $R_x(\tau)$ estimates the similarity in the statistics of a single waveform, as a function of τ. For $\tau = 0$, the similarity measured by $R_x(\tau)$ becomes a waveform identity, since matching of any function with itself must be perfect:

$$R_x(0) = \lim_{T \to \infty} \frac{1}{T} \int_{-T/2}^{T/2} v_x^2(t) \, dt. \tag{A.39}$$

Equation (A.39) shows that the autocorrelation function of $v_x(t)$ equals, for $\tau = 0$, the mean square value of the voltage or the normalized power $\bar{W}_x$ dissipated by that voltage in a resistance of $1\,\Omega$.

Correlation functions not only describe the statistical properties of random signals, but also those of deterministic ones. Hence, they are a useful tool in the classification of the latter. Because of the difference in the autocorrelation functions for various signals and also for noise, they provide a powerful tool for detecting deterministic data in the presence of noise. Chapters 9 and 10 deal with these aspects.

As an example of a practical application of (A.38) and (A.39) consider two partly correlated noise voltages, defined as containing some noise that arises from a common phenomenon as well as some independently generated noise. The normalized power in the sum of the two voltages is calculated to be

$$\begin{aligned} \bar{W} &= \lim_{T \to \infty} \frac{1}{T} \int_{-T/2}^{T/2} [v_x(t) + v_y(t)]^2 \, dt \\ &= R_x(0) + R_y(0) + R_{xy}(0) + R_{yx}(0) \\ &= \bar{W}_x + \bar{W}_y + [\rho_{xy}(0) + \rho_{yx}(0)](\bar{W}_x \bar{W}_y)^{1/2}. \end{aligned} \tag{A.40}$$

The time domain response of a linear time-invariant filter to noise

The output–input relationship of a linear time-invariant filter is given by the convolution integral

$$y(t) = \int_{-\infty}^{\infty} x(u)h(t-u) \, du = x(t) * h(t). \tag{A.41}$$

We call $h(t)$ the system weighting function of the filter, and the time invariance implies that if $y(t)$ is the filter output in response to an input $x(t)$, then $y(t-t')$ is its output in response to the time-shifted input $x(t-t')$.

A heuristic way of obtaining the transmission of noise through a linear time-invariant filter is to substitute its autocorrelation function $R_x(\tau)$ for $x(t)$ and the autocorrelation function $R_h(\tau)$ of the filter for $h(t)$:

$$R_y(\tau) = R_x(\tau) * R_h(\tau), \tag{A.42}$$

where

$$R_h(\tau) = \int_{-\infty}^{\infty} h(t) h(t-\tau) \, dt \qquad (A.43)$$

and $R_y(\tau)$ is the autocorrelation function of the noise at the output.

Noise power spectra

Convolution in the time domain transforms into multiplication in the frequency domain. Due to the algebraic convenience of multiplication it is common engineering practice to calculate linear filter responses in the frequency domain.

The relationship between the two domains is given by the Fourier integral:

$$F(\omega) = \int_{-\infty}^{\infty} f(t) \exp(-j\omega t) \, dt. \qquad (A.44)$$

Equation (A.44) reveals that the Fourier transform is a crosscorrelation of the signal with a complex sinusoid of each of the possible harmonic signal frequencies, respectively. The result is a complex number for each ω, whose modulus and argument equal that of the corresponding component in the signal.

The conventional Fourier integral cannot be applied to a noise wave, since it does not converge for a typical time history record from a noise ensemble. It can be shown,[2] however, that the integral does converge if it is applied to mean square powers in infinitesimal frequency intervals, instead of first powers. The resulting power density spectrum $W(f)$ is a second order statistic in the frequency domain and can therefore be related to the autocorrelation, which is a second order statistic in the time domain. In fact, the two form a Fourier transform pair:

$$W_x(f) = \int_{-\infty}^{\infty} R_x(\tau) \exp(-j2\pi f\tau) \, d\tau \qquad (A.45)$$

where $W_x(f)$ is defined as the power spectral density of $\overline{v_x^2}$. Conversely,

$$R_x(\tau) = \frac{1}{2\pi} \int_{-\infty}^{\infty} W_x(f) \exp(j2\pi f\tau) \, d\omega = \int_{-\infty}^{\infty} W_x(f) \exp(j2\pi f\tau) \, df. \qquad (A.46)$$

Equation (A.46) shows that the proper scale for power density spectra is frequency and not angular frequency.

General properties of the spectral density function
Some general properties of $W(f)$ can be derived from (A.45) and (A.46).
Equation (A.45) yields, for $f = 0$,

$$W_x(0) = \int_{-\infty}^{\infty} R_x(\tau) \, d\tau, \qquad (A.47)$$

i.e. the spectral density at $f = 0$ equals the total area under the graph of
the autocorrelation function. Similarly, (A.46) yields, for $\tau = 0$,

$$R_x(0) = \int_{-\infty}^{\infty} W(f) \, df. \qquad (A.48)$$

A comparison between (A.48) and (A.39) shows that

$$\lim_{T \to \infty} \frac{1}{T} \int_{-T/2}^{T/2} v_x^2(t) \, dt = \int_{-\infty}^{\infty} W_x(f) \, df = \bar{W}_x. \qquad (A.48a)$$

Hence, the total power in the time and in the frequency domains is the
same, a theorem established by Parseval. Equation (A.48a) also
establishes the fact that $W_x(f)$ represents the mean square value of $v_x(t)$
lying in the frequency interval $(f, f + df)$.

If we expand the exponential term in (A.45) as $\cos(2\pi f\tau) - j \sin(2\pi f\tau)$,
and if we further note that $W(f)$ is a real function of frequency, then it
follows that the imaginary term in (A.45) must vanish. Hence,

$$W(f) = \int_{-\infty}^{\infty} R_x(\tau) \cos(2\pi f\tau) \, d\tau. \qquad (A.49)$$

But since the cosine is an even function of f, $W_x(f)$ must be even as well:

$$W_x(f) = W_x(-f). \qquad (A.50)$$

Finally, substituting (A.50) in (A.46), we may write

$$R_x(\tau) = \int_{-\infty}^{\infty} W_x(f) \cos(2\pi f\tau) \, df$$

$$= 2 \int_{0}^{\infty} W_x(f) \cos(2\pi f\tau) \, df$$

$$= \int_{0}^{\infty} S_x(f) \cos(2\pi f\tau) \, df \qquad (A.51)$$

with $S_x(f) = 2W(f)$.

This result deserves the following comment. For mathematical convenience we define the spectral density function $W(f)$ for both positive

and negative frequencies. In physical apparatus, however, the mean power contributions from positive and negative frequencies add to give a single contribution. Hence $W(f)$ is defined as the *mathematical* or *two-sided power spectral density* and $S(f)$ as the *physical* or *one-sided power spectral density*.

The frequency domain response of a linear time-invariant filter to noise

The output–input relationship of a linear system for noise has been given in (A.42) for the time domain. The corresponding relationship in the frequency domain is its transform pair. Considerations of both the physical process and mathematical formality show that convolution between the two signals $v_x(\tau)$ and $v_y(\tau)$ is the same as their crosscorrelation with the time scale of one signal reversed and the variables t and τ, employed in the convolution, exchanged. Formally, $R_{xy}(\tau) = v_x(\tau) * v_y(-\tau)$. But convolution transforms into the frequency domain as multiplication. Hence, we have the following transform pairs:

$$v_x(t) \leftrightarrow V_x(j\omega),$$

$$v_x(-t) \leftrightarrow V_x(-j\omega),$$

$$v_x(t) * v_y(t) \leftrightarrow V_x(j\omega) V_y(j\omega),$$

$$R_{xy}(\tau) \leftrightarrow V_x(j\omega) V_y(-j\omega)$$

$$R_x(\tau) \leftrightarrow V_x(j\omega) V_x(-j\omega) = |V_x(j\omega)|^2.$$

Hence, the output–input relationship for a linear system or a filter is, in accordance with (A.43),

$$R_h(\tau) = h(\tau) * h(-\tau) \leftrightarrow H(j\omega) H(-j\omega) = |H(j\omega)|^2. \qquad \text{(A.52)}$$

It follows that, with $S_x(f)$ and $S_y(f)$ representing the noise spectral densities at the input and the output respectively, we have

$$S_y(f) = S_x(f) |H(j\omega)|^2. \qquad \text{(A.53)}$$

Cross-spectral density

The cross-spectral density function $S_{xy}(f)$ for a pair of time history records $x(t)$ and $y(t)$, or noise signals, is the Fourier transform of their crosscorrelation function $R_{xy}(\tau)$. Being a complex quantity, its amplitude at a given frequency equals the average product between the amplitudes of the corresponding components contained in the two signals, at that frequency; similarly, its phase equals the average difference between the phases of the same components. Hence, unlike the power spectral density function $S_x(f)$, it estimates both amplitude and phase.

As a demonstration of the practical application of this property, consider the complex spectrum of a filter $H(j\omega)$, which can be measured by the ratio between its response $H_y(j\omega)$ to an input signal of spectrum $H_x(j\omega)$:

$$H(j\omega) = \frac{H_y(j\omega)}{H_x(j\omega)} = \frac{H_y(j\omega)H_x^*(j\omega)}{|H_x(j\omega)|^2} \tag{A.54}$$

If we identify $|H_x(f)|^2$ with the power spectral density $S_x(f)$ of a noise voltage $v_x(t)$ applied to the input of the filter, then $H_y(j\omega)H_x^*(j\omega)$ corresponds to the output–input cross-spectral density $S_{yx}(f)$, i.e. $H(f)$ can be obtained from $S_{yx}(f)/S_x(f)$. This is a practical way of monitoring a system transfer function without interfering with its normal operation, employing noise as a test signal (see problem A.2 and Chapter 9). In practice, $S_{yx}(f)$ and $S_x(f)$ are not measured directly, but computed from the measured functions $R_{yx}(\tau)$ and $R_x(\tau)$ in accordance with the transform pairs

$$R_{yx}(\tau) \leftrightarrow S_{yx}(f),$$

and

$$R_x(\tau) \leftrightarrow S_x(f). \tag{A.55}$$

A useful concept is the *coherence function*, which is defined as

$$\gamma_{xy}^2(f) = \frac{|S_{xy}(f)|^2}{S_x(f)S_y(f)}. \tag{A.56}$$

It yields the fraction of the power in $v_y(t)$ which is coherent with that in $v_x(t)$ independent of the respective power levels in each signal. Note that $\gamma_{xy}^2(f)$ differs significantly from a normalised crosscorrelation function since it is normalised at each frequency, whereas the correlation function can be normalised only for the total power. Furthermore, unlike the latter, the coherence function between the output and input of a linear time-invariant system is not a function of the transmission gain between them.

There is a cross-spectrum inequality analogous to the crosscorrelation inequality of (A.33).

$$|S_{xy}(f)|^2 \leqslant S_x(f)S_y(f). \tag{A.57}$$

From (A.57) it follows that

$$0 \leqslant \gamma_{xy}^2(f) \leqslant 1. \tag{A.58}$$

If applied to a time-invariant linear system with a single input and output, the value of the output–input coherence function is unity. If it is less than unity, then either there are other inputs to the system with different transfer functions, or the system is not linear. Actually these two conditions are corollaries, since non-linearity contributes components in addition to the linear output.

In practical applications where $S_x(f)$ relates to a measured variable, the inherent advantage of $\gamma^2_{xy}(f)$ over the unnormalized cross-spectral density is because the conversion factor of the transducer converting the measured quantity into an electric signal, and the gain of the following signal processing channel, appear as a factor in the numerator and the denominator of (A.56). Since $S_x(f)$ and $S_y(f)$, multiplied by the corresponding gain factors, are obtained from the same data as $S_{xy}(f)$, the corresponding factors in the numerator and in the denominator are equal and hence cancel out. This is essential in a practical measurement, where calibration of the various gain factors is impractical.

The response of gated filters to stationary noise

Circuits such as TACs or integrating A/D converters measure voltages or time intervals, respectively. Each measurement is performed by integrating over a limited interval of time, and the result is the voltage at the end of a certain integrating period, or the period of integration required to reach a certain voltage. An important performance parameter for such gated integrators, or gated filters in the general case, is the variance of this series of measurements.

Although gated filters are time variant networks, they can be considered as dealing with a noise source switched onto the input of a filter with time invariant parameters, if the switch is repeatedly opened during a fixed time interval T and closed in synchronization with the instant of observation.

We shall first consider the response of a gated filter to a deterministic input. Since we are dealing with a physical – perforce causal – system and t stands for real time, the system impulse response satisfies $h(t) = 0$ for $t < 0$. Hence, (A.41) may be written as

$$y(t_o) = \int_{-\infty}^{t_o} x(t)h(t_o - t)\,\mathrm{d}t, \tag{A.59}$$

where t_o is defined as the observation instant and $x(t)$ represents the input.

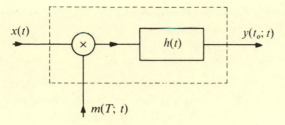

Fig. A.2. Gated filter.

Now assume that $h(t)$ is modulated by $m(t)$ as shown in Fig. A.2. As an example, $m(t)$ could simulate a switch opening during a time interval T before t_o, in which case it is defined by

$$m(T; t) \triangleq \begin{cases} 1 & \text{for } (t_o - T) \le t \le t_o, \\ 0 & \text{otherwise.} \end{cases} \tag{A.60}$$

Then we obtain

$$y(t_o) = \int_{t_o - T}^{t_o} x(t) h(t_o - t) \, dt. \tag{A.61}$$

By choosing $t_o = T$, (A.61) becomes

$$y(T) = \int_0^T x(t) h(T - t) \, dt. \tag{A.62}$$

The transmission of noise through a gated filter has been derived from (A.62) by Lampard,[5] who shows that the normalized output noise power of the filter, with stationary random noise applied to its input, is

$$R_y(T; 0) = \int_0^T R_h(T - u; u) R_x(u) \, du, \tag{A.63}$$

where

$$R_h(T; v) = \int_0^T h(t) h(t - v) \, dt \tag{A.64}$$

is the time-dependent autocorrelation function of the filter introduced by Lampard, and $R_y(T; 0)$ the autocorrelation at the output, both for $\tau = 0$. The time dependency due to the switching is reflected by the integration over the finite time interval T.

Equation (A.63) may be simplified if we assume that the noise applied to the input is white, i.e. the input spectral noise density is constant over the range of passband frequencies of the following filter. In that case, in accordance with Parseval's theorem and (A.53), (A.63) can be written as

$$R_y(T; 0) = W_x(0) \int_0^T h^2(t) \, dt, \tag{A.65}$$

where $W_x(0)$ is the mathematical power density of the white noise source.

The validity of (A.65) may be extended to the general case in which the input noise is not white. For this case, the response $R_y(T; 0)$ may be considered as being due to a white noise source whose noise impulses are filtered by a fictive noise filter of impulse response $h_n(t)$, which is cascaded with $h(t)$ and modulated by $m(t)$. Hence, the impulse response of this fictive noise channel is

$$w(T; t) = m(T; t)[h_n(t) * h(t)]. \tag{A.66}$$

We define $w(T; t)$ as the gated weighting function for the input noise impulses, which is the response of the noise channel to a unit impulse sliding backwards in time, starting from the instant of observation towards minus infinity. According to this definition, the ungated weighting function

$$w(t) = h_n(t) * h(t) \tag{A.67}$$

of the noise channel is identical with its impulse response.

Applying the concept of the weighting function to the impulses emitted by a white noise source, (A.63) may be written as

$$R_y(T; 0) = W_x(0) \int_0^\infty w^2(T; t) \, dt = W_x(0) \int_0^T w^2(t) \, dt. \tag{A.68}$$

In a practical evaluation, the convolution between $h_n(t)$ and $h(t)$ in (A.66) is preferably computed in the frequency domain and then transformed back into the time domain, where $w(t)$ is squared and integrated over the time interval T as indicated in (A.68). Note that algebraic integration of (A.68) is more easily performed in the time domain than in the frequency, where multiplication by $m(t)$ transforms into convolution. In fact, the modulating function $m(t)$ is not restricted to the two values 1 and 0, as has been assumed in (A.60), but may represent any kind of modulation synchronized with t_o. Furthermore, graphical integration yields excellent approximations in cases where algebraic computations become cumbersome.

Application of (A.68) is demonstrated in Chapter 3, in the computation of a gated integrator. However, its use is not restricted to gated integrators since it is applicable to any kind of gated filter satisfying the conditions defined in this section. Further applications and the use of graphical integration for $w^2(t)$ are discussed in references [6] and [7].

Problems

A.1. Show that (A.19) can also be written as

$$C_{xy} = E[x(k)y(k)] - E[x(k)]E[y(k)].$$

A.2. Consider white noise applied to the input of a system of impulse response $h(t)$. The autocorrelation function $R_x(\tau)$ of white noise is an impulse $R_x(0)$ at $\tau = 0$ and zero at $\tau \neq 0$. Show in the time domain that

$$R_{yx}(\tau) = W(0)h(\tau), \qquad \text{(A.69)}$$

where $R_{yx}(\tau)$ is the crosscorrelation between the output of the filter and the noise, and $W(0)$ the mathematical power spectral density of the noise, whose value is constant over the frequency spectrum of $h(t)$.

A.3. (a) Show that for a time-invariant linear system, the coherence function between the output and the input equals unity provided no other input is present.
   (b) If there is more than one input, show that

$$0 < \gamma_{xy}^2(f) < 1.$$

Note that in both cases the coherence function is not a function of the transmission gain between input and output.

A.4. A white noise voltage source of power spectral density $S(f) = 2W(f)$ is connected to two signal processing channels as shown in Fig. A.3. $H_1(j\omega)$ and $H_2(j\omega)$ model the transfer functions of the two channels, and τ_D the delay in channel 1.
   (a) Find $S_{vo}(f)$ in terms of the given parameters.
   (b) Identify $\rho_{12}(\tau)$ in the result of (a).

A.3 Noise sources

The ultimate limit to the sensitivity of electronic instrumentation is set by noise. Two kinds of noise widely encountered are Shot noise and Thermal or Johnson noise.

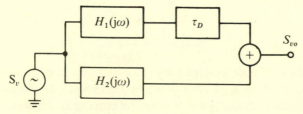

Fig. A.3. Addition of noise voltage transmitted through two different filters.

Shot noise is characterized by discrete and randomly distributed charge carriers crossing a region, a typical example being the saturation current of a reverse-biased junction. The current pulses due to the individual electrons, passing through a circuit, constitute statistically independent events following the Poisson distribution, which for a sufficiently large number of electrons becomes a Gaussian distribution. Because of the very short duration of these pulses and the limited bandwidth of practical circuits, this 'shot effect' cannot be observed directly, but only by the response of a circuit to the current pulses.

Since the resulting fluctuations are statistical in nature, shot noise is modelled by an equivalent mean square noise current generator whose spectral density can be shown to equal[3]

$$S_i = \mathrm{d}(\overline{i_n^2})/\mathrm{d}f = 2qI, \tag{A.70}$$

with I representing the value of the dc current producing the shot noise.

It is convenient to express the effect of shot noise as being due to a fictive parallel noise resistance R_p replacing the noise current source: $S_i = 2qI = 4k\mathcal{T}/R_p$, from which the value of the fictitious resistance is

$$R_p = 2k\mathcal{T}/qI \approx 0.052/I \tag{A.71}$$

at room temperature.

Johnson noise occurs in resistive conductors due to the conduction electrons being subject to 'Brownian motion'. These electrons have a thermal energy which is in equilibrium with the thermal energy of the surrounding atoms, and their fluctuations (whose average value is zero) can be represented by a Norton current source in parallel or a Thevenin voltage source in series with the resistor. The intensity of an equivalent mean square noise voltage source connected in series with a resistor has been derived by Nyquist[8] from thermodynamic laws as

$$\mathrm{d}(\overline{v_n^2})/\mathrm{d}f = 4hfR/[\exp{(hf/k\mathcal{T})} - 1] \tag{A.72}$$

which can be simplified for frequencies sufficiently low that

$$hf/k\mathcal{T} \ll 1:$$
$$\mathrm{d}(\overline{v_n^2})/\mathrm{d}f = 4k\mathcal{T}R. \tag{A.73}$$

Fig. A.4 shows the equivalent diagram of a resistor, with a series noise voltage source S_v or a parallel noise current source S_i representing the respective mean square noise voltage or current spectral density, with $S_i = 4k\mathcal{T}/R$.

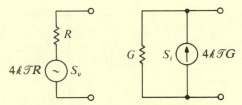

Fig. A.4. Equivalent Thevenin and Norton noise source of resistor.

Equations (A.70) and (A.73) assume ideal noise impulses. Hence, their frequency spectrum is flat, which would lead to an infinite noise power if the frequency band of the cascaded filter were extended without upper limit. Since actual current impulses due to single electrons have a finite duration, they are frequency limited, which would theoretically require a quantum correction as given by (A.72) for Johnson noise. In practice, however, the band limitation of electrical noise sources (both shot noise and Johnson noise) occurs at such high frequencies that, over the useful frequency band of practical circuits, their spectral noise density can be considered as independent of frequency, and the noise as 'white'.

Other kinds of noise sources
Shot and thermal noise sources are not the only ones encountered in electrical circuits. There is a whole group of sources which may be lumped together as 'low frequency noise' or 'semiconductor noise', all of them having the characteristic that their spectral density is inversely proportional to frequency. 'Flicker effect' and 'popcorn noise' belong to this category. Also, the noise in carbon-composition resistors increases with current above the value indicated by (A.73) (excess noise).

A.4 Useful notation

Equivalent noise sources

Dissipative networks produce noise which is frequency dependent in accordance with the transfer function between the internal noise sources and the output terminals. Nyquist has shown[8] that all internal noise sources of a complex immittance can be lumped into a single equivalent mean squared noise voltage spectral density source in series or a noise current spectral density source in parallel with the noiseless network, as shown in Fig. A.5, with

$$S_v = 4k\mathcal{T}\,\mathrm{Re}\,[Z(\mathrm{j}\omega)], \quad S_i = 4k\mathcal{T}\,\mathrm{Re}\,[Y(\mathrm{j}\omega)]. \tag{A.74}$$

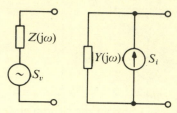

Fig. A.5. Equivalent noise source of complex immittances.

Equivalent noise bandwidth

Noise performance evaluation of amplifiers requires the computation of amplifier noise referred to its input, where it can be conveniently compared with the noise of the signal source. The effective noise output voltage can be referred to the input by using the concept of the ENB (*equivalent noise bandwidth*).

Consider a white noise voltage source of intensity S_v connected to a network as shown in Fig. A.6, and assume that

$$H(j\omega) = A(0)/(1+j\omega\tau).$$

Then

$$\overline{v_{no}^2} = S_v A^2(0) \int_0^\infty (1+\omega^2\tau^2)^{-1}\,df = S_v A^2(0)/4\tau. \qquad (A.75)$$

In accordance with Fig. A.7 we may replace $H(j\omega)$ by an imaginary filter whose normalized transmission over the passband is $A(0)$ and whose bandwidth Δf is chosen such that the noise voltage at the output becomes the same as in (A.75):

$$\overline{v_{no}^2} = S_v A^2(0)\Delta f. \qquad (A.76)$$

This yields an equal area under the two responses shown in Fig. A.7. Comparison of (A.76) with (A.75) defines

$$\Delta f = \int_0^\infty |H(j\omega)|^2\,df/A^2(0) = 1/4\tau = \pi f_h/2 \qquad (A.77)$$

for that particular case.

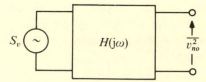

Fig. A.6. ENB computation.

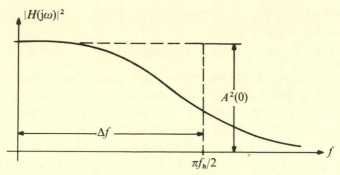

Fig. A.7. ENB of single pole low-pass filter.

Finally, the output noise voltage is referred to the input, dividing by the gain:

$$\overline{v_{ni}^2} = S_v \, \Delta f. \tag{A.78}$$

In the case of a narrow bandpass filter, whose characteristic is shown in Fig. A.8, the normalized transmission or midband gain would be $H(f_d)$, with f_d being the frequency of maximum response.

If the ENB is normalized with respect to f_h of a low-pass filter to be investigated, we obtain $\Delta f / f_h$. This is a useful figure of merit,[9] defined as the ratio between the noise power passed by the real filter and the noise power passed by an ideal square-sided filter, if the width of the latter equals f_h. Hence, $\Delta f / f_h$ is preferably employed in the comparison of the noise performance of various filters.

Equivalent series noise resistance

If $H(j\omega)$ in Fig. A.6 represents a network comprising several noise sources in an arbitrary configuration, then the ENB is useful in defining

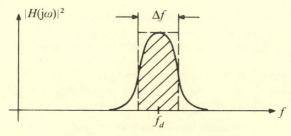

Fig. A.8. ENB of bandpass filter.

the *equivalent series noise resistance* R_{eq} of $H(j\omega)$. Substituting $S_v = 4 k \mathcal{T} R_{eq}$ in (A.78), we obtain

$$R_{eq} = \overline{v_{ni}^2}/4 k \mathcal{T} \, \Delta f. \tag{A.79}$$

As defined by (A.79) R_{eq} defines the value of a fictitious resistor, connected in the input loop of the noiseless network in series with the signal voltage generator, which produces the same rms output noise as the actual internal noise sources of the network.

The equivalent series noise resistance is useful in noise figure computations of amplifiers (see Appendix A.5) and complete systems (see Chapter 11).

Noise corner frequency

In the general case, R_{eq} is a function of frequency. Consider, as an oversimplified example, a white noise source S_n connected to the output of an amplifier whose gain equals $A(0)/(1 + j\omega/\omega_h)$. The spectral noise density of this source, referred to the input of the amplifier, is

$$S_i = 4 k \mathcal{T} R_{eq}(f), \tag{A.80}$$

with

$$R_{eq}(f) = S_n (1 + f/f_h^2)/4 k \mathcal{T} A^2(0)$$

i.e. the assymptotic frequency response of the spectral noise density is flat below, and rises by 40 dB/ decade above ω_h. Hence, if S_n were the only noise source of the amplifier, the amplifier's noise corner frequency f_c would equal f_h. A more realistic example is given in problem A.6.

Qualitatively, the corner frequency f_c is defined as that frequency beyond which the noise spectral density referred to the input of an amplifier or a gain element becomes frequency dependent. Below f_c, the noise spectral density is independent of frequency.

Problems

A.5. Consider the network shown in Fig. A.9. The resistors R_1, R_2 and R_3 are at temperatures $\mathcal{T}_1$, $\mathcal{T}_2$ and $\mathcal{T}_3$, respectively. Assume $\mathcal{T}_1 > \mathcal{T}_2 > \mathcal{T}_3$.

(a) Compute S_a, the noise power density flowing through node a, from left to right.

(b) Repeat (a) for S_b flowing from right to left.

(c) Show that $S_a > S_b$ for the temperature relationship as given.

(d) Obtain an expression for the total power flowing from left to right.

A.6. The noise performance of an amplifier is modelled by two white noise sources S_1 and S_2, connected to its input and output, respectively. The gain of the

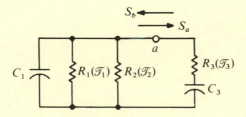

Fig. A.9. Problem A.5.

amplifier is

$$A(s) = A(0)/[1 + sA(0)\tau_0].$$

Show that, for $S_2 \ll A^2(0)S$, the noise corner frequency equals $\omega_c^2 \approx S_1/S_2\tau_0^2$.

A.5 The noise figure

Amplifiers connected to low-level signal sources should contribute as little noise as possible. A convenient measure for the noise contribution of an amplifier is its *noise figure F*, defined as its $(\text{SNR})^2$ or W_{sg}/W_{ng} at the input, due to the source only, divided by its $(\text{SNR})^2$ or $W_{sg}/(W_{ng} + W_{na})$ at the output, with W_{sg} and W_{ng} being the available signal power and noise power from the source, and W_{na} that from the amplifier referred to its input, respectively. According to Fig. A.10,

$$F = (W_{sg}/W_{ng})/[P_a W_{sg}/P_a(W_{ng} + W_{na})] = 1 + W_{na}/W_{ng}, \quad (A.81)$$

where P_a is the power gain of the amplifier.

It follows that, although F is defined in terms of power ratios between signal and noise, the signal actually cancels out and so does the power gain of the amplifier. Hence, F is a function only of W_{ng} and of W_{na}, since W_{na} is the only quantity added by the amplifier, whereas the signal is merely amplified by the same power gain as the sum of W_{ng} and W_{na} at the input.

Next, we consider the noise figure of two cascaded amplifiers, as shown in Fig. A.11:

$$F_{12} = 1 + (W_{na1} + W_{na2}/P_{a1})/W_{ng} = F_1 + (F_2 - 1)/P_{a1}, \quad (A.82)$$

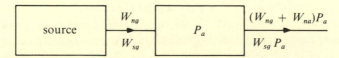

Fig. A.10. Noise figure computation for single amplifier.

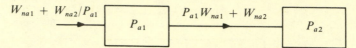

Fig. A.11. Noise figure computation for cascaded amplifiers.

with $F_1 = 1 + W_{na1}/W_{ng}$ and $F_2 = 1 + W_{na2}/W_{ng}$.   For n cascaded amplifiers we have

$$F = F_1 + \cdots + \frac{F_i - 1}{P_{a1}P_{a2} \cdots P_{a(i-1)}} + \cdots + \frac{F_n - 1}{\prod\limits^{n-1} P_{ai}} \qquad (A.83)$$

Equation (A.82) brings out the importance of the power gain P_a of an amplifier, which, although not affecting its own noise figure, should make the noise contribution of the next amplifier as small as possible.

In noise figure computations for mismatch-designed amplifiers, it is essential to replace the power ratios by ratios between the corresponding squared voltages or currents, as the case may be:

$$F = 1 + W_{na}/W_{ng} = 1 + \overline{v_{na}^2}/\overline{v_{ng}^2} = 1 + \overline{i_{na}^2}/\overline{i_{ng}^2}. \qquad (A.84)$$

This is permissible, provided that the voltages are referred to the same loop, or the currents to the same node at the amplifier input.

The use of voltage or current ratios in (A.84) implies the use of voltage or current as the input signal of the amplifier. In the corresponding relationship for two cascaded amplifiers, in (A.82), P_{a1} will be replaced by the squared voltage or current gain $|A_{v1}(j\omega)|^2$ or $|A_{i1}(j\omega)|^2$ of the first stage, whose definition is straightforward for voltage or current amplifiers.

Usually, the noise figures is given in decibels. If a source of given SNR is connected to an amplifier whose noise figure is F, the $(SNR)^2$ at the output, in dB, is obtained from (A.81) as

$$10 \log (SNR)^2_{output} = 10 \log (SNR)^2_{source} - 10 \log F. \qquad (A.85)$$

A.6   The matched filter

For a rigorous derivation of matched filter theory, the reader is referred to one of the standard texts.[10] Here we shall content ourselves with intuitive reasoning, which provides us with a simple explanation.

To maximize the SNR, two conditions should be met. First, we choose the relationship between the noise and signal spectra at the input of the filter to be maximally random, or the noise to be white. Hence, the matched filter should be preceded by a noise whitening filter. Second, to obtain at the output the total power due to the signal $f_a(t)$ at the input, we choose the filter response to $f_a(t)$ to yield the autocorrelation function of $f_a(t)$. Hence, if $H_a(j\omega)$ is the signal spectrum at the input of the filter, and $H_b(j\omega)$ the filter's frequency response, then this condition defines the transfer function of the matched filter as $H_b(j\omega) = H_a(-j\omega)$.

The Fourier transform pairs $H(j\omega) \leftrightarrow f(t)$ and $H(-j\omega) \leftrightarrow f(-t)$ provide us with an interpretation of the matched filter impulse response in the time domain, which should be the mirror image of the signal shape at the output of a noise whitening filter.

In the frequency domain, we may draw a parallel with matching a load for maximum power transfer from the source, which is obtained by choosing the load impedance as the complex conjugate of the source.

If the signal is a causal function, i.e. $f_a(t) = 0$ for $t < 0$, then the matched filter response is defined by $f_a(t)$, reversed in time and delayed by a sufficiently long time to be causal. In cases where such a response is not realizable, the computed optimal SNR using the matched filter merely provides a lower bound, to which the SNR obtained with a practical filter can be compared.

A matched filter may, of course, also be defined for the case in which the noise spectrum at its input is not white. A practical example is given in Section 3.1, equation (3.57).

Appendix B

Analysis of feedback amplifiers

The properties of a feedback amplifier can be described by its transfer function

$$H = \frac{1}{B} \frac{(-LT)}{(1-LT)} \tag{B.1}$$

and its closed-loop input (or output) immittance (impedance or admittance) I_{in} (or I_{out}) by

$$x_j / r_j = [I_{jj}(1 - LT_{jj})]^{-1} \tag{B.2}$$

where r is the independent variable and x the dependent variable. The flow graph for (B.2) is shown in Fig. B.1. We define

$$1/B = \lim_{|LT| \to \infty} H \tag{B.3}$$

as the ideal transfer function with infinite LT (*loop transmission*). The physical significance for the input immittance of port j is obtained, comparing the flow graph in Fig. B.1 with Figs. B.2(a), (b), (c) and (d), with I_{jj} being defined as the open-loop self-immittance of node or loop j. On the basis of (B.2) we conclude that feedback increases the open-loop input and output immittance I_{jj} of a feedback amplifier – i.e. it modifies its properties in a unified way. That the self-immittance I_{jj} is an impedance in the case of a loop (series feedback) or an admittance in the case of a node (shunt feedback) manifests itself merely in the choice of the variables, but is basically irrelevant.

Note that (B.2) defines the closed-loop input or output immittance (I_{in} or I_{out}) as the *inverse* of the closed-loop self-immittance. This definition stems from the topological derivation of the feedback parameters presented in Appendix B.4, where I_{in} and I_{out} are identified with the corresponding term in the parameter set equations.

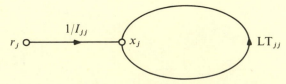

Fig. B.1. Modification of input immittance by feedback.

Knowledge of B, LT, and I_{jj} completely defines the properties of a feedback amplifier. Their computation will be demonstrated in the following paragraphs for the four different kinds of feedback amplifiers shown in Fig. B.2, which demonstrates the connection between a basic amplifier and the feedback network for the four possible combinations

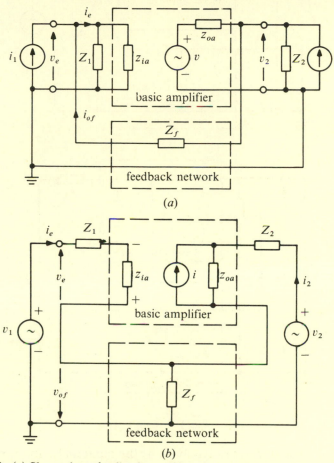

Fig. B.2. (a) Shunt-shunt feedback amplifier; for a definition of v see text.
(b) Series-series feedback amplifier:

$$i = -v_e Y_T z_{ia}/(Z_1 + z_{ia}).$$

(c) (*Overleaf*) Shunt-series feedback amplifier:

$$v = v_e A_v z_{ia}/(Z_1 + z_{ia}).$$

(d) (*Overleaf*) Series-shunt feedback amplifier:

$$i = i_e A_i Z_1/(Z_1 + z_{ia}).$$

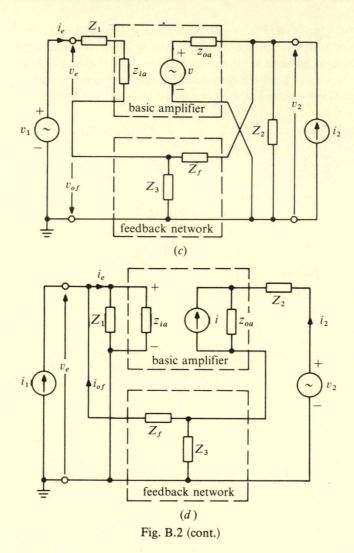

(c)

(d)

Fig. B.2 (cont.)

between parallel and series feedback at the input and output, respectively, of the circuit.

B.1   Identification of the ideal response by inspection

The ideal response $1/B$ as defined in (B.3) can be obtained if the gain value of the dependent source is assumed to be infinitely large. Applying this to each part of Fig. B.2, an infinite gain value of the corresponding active elements causes both the input voltage and current of the basic amplifiers to become zero. The ideal transfer function is therefore obtained when the response to the driving function satisfies this condition. Applied to Fig. B.2, (B.3) becomes $v_2/i_1 = -Z_f$ for part (a), $i_2/v_1 = 1/Z_f$ for part (b), $v_2/v_1 = (Z_3 + Z_f)/Z_3$ for part (c) and $i_2/i_1 = -(Z_f + Z_3)/Z_3$ for part (d).

It is important to recognize that *the nature of the ideal response $1/B$ is determined solely by the kind of feedback applied.* For example, the basic amplifier in Fig. B.2(a) may be a voltage or a transimpedance amplifier, but $1/B$ will still be a transimpedance because of the shunt-shunt feedback applied.

B.2 Identification of LT by source splitting

In control theory, the closed-loop response is usually described as

$$H = a/(1 - af) \tag{B.4}$$

with the open-loop gain a and the LT af being easily obtained by inspection. For electronic feedback circuits, however, recognition of LT is not straightforward, because two signal parameters – voltage and current – are present.

Most textbooks[1,2] employ a set of rules by means of which a, f and the return ratio af in (B.4) are determined subject to circuit modifications varying with the topology of the feedback network.

Another method,[3] employed throughout this book, leaves the circuit intact, but splits the dependent source into two, one driving the circuit with a signal of unit value and the other being driven by the circuit, its value representing the LT. Both methods yield the same result, but the latter is applicable to all circuit topologies without modifications and yields the LTs 'by inspection'.

The four parts of Fig. B.2 may be used to demonstrate the computation of LT by the *source splitting* method for the four possible combinations between shunt and series feedback applied to the input and output of an amplifier. They have been drawn with the source and load impedances Z_1 and Z_2 considered as part of a two-port driven by two independent sources v_1 or i_1 at the input and v_2 or i_2 at the output. One of the two error parameters (v_e or i_e) is shown as the sum of the input and the feedback parameters, i.e. $v_e = v_1 + v_{of}$ or $i_e = i_1 + i_{of}$.

The source splitting method will be demonstrated considering shunt-shunt feedback applied to a basic voltage or transimpedance amplifier, as shown in Fig. B.2(a). Both kinds of amplifier may be employed with this type of feedback, but the dependent voltage source driving the amplifier

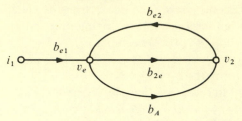

Fig. B.3. Flow graph representing Kirchhoff equations for Fig. B.2(a), with $i_2 = 0$.

output is $v = -A_v v_e$ for a basic voltage amplifier and $v = -Z_T v_e / z_{ia}$ for a basic transimpedance one, since the error parameter is v_e for the first but i_e for the second. The resulting LT is, however, the same in both cases, unless the appropriate mismatch conditions are taken into account (Appendix B.5). Here we shall choose a voltage amplifier, for which the relevant Kirchhoff equations for v_e and v_2 yield the flow graph of Fig. B.3, with $b_{e1} = 1/Y_{11}$, $b_A = -A_v y_{oa}/Y_{22}$, $b_{2e} = Y_f/Y_{22}$ and $b_{e2} = Y_f/Y_{11}$, where $Y_{11} = Y_1 + Y_f + Y_{ia}$ and $Y_{22} = Y_2 + Y_f + y_{oa}$.

Splitting the dependent voltage source into two yields the circuit diagram of Fig. B.4 and the corresponding flow graph of Fig. B.5, from which $\mathrm{LT_s}$ is identified as

$$\mathrm{LT_s} = -A_v (z_{ia} \| Z_1) Z_2 / (z_{oa} + Z_2)[(z_{oa} \| Z_2) + Z_f + (z_{ia} \| Z_1)]$$

$$= b_A b_{e2} / (1 - b_{e2} b_{2e}), \tag{B.5}$$

where $\mathrm{LT_s}$ signifies that the LT has been obtained by source splitting.

The inherent error in the source splitting method

Although these methods – i.e. the one referenced in[1,2] and, in particular, the source splitting one – allow the path of the LT to be traced throughout

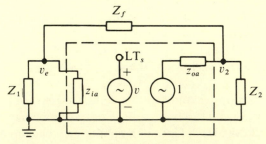

Fig. B.4. Source splitting for circuit of Fig. B.2(a).

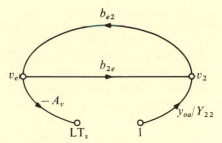

Fig. B.5. Flow graph yielding LT_s by source splitting.

the circuit, an inspection of Fig. B.3 and B.5 reveals that the LT_s thus obtained is in error, since the signal path b_{2e} which is part of the forward transmission should have been interrupted as well. Hence, the correct expression for LT is obtained from Fig. B.6 as

$$LT = (b_A + b_{2e})b_{e2} = -A_v y_{oa}(1 - Y_f/A_v y_{oa})Y_f/Y_{11}Y_{22}, \qquad (B.6)$$

which differs from (B.5).

Computation of LT by the source splitting method or in accordance with [1,2] for the remaining configurations shown in Fig. B.2 will be left to the reader as a problem.

Digression to feedback terminology

We note that the LT is commonly termed the *return ratio*, the ratio between the parameter returned by the opened loop and the one driving it (both either voltage or current). Similarly, the difference between the driving parameter of unit value and the one returned by the loop, $1 - LT$, is termed *return difference* $\mathscr{F}$ (see problem 1.9).

Related to the flow graph in Fig. B.6, $\mathscr{F}$ is defined as the vector drawn from the output of the open loop (LT) to the input (1).

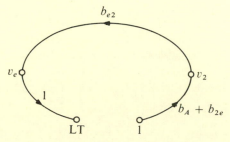

Fig. B.6. Flow graph yielding topologically correct LT.

B.3   Impedance computations

After the LT has been obtained, impedance calculations from (B.2) are a simple matter. We note that in each part of Fig. B.2 the input and output driving functions have been chosen as voltages for series feedback and currents for parallel. Accordingly, we shall be dealing with a loop or a node and I_{jj} will be its self-impedance or admittance, respectively.

We recall that to evaluate a self-immittance all adjacent loops should be opened and all adjacent nodes short-circuited. Hence, as a sample computation for Fig. B.2(c), (B.2) becomes:

$$I_{\text{in}} = i_e/v_1 = [Z_{11}(1 - \text{LT}_s)]^{-1}, \tag{B.7}$$

$$I_{\text{out}} = v_2/i_2 = [Y_{22}(1 - \text{LT}_s)]^{-1}, \tag{B.8}$$

$$Z_{11} = Z_1 + z_{ia} + (Z_f \| Z_3), \tag{B.9}$$

$$Y_{22} = Y_2 + y_{oa} + 1/(Z_3 + Z_f). \tag{B.10}$$

Keeping in mind the dual relationship between impedance (associated with loop equations and series feedback) and admittance (associated with node equations and parallel feedback), interpretation of the flow graph in Fig. B.1 for Figs. B.2(a), (b) and (d) should not present any difficulty.

B.4   Identification of LT from the Kirchhoff equations

In most practical cases, obtaining the LT by source splitting or equivalent techniques is quite satisfactory in spite of the inaccuracies involved. However, a formal approach based upon topological considerations[3,4] is found to yield not only the exact relationships, but also simpler ones than the source splitting approach. Furthermore, it is also applicable to multi-variable feedback circuits.

Consider again the amplifier of Fig. B.2(a). This uses a basic voltage amplifier whose Kirchhoff equations for the input and output node are obtained in matrix form as

$$\underbrace{\begin{matrix} \text{independent} \\ \text{variables} \end{matrix}}_{} \qquad \underbrace{\begin{matrix} \text{dependent} \\ \text{variables} \end{matrix}}_{}$$

$$\begin{matrix} \text{input} \\ \text{output} \end{matrix} \begin{bmatrix} r \\ n \end{bmatrix} = \begin{bmatrix} I_{11} & B \\ A & I_{22} \end{bmatrix} \begin{bmatrix} e \\ y \end{bmatrix} \begin{matrix} \text{input} \\ \text{output} \end{matrix} \tag{B.11}$$

where $r = i_1$, $n = i_2$, $e = v_e$, $y = v_2$, $I_{11} = Y_{11}$, $I_{22} = Y_{22}$, $B = -Y_f$, and $A = A_v y_{oa} - Y_f = A_v y_{oa}(1 - Y_f/A_v y_{oa})$.

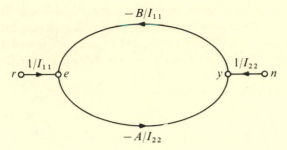

Fig. B.7. Flow graph for feedback amplifier based on Kirchhoff equations.

Equation (B.11) is represented by the flow graph of Fig. B.7, which yields the 'topologically correct' LT given by (B.6). Solving the flow graph for the dependent variables e and y in terms of LT, we obtain

$$\begin{bmatrix} e \\ y \end{bmatrix} = \begin{bmatrix} I_{\text{in}} & H_r \\ H_f & I_{\text{out}} \end{bmatrix} \begin{bmatrix} r \\ n \end{bmatrix} \tag{B.12}$$

with

$$I_{\text{in}} = [I_{11}(1 - \text{LT})]^{-1},$$

$$I_{\text{out}} = [I_{22}(1 - \text{LT})]^{-1},$$

$$H_f = (-\text{LT})/B(1 - \text{LT}),$$

$$H_r = (-\text{LT})/A(1 - \text{LT}).$$

Finally, $\text{LT} = AB/I_{11}I_{22}$, with $-A/I_{22}$ denoting the forward gain, and

$$-B/I_{11} = B', \quad \text{the feedback attenuation.}$$

Equation (B.12) corresponds to the two-port parameter set (known as y, z, g and h parameters),[5] whose coefficients have here been obtained in terms of LT which in turn is directly derived from (B.11). This inversion of the Kirchoff matrix in (B.11) in terms of LT reveals the physical significance of its coefficients: I_{11} and I_{22} are the open-loop input and output immittances, $1/B$ is the ideal transfer function for $\lim |\text{LT}| \to \infty$, and $1/A$ is the reverse transmission for $\lim |\text{LT}| \to \infty$. In the process we have formally derived H in (B.1), which is identical to H_f.

Summarizing the rules for obtaining the feedback parameters of an active two-port network from the Kirchhoff equations, it is essential to arrange the independent and dependent variables as shown in (B.11). Specifically, the error parameter e should be chosen as the input parameter of the amplifier or of the gain element, and y as the feedback-stabilized output parameter of the two-port. A criterion for the correct choice of e is that it vanishes if the gain becomes infinitely large, whereas y/r becomes a function of the feedback network only, for the same condition.

The corresponding equations are:

$$\begin{bmatrix} i_1 \\ i_2 \end{bmatrix} = \begin{bmatrix} (Y_1 + Y_f + y_{ia}) & -Y_f \\ (A_v y_{oa} - Y_f) & (Y_2 + Y_f + y_{oa}) \end{bmatrix} \begin{bmatrix} v_e \\ v_2 \end{bmatrix} \qquad \text{(B.13)}$$

for shunt-shunt feedback (Fig. B.2(a));

$$\begin{bmatrix} v_1 \\ v_2 \end{bmatrix} = \begin{bmatrix} (z_{ia} + Z_f + Z_1) & Z_f \\ (Z_f - Y_T z_{ia} z_{oa}) & (Z_2 + z_{oa} + Z_f) \end{bmatrix} \begin{bmatrix} i_e \\ i_2 \end{bmatrix} \qquad \text{(B.14)}$$

for series-series feedback (Fig. B.2(b));

$$\begin{bmatrix} v_1 \\ i_2 \end{bmatrix} = \begin{bmatrix} [Z_1 + z_{ia} + (Z_3 \| Z_f)] & [Y_3/(Y_f + Y_3)] \\ [Y_3/(Y_f + Y_3) - A_v z_{ia} y_{oa}] & [Y_{oa} + Y_2 + 1/(Z_3 + Z_f)] \end{bmatrix} \begin{bmatrix} i_e \\ v_2 \end{bmatrix}$$
$$\text{(B.15)}$$

for shunt-series feedback (Fig. B.2(c)); and

$$\begin{bmatrix} i_1 \\ v_2 \end{bmatrix} = \begin{bmatrix} [y_{ia} + Y_1 + 1/(Z_3 + Z_f)] & -[Y_f/(Y_f + Y_3)] \\ [A_i y_{ia} z_{oa} - Y_f/(Y_f + Y_3)] & [z_{oa} + Z_2 + (Z_f \| Z_3)] \end{bmatrix} \begin{bmatrix} v_e \\ i_2 \end{bmatrix}$$
$$\text{(B.16)}$$

for series-shunt feedback (Fig B.2(d)).

At this point, two observations should be made. (a) Frequently, a two-port may be described by more than two equations, in which case (B.11) is obtained through suitable reduction and elimination of undesirable variables. (b) The choice of e is not unique. Referring to (B.15) as an example, i_e has been chosen as error parameter, in which case the parameter v_e of the dependent source will be replaced by $i_e z_{ia}$. Alternatively, v_e could have been chosen as error parameter, employing the relationship $v_e = v_1 + v_{of}$ and eliminating v_{of}.

Verification of equations (B.13) to (B.16) is left to the reader.

B.5   Mismatch oriented design of feedback amplifiers

Mismatch is essential in the design of broadband amplifiers. It is widely employed in the design of feedback amplifiers, and its application to the expressions derived in the preceding section provides the conditions for using the simplified design equations employed throughout the book.

Consider again the shunt-shunt feedback amplifier shown in Fig. B.2(a), employing a basic voltage amplifier. Our design target is that the LT should not be affected by the input and output impedances of the basic amplifier, since these parameters may exhibit a large spread even for ICs having identical model numbers. Recalling that a voltage amplifier exhibits 'high input' and 'low output' impedances, we apply the corresponding mismatch conditions to I_{11} and I_{22} in (B.13): $y_{ia} \ll Y_1 + Y_f$, $y_{oa} \gg Y_2 + Y_f$. This yields $\overline{I_{11}} \simeq Y_1 + Y_f$, $\overline{I_{22}} \simeq y_{oa}$, and $Y_f \ll A_v y_{oa}$, the last inequality indicating negligible direct signal transmission from v_e to v_2. The bar above I_{jj} indicates that mismatch conditions have been applied.

Finally,

$$\mathrm{LT} \simeq -A_v Y_f / (Y_1 + Y_f) = -A_v Z_1 / (Z_1 + Z_f) \qquad (B.17)$$

is the simplified design condition for stability, if the mismatch conditions are satisfied. The corresponding mismatch conditions for the remaining three amplifiers are: for series-series feedback (Fig. B.2(b) and (B.14))

$$z_{ia} \gg Z_1 + Z_f, \qquad z_{oa} \gg Z_f + Z_2, \quad \text{and} \quad Y_T z_{ia} z_{oa} \gg Z_f,$$

hence

$$\mathrm{LT} \simeq -Y_T Z_f; \qquad (B.18)$$

for shunt-series feedback (Fig. B.2(c) and (B.15))

$$z_{ia} \gg Z_1 + (Z_3 \| Z_f), \qquad y_{oa} \gg Y_2 + 1/(Z_3 + Z_f),$$

$$A_v z_{ia} y_{oa} \gg Y_3 / (Y_f + Y_3),$$

hence

$$\mathrm{LT} \simeq -A_v Y_3 / (Y_3 + Y_f); \qquad (B.19)$$

and finally, for series-shunt feedback (Fig. B.2(d) and (B.16))

$$y_{ia} \gg Y_1 + 1/(Z_3 + Z_f), \qquad z_{oa} \gg Z_2 + (Z_f \| Z_3),$$

$$A_i y_{ia} z_{oa} \gg Y_f / (Y_f + Y_3),$$

hence

$$\mathrm{LT} \simeq -A_i Y_f / (Y_f + Y_3). \qquad (B.20)$$

At this point we may summarize the benefits derived from topological identification of the feedback parameters, comparing (B.6) with (B.5).

First, the topological method provides a rigorous basis for the derivation of mismatch conditions. Second, it reveals that, because of direct transmission between input and output, at least one zero is present among the singularities of LT, whose value can be obtained in the case of (B.6) by factorizing the term $(1 - Y_f/A_v y_{oa})$. The resulting singularities are located either in the RHS of the complex frequency plane as in (B.6) for inverting amplifiers, or in the LHS in the case of non-inverting amplifiers (see Sections C.1 and 1.7.1). This has implications on stability, which is enhanced by LHS zeros and adversely affected by RHS zeros. Last but not least, the LT thus obtained is composed of four physically significant factors $(A, B, I_{11}$ and $I_{22})$, which is far more revealing than the rather cumbersome expression of (B.5).

Problems

B.1. For a basic transimpedance amplifier employed in the circuit of Fig. B.2(a), find conditions for mismatch, and the simplified expression for LT under mismatch conditions.

B.2. Compute the closed-loop input and output immittances I_{in} and I_{out} of the circuits shown in Figs. B.2(a), (b) and (d). Employ mismatch conditions.

B.3. (a) Identify I_{11}, I_{22}, A and B for the feedback amplifier shown in Fig. B.8. (b) Find the appropriate mismatch conditions and the simplified LT for this case.

B.4. A basic transadmittance amplifier is externally connected as shown in Fig. B.9. For this amplifier, three feedback-stabilized transmittances can be identified: $H_1 = v_3/v_1$ (voltage follower), $H_2 = -i_2/i_3$ (current follower), and $H_3 = i_2/v_1$ (transadmittance amplifier).

(a) Write the 3×3 Kirchhoff matrix equation for the three-port network, arranging the coefficients as shown in (B.11).

(b) Find the reduced Kirchhoff matrix equations corresponding to H_1, H_2 and H_3, respectively. This can be done in a formal way, reducing the 3×3 Kirchhoff matrix to the proper 2×2 matrix for each case, or eliminating the corresponding dependent variable.

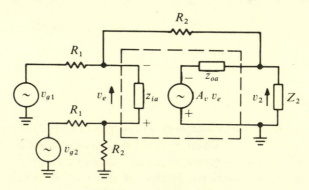

Fig. B.8. Problem B.3.

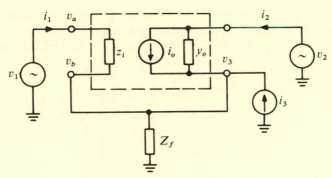

Fig. B.9. Problem B.4; $i_o = (v_b - v_a)Y_T$.

(c) For each transmittance, find the appropriate mismatch conditions and identify the coefficients of the corresponding Kirchhoff matrix. In particular, identify LT and $1/B$ for each case.

[Hint: Convert i and y_0 into an equivalent voltage source and choose i_1 as error parameter.]

B.6 Multivariable feedback circuits

The technique of analysing feedback circuits discussed in Appendix B.4 can be extended to multivariable feedback circuits – i.e. to circuits employing more than a single amplifier.[4] Possible applications are filters, fast-slow amplifiers, and circuits in which an additional amplifier is employed to improve a particular property of the circuit such as the CMRR (*common mode rejection ratio*) (see problem 1.11).

Consider the active filter[7] shown in Fig. B.10. This circuit can be represented by the multidimensional counterpart of Fig. B.7 as shown in Fig. B.11 in which the variables are replaced by 'node groups' and the branch transmissions by 'flow matrices'. Hence

$$\begin{bmatrix} \mathbf{R} \\ \mathbf{N} \end{bmatrix} = \begin{bmatrix} \mathbf{I}_{11} & \mathbf{B} \\ \mathbf{A} & \mathbf{I}_{22} \end{bmatrix} \begin{bmatrix} \mathbf{E} \\ \mathbf{Y} \end{bmatrix} \tag{B.21}$$

in which the corresponding transform vectors and submatrices are identified as

$$\mathbf{R} = \begin{bmatrix} v_g \\ v_g \end{bmatrix}, \qquad \mathbf{N} = \begin{bmatrix} i_{o1} \\ i_{o2} \end{bmatrix}, \qquad \mathbf{E} = \begin{bmatrix} e_1 \\ e_2 \end{bmatrix}, \qquad \mathbf{Y} = \begin{bmatrix} v_{o1} \\ v_{o2} \end{bmatrix},$$

$$\mathbf{I}_{11} = \begin{bmatrix} Y_{i1}/Y_1 & 0 \\ 0 & Y_{i2}/Y_4 \end{bmatrix}, \qquad \mathbf{B} = \begin{bmatrix} -Y_2/Y_1 & -Y_6/Y_1 \\ -Y_3/Y_4 & -Y_5/Y_4 \end{bmatrix},$$

$$\mathbf{A} = \begin{bmatrix} (a_1 y_{o1} - Y_2) & -Y_3 \\ -Y_6 & (a_2 y_{o2} - Y_5) \end{bmatrix}, \qquad \mathbf{I}_{22} = \begin{bmatrix} Y_{o1} & 0 \\ 0 & Y_{o2} \end{bmatrix}$$

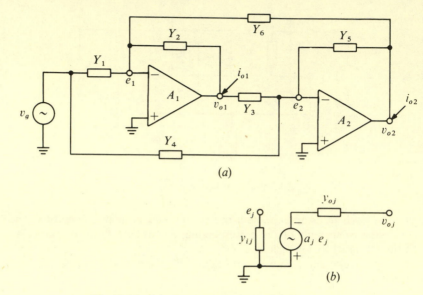

Fig. B.10. (a) Active filter. (b) Amplifier model

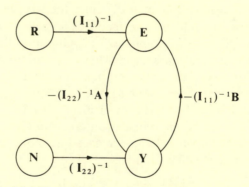

Fig. B.11. Multidimensional equivalent of flow graph of Fig. B.7.

where

$$Y_{i1} = Y_1 + Y_2 + Y_6 + y_{i1}, \qquad Y_{i2} = Y_3 + Y_4 + Y_5 + y_{i2},$$

$$Y_{o1} = Y_2 + Y_3 + y_{o1}, \qquad Y_{o2} = Y_5 + Y_6 + y_{o2}.$$

Applying the appropriate mismatch conditions

$$y_{i1} \ll Y_1 + Y_2 + Y_6, \qquad y_{i2} \ll Y_3 + Y_4 + Y_5,$$

$$y_{o1} \gg Y_2 + Y_3 \quad \text{and} \quad y_{o2} \gg Y_5 + Y_6$$

for the voltage amplifiers A_1 and A_2, we obtain

$$\bar{Y}_{i1} \simeq Y_1 + Y_2 + Y_6, \qquad \bar{Y}_{i2} \simeq Y_3 + Y_4 + Y_5;$$

$$\bar{Y}_{o1} \simeq y_{o1}, \quad \text{and} \quad \bar{Y}_{o2} \simeq y_{o2}.$$

From Fig. B.11 we can identify the two return ratio matrices

$$\mathbf{E} = \mathbf{LT}_A \mathbf{1} \tag{B.22}$$

and

$$\mathbf{Y} = \mathbf{LT}_B \mathbf{1}, \tag{B.23}$$

which are obtained by interrupting either of the feedback flow matrices, $\mathbf{I}_{22}^{-1}\mathbf{A}$ or $\mathbf{I}_{11}^{-1}\mathbf{B}_f$, and feeding the open input from a unity matrix:

$$\mathbf{LT}_A = \mathbf{I}_{11}^{-1}\mathbf{B}\,\mathbf{I}_{22}^{-1}\mathbf{A}, \qquad \mathbf{LT}_B = \mathbf{I}_{22}^{-1}\mathbf{A}\mathbf{I}_{11}^{-1}\mathbf{B}.$$

The circuit diagram corresponding to (B.22) is shown in Fig. B.12, from which the physical significance of the elements of $\mathbf{LT}_A$ emerges on inspection.

Inversion of (B.21) yields the multidimensional parameter set

$$\begin{bmatrix} \mathbf{E} \\ \mathbf{Y} \end{bmatrix} = \begin{bmatrix} \mathscr{F}_A^{-1}\mathbf{I}_{11}^{-1} & \mathscr{F}_A^{-1}(-\mathbf{LT}_A)\mathbf{A}^{-1} \\ \mathscr{F}_B^{-1}(-\mathbf{LT}_B)\mathbf{B}^{-1} & \mathscr{F}_B^{-1}\mathbf{I}_{22}^{-1} \end{bmatrix} \begin{bmatrix} \mathbf{R} \\ \mathbf{N} \end{bmatrix}, \tag{B.24}$$

with the return difference matrices given by $\mathscr{F}_A = (\mathbf{1} - \mathbf{LT}_A)$ and $\mathscr{F}_B = (\mathbf{1} - \mathbf{LT}_B)$.

B.6.1. Simplifications for high-gain local feedback loops

In an active filter of the type shown in Fig. B.10, distinction should be made between the local feedback loops whose LT is high but may vary considerably without substantial effect on filter response, and the principal loop whose LT is low and determines the exact location of the poles of the filter. In our example, A_1 and A_2 and their associated local feedback networks may be considered as transimpedance amplifiers with local return ratios $|\mathrm{LT}_{11}| \gg 1$ and $|\mathrm{LT}_{22}| \gg 1$, and with feedback-stabilized transimpedances equal to $-1/Y_2$ and $-1/Y_5$, respectively. The principal feedback loop then consists of these two transimpedances in series with the two transadmittances Y_3 and Y_6. For the case of high local return ratios, $\mathscr{F}_A \simeq -\mathbf{LT}_A$, $\mathscr{F}_B \simeq -\mathbf{LT}_B$, and the transfer function in terms of (B.24) yields

$$\lim_{|\mathrm{LT}_{ii}| \to \infty} \mathbf{Y}\mathbf{R}^{-1} = \mathscr{F}_B^{-1}(-\mathbf{LT}_B)\mathbf{B}^{-1} \simeq \mathbf{B}^{-1} \tag{B.25}$$

which is the multidimensional equivalent of (B.3).

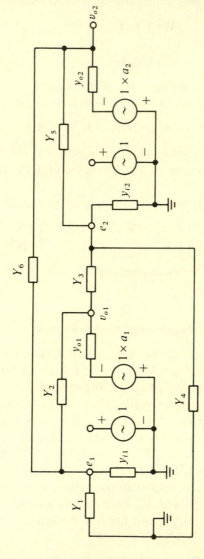

Fig. B.12. Circuit diagram for $\mathbf{E} = \mathbf{LT_A} 1$ of active filter, Fig. B.10.

Hence, substituting the appropriate terms from (B.21) in (B.25), the ideal response in matrix form is

$$\begin{bmatrix} v_{o1} \\ v_{o2} \end{bmatrix} = \begin{bmatrix} -Y_2/Y_1 & -Y_6/Y_1 \\ -Y_3/Y_4 & -Y_5/Y_4 \end{bmatrix}^{-1} \begin{bmatrix} v_g \\ v_g \end{bmatrix} \tag{B.26}$$

Similarly, a simplified expression for the return difference for the gain element a_i in the presence of high local LTs is obtainable with the aid of Bode's formula $\mathscr{F}(a_i) = \Delta/\Delta_i^0$,[8] where Δ is the system determinant and Δ_i^0 its particular value with $a_i = 0$. To this end, Δ/Δ_i^0 is expressed in terms of the return difference:

$$\Delta/\Delta_i^0 = \Delta\mathscr{F}/\Delta\mathscr{F}_i^0, \tag{B.27}$$

which is proved at the end of this chapter. The subscript A or B has been omitted in (B.27), since it is irrelevant in this relationship.

For the condition of high local LTs, the amplification factors a_i are assumed to be sufficiently high, to make all terms in LT not containing a_i negligible. Applied to the circuit of Fig. B.10, the complete expression for LT_A becomes for high local LTs

$$LT_A = \begin{bmatrix} \left[\dfrac{-Y_2(a_1y_{o1} - \cancel{Y_2})}{Y_{i1}Y_{o1}} + \dfrac{\cancel{Y_6^2}}{\cancel{Y_{i1}}Y_{o2}} \right] & \left[\dfrac{-Y_6(a_2y_{o2} - \cancel{Y_5})}{Y_{i1}Y_{o2}} + \dfrac{Y_2\cancel{Y_3}}{\cancel{Y_{i1}}Y_{o1}} \right] \\ \left[\dfrac{-Y_3(a_1y_{o1} - \cancel{Y_2})}{Y_{i2}Y_{o1}} + \dfrac{Y_5\cancel{Y_6}}{\cancel{Y_{i2}}Y_{o2}} \right] & \left[\dfrac{-Y_5(a_2y_{o2} - \cancel{Y_5})}{Y_{i2}Y_{o1}} + \dfrac{\cancel{Y_3^2}}{\cancel{Y_{i2}}Y_{o1}} \right] \end{bmatrix} .$$

Applying mismatch conditions, the return difference $\mathscr{F}$ for high local LTs is finally obtained as

$$\bar{\bar{\mathscr{F}}} = \begin{bmatrix} (1 - \overline{LT}_{11}) & -\overline{LT}_{12} \\ -\overline{LT}_{21} & (1 - \overline{LT}_{22}) \end{bmatrix} \tag{B.28}$$

$$\overline{LT}_{11} \simeq -a_1 Y_2/\bar{Y}_{i1}, \qquad \overline{LT}_{12} \simeq -a_2 Y_6/\bar{Y}_{i1},$$

$$\overline{LT}_{21} \simeq -a_1 Y_3/\bar{Y}_{i2} \quad \text{and} \quad \overline{LT}_{22} \simeq -a_2 Y_5/\bar{Y}_{i2},$$

where the bar indicates that mismatch conditions have been applied. The ratio $\Delta\bar{\bar{\mathscr{F}}}/\Delta\bar{\bar{\mathscr{F}}}_i^0$ can be further simplified, if we expand the numerator and the denominator in terms of the elements of the ith column, multiplied by their corresponding cofactors. In the general case,

$$\Delta\bar{\bar{\mathscr{F}}} = (1 - \overline{LT}_{ii})\Delta_{ii} - \sum_{j \neq i}^{n} \overline{LT}_{ji}\Delta_{ji}, \tag{B.29}$$

where Δ_{ji} is the cofactor of element f_{ji}. Since all elements in column i of $\bar{\bar{\mathscr{F}}}$ are multiplied by the factor a_i, all elements of $\bar{\bar{\mathscr{F}}}_i^0$ in column i vanish except for the element f_{ii}, which equals unity. Hence, $\Delta \bar{\bar{\mathscr{F}}}^0 = \Delta_{ii}$, and finally, for high local LTs,

$$\frac{\Delta \bar{\bar{\mathscr{F}}}}{\Delta \bar{\bar{\mathscr{F}}}^0_i} = -\overline{LT}_{ii} \left[1 + \frac{\sum\limits_{\substack{j=1 \\ j \neq i}}^{n} \overline{LT}_{ji} \Delta_{ji}}{\overline{LT}_{ii} \Delta_{ii}} \right] \tag{B.30}$$

All gain factors except a_i cancel in (B30), and the return difference for the gain element a_i in the presence of high local LTs is a function of a_i and passive components only. Applied to (B.28)

$$\bar{\mathscr{F}}(a_1) \simeq -\overline{LT}_{11}(1 - \overline{LT}_{12}\overline{LT}_{21}/\overline{LT}_{11}\overline{LT}_{22})$$

$$= a_1 y_{o1} Y_2 / \bar{Y}_{i1} \bar{Y}_{o1}(1 - Y_6 Y_3 / Y_2 Y_5). \tag{B.31}$$

Note that $\bar{Y}_{o1} = y_{o1}$ and hence cancels.

Arriving at such a simple expression after extensive calculations makes one suspect that there might be a simpler way of obtaining the same result. Indeed, there is (see problem B.5). Then why did we bother making such an effort? The reason is that feedback loops in electronic circuits cannot always be as neatly localized as in our example, and, where simplifying approaches fail, a generalized solution is required.

Proof that $\Delta / \Delta^0_i = \Delta \mathscr{F} / \Delta \mathscr{F}^0_i$.

The system determinant of (B.21) can be brought into the form

$$\Delta \begin{bmatrix} (\mathbf{I}_{11} - \mathbf{BI}^{-1}_{22}\mathbf{A}) & \mathbf{0} \\ \mathbf{A} & \mathbf{I}_{22} \end{bmatrix} = \Delta \mathbf{I}_{22} \Delta(\mathbf{I}_{11} - \mathbf{BI}^{-1}_{22}\mathbf{A})$$

by premultiplying the second row by $\mathbf{BI}^{-1}_{22}$, subtracting it from the first one and using the identity $\Delta(\mathbf{XY}) = \Delta\mathbf{X} \cdot \Delta\mathbf{Y}$. Further manipulation yields, finally,

$$\Delta = \Delta\mathbf{I}_{22}\Delta\mathbf{I}_{11}\Delta(1 - \mathbf{I}^{-1}_{11}\mathbf{BI}^{-1}_{22}\mathbf{A}) = \Delta\mathbf{I}_{22}\Delta\mathbf{I}_{11}\Delta\mathscr{F}_A \tag{B.32}$$

Since in (B.32) only $\mathscr{F}_A$ is a function of a_i, $\Delta^0_i = \Delta\mathbf{I}_{22}\Delta\mathbf{I}_{11}\Delta\mathscr{F}^0_{Ai}$. But $\Delta(1 - \mathbf{XY}) = \Delta(1 - \mathbf{YX})$, so that $\Delta\mathscr{F}_A = \Delta\mathscr{F}_B = \Delta\mathscr{F}$, and finally

$$\Delta / \Delta^0_i = \Delta\mathscr{F} / \Delta\mathscr{F}^0_i, \tag{B.27}$$

QED.

Problem

B.5. (a) Verify the branch transmittances of Fig. B.13, which shows the flow graph for the circuit of Fig. B.10 for the case of mismatch, assuming the basic amplifiers have infinite input and zero output impedances.

(b) Assuming high local LTs, verify (B.26) from the flow graph.

(c) Employing source splitting, verify the return difference for a_1 as given by (B.31). Assume the local LT of a_2 to be high enough for substituting v_{o2}/v_{o1} by the corresponding feedback-stabilized transmission $1/B$.

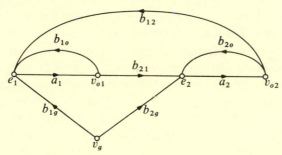

Fig. B.13. Flow graph for circuit of Fig. B.10; $b_{1o} = Y_2/Y_{i1}$, $b_{2o} = Y_5/Y_{i2}$, $b_{21} = Y_3/Y_{i2}$, $b_{12} = Y_6/Y_{i1}$, $b_{1g} = Y_1/Y_{i1}$, $b_{2g} = Y_4/Y_{i2}$.

Appendix C

Analysis and design techniques for LT

A short settling time in signal processing systems for instrumentation is highly desirable. However, feedback, widely used to stabilize transfer functions, always contains the danger of an oscillatory response, which is anathema to short settling times. In this appendix, a practical design procedure for feedback systems will be presented which consists of two stages. The first stage employs a simplified model whose closed-loop transfer function approaches sufficiently the desired response to supply the initial data for the second stage. The latter employs a physical model of the actual system or CAD which also includes parameters of secondary importance not present in the simplified model. The resulting response may then be finally adjusted by trial and error or by employing an optimization program.

A desirable design target for the simplified model would be monotonicity of response. A transfer function is defined as being monotonic if its impulse response does not change sign. Unfortunately, this criterion does not lend itself to a simple design procedure.[1]

The concept of 'real poles only'

A criterion which lends itself more readily to a simplified evaluation is the condition that all closed-loop poles of the system function must be real (located on the real axis). This condition is very useful, since the error $\varepsilon(t)$ (see Section 1.7.8) of such a system is composed of purely exponential terms, and, whilst its response may not be monotonic by definition, it will not contain any oscillatory terms. As a matter of fact, some pole-zero patterns having real poles only are indeed monotonic, although this is neither a necessary nor a sufficient condition for monotonicity. For example, the case in which all singularities (excluding those at infinity) are negative real poles only is truly monotonic, as is that in which poles and zeros on the negative real axis alternate and the singularity next to the origin is a pole. In the latter case, a zero next to the origin interferes with monotonicity.

The design procedure to be followed employs the well-established root locus technique, a method of plotting the path of the migrating closed-loop poles as a function of k_0, a multiplicative gain factor of LT. If the open-loop poles are all real, they will migrate with increasing k_0 along the real axis and may finally break away into the complex frequency plane, if the value k_0 exceeds a certain critical value. Since a large value of k_0 will desensitize the system with respect to parameter changes, it will be our design target to find the maximum value of k_0 for which the closed-loop poles are still lying on the real axis. This value will be designated as $k_0(\sigma_b)$ (the *breakaway LT factor*), since for this value of LT there will always be a pole of second or higher order lying on the real frequency axis, with the poles breaking away (branching out) from the real axis if k_0 deviates from its critical value $k_0(\sigma_b)$. This deviation implies an increase or decrease in k_0, or both in certain special cases to be discussed in Appendix C.4.

The present section makes extensive use of the root locus technique, with which the reader is assumed to be familiar. Trivial concepts are introduced mainly to define the notation, and to serve as a starting point for more sophisticated concepts to be developed later on.

C.1 The breakaway LT factor

The root locus

The root locus technique[2,3] is a convenient method for evaluating the zeros s_ρ of the characteristic equation

$$[1 - LT(s_\rho)] = 0 \qquad (C.1)$$

as a function of k_0, a multiplicative gain factor of LT.

The root locus plot depicts the paths of the poles of

$$[-LT(s_\rho)]/[1 - LT(s_\rho)] \qquad (C.2)$$

which, starting at the open-loop poles of LT(s) for $k_0 = 0$, migrate towards the zeros as k_0 increases towards infinity. For negative feedback we define the LT as

$$LT(s) = -LT(0) \prod^{a} (1 + s\tau_{zi})/\prod^{b} (1 + s\tau_{pj}) = (-1)^{m+1} k_0 z(s)/p(s),$$

$$\qquad (C.3)$$

with

$$k_0 = [-LT(0)] \prod^{a} \tau_{zi}/\prod^{b} \tau_{pj}; \qquad (C.4)$$

where

$$z(s)/p(s) = \prod^{a} (s - z_i)/\prod^{b} (s - p_j)$$

and

$$z_i = -1/\tau_{zi}, \; p_j = -1/\tau_{pj}.$$

The symbol m denotes the number of singularities located in the RHS of the complex frequency plane. It determines the angle of $LT(s_\rho)$, which is $\pi \pm 2n\pi$ for an even and $\pm 2n\pi$ for an odd number of singularities in the RHS (n is an integer). Since we shall deal here only with open-loop stable systems, m relates to RHS zeros only.

Solving for the zeros of (C.1) yields the angle and the absolute value of $LT(s_\rho)$. The angle is obtained from the condition

$$\measuredangle LT(s_\rho) = \pm 2\pi n, \qquad n = 0, 1, 2, \ldots,$$

or in accordance with (C.3)

$$\measuredangle k_0 z(s_\rho)/p(s_\rho) = (m+1)\pi \pm 2\pi n \qquad (C.5)$$

which defines the paths of the root locus. The second condition satisfying (C.1) is

$$|LT(s_\rho)| = |k_0 z(s_\rho)/p(s_\rho)| = 1 \qquad (C.6)$$

which yields the value of k_0 as a function of s_ρ. In accordance with (C.6), the root locus starts for $k_0 = 0$ at the poles and terminates for $k_0 = \infty$ at the zeros of $z(s_\rho)/p(s_\rho)$.

The root locus on the real axis

In accordance with the introduction we shall employ the criterion that all singularities of the closed-loop system function lying within the finite region of the complex frequency plane are negative real (located on the real axis). In order to derive the breakaway LT factor $k_0(\sigma_b)$, we shall first examine the properties of the root locus along the real frequency axis.

The sections of the root locus along the real axis are obtained from the angle defined by (C.5), noting that the only contribution comes from real singularities lying to the right of the root locus, each contributing 180°. Complex singularities do not contribute since they occur in conjugate pairs, which makes the angles of the contributing vectors $(\sigma_\rho - s_j)$ and $(\sigma_\rho - s_j^*)$ complementary with respect to 2π (s_j^* denotes the complex conjugate value of s_j). Neither do real singularities lying to the left of the root locus contribute, since their contributions $(\sigma_\rho - \sigma_j)$ are zero. Finally, for even m the number of singularities to the right of the root locus along the real axis must be odd, in order to contribute $\pi \pm 2n\pi$ radians, and for odd m it must be even.

Computation of the breakaway LT factor

Next we shall consider the behavior of $k_0(\sigma_\rho)$ along the real axis. In particular, we shall evaluate its extreme values $k_0(\sigma_b)$ and their physical significance. To this end we shall (a) solve the characteristic equation (C.1) for k_0 as a function of real poles only:

$$k_0(\sigma_\rho) = (-1)^{(m+1)} p(\sigma_\rho)/z(\sigma_\rho); \tag{C.7}$$

(b) set the derivative of $k_0(\sigma_\rho)$ with respect to σ_ρ equal to zero and solve for σ_{bi}; and (c) substitute σ_{bi} in (C.7) yielding i extreme values of k_0 for the root locus on the real frequency axis. Each $k_{0i}(\sigma_{bi})$ is an extreme value (maximum or minimum) yielding a double pole on the real frequency axis, from where the migrating poles break away (hence σ_b) into the complex frequency plane if k_0 is changed any further (increased or decreased). The order i depends on the number of real roots of the derivative of (C.7).

The maximum $k_{0i}(\sigma_{bi})$ yielding real poles may then be chosen as initial data for CAD or for a physical model to be optimized by trial and error.

A simple example will illustrate the evaluation of $k_0(\sigma_b)$. Consider

$$LT(s) = k_0[s(s+1/\tau_1)]^{-1}; \qquad k_0 = -1/\tau_0\tau_1. \tag{C.8}$$

From $[1 - LT(\sigma_\rho)] = 0$ we obtain

$$k(\sigma_\rho) = \sigma_\rho(\sigma_\rho + 1/\tau_1). \tag{C.9}$$

Differentiation of (C.9) yields $\sigma_b = -1/2\tau_1$; $k_0(\sigma_b) = -1/4\tau_1^2 = k_{0\text{max}}$.

Hence, the condition for real poles is $k_0 \leqslant k_{0\text{max}}$ or $\tau_0 \geqslant 4\tau_1$. The root locus of (C.8) and (C.9) are shown in Fig. C.1. Since no zeros are included in the RHS of the complex frequency plane, $m = 0$ and there is an odd number of singularities between the root locus and positive infinity along the real axis.

The effect of a RHS zero on the root locus

It is instructive to observe the effect of a positive real zero τ_2 on the root locus of (C.8). Consider

$$LT = k_0(s - 1/\tau_2)/[s(s+1/\tau_1)]; \qquad k_0 = \tau_2/\tau_0\tau_1. \tag{C.10}$$

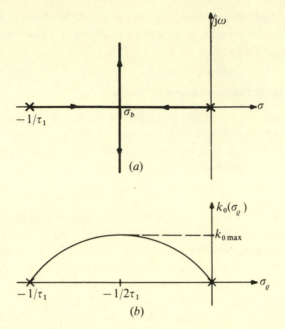

Fig. C.1. (a) Root locus of (C.8). (b) k_0 against σ_ρ of (C.8).

We notice the change in the sign of k_0, compared with (C.8), owing to $m = 1$. Accordingly, the number of singularities to the right of the root locus on the real axis is even, as shown in Fig. C.2. This root locus demonstrates the effect of a positive zero, which 'draws' the migrating poles towards the RHS and thereby converts an unconditionally stable system, such as the one defined by (C.8), into a conditionally stable one.

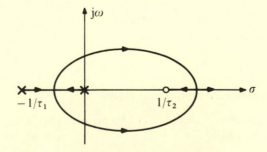

Fig. C.2. Effect of RHS zero on root locus.

C.2 The effect of delay on the breakaway LT factor

Signal delay by τ_D is described by the transform pair

$$\mathscr{L}\{f(t-\tau_D)\} = F(s) \exp(-s\tau_D). \qquad (C.11)$$

Hence, (C.7) including delay becomes

$$k_0(\sigma_\rho) = (-1)^{(m+1)} \exp(+\sigma_\rho\tau_D)p(\sigma_\rho)/z(\sigma_\rho). \qquad (C.12)$$

Setting the derivative of (C.12) with respect to σ_ρ equal to zero yields:

$$\tau_D p(\sigma_\rho) + \dot{p}(\sigma_\rho) - p(\sigma_\rho)\dot{z}(\sigma_\rho)/z(\sigma_\rho) = 0, \qquad (C.13)$$

with $\dot{p}(\sigma_\rho)$ denoting the derivative of $p(\sigma_\rho)$ with respect to σ_ρ, and $\dot{z}(\sigma_\rho)$ being defined similarly. Substituting the values of σ_{bj} satisfying (C.13) into (C.12) yields the extreme values of k_{0j}, each producing a particular σ_b on the real axis:

$$k_0(\sigma_b) = \exp(\sigma_b\tau_D)p(\sigma_b)/z(\sigma_b). \qquad (C.14)$$

The last equation shows that presence of delay reduces the value of $k_0(\sigma_b)$, since $\sigma_b < 0$ and hence $\exp(\sigma_b\tau_D) < 1$.

Drawing the root locus in the presence of delay

The term $\exp(-s\tau_D)$ exhibits an infinite number of poles and zeros at infinity. Those poles due to the delay which migrate within the region of interest in the complex frequency plane are an essential part of the root locus. In order to find the open-loop singularities due to $\exp[-\sigma - j(\omega\tau_D \pm 2n\pi)]$, we note that the real value of its poles is located at $\sigma = -\infty$ and that of its zeros at $\sigma = +\infty$. The angle of these singularities is a multivalued function, i.e. there is an infinite number of poles and zeros (essential singularity), because of the cyclic nature of the angle.

The location of these singularities is indicated in the complex frequency plane by their asymptotes, which cross the $j\omega$ axis at angles satisfying the characteristic equation. Including the delay factor in (C.5) and substituting the complex variable $\sigma + j\omega$, the angle of the delay factor is found to be

$$\measuredangle\{\exp[-\sigma - j(\omega_\rho\tau_D \pm 2n\pi)]\} = -\omega_\rho\tau_D \pm 2n\pi$$

$$= (m+1)\pi - \measuredangle[z(s_\rho)/p(s_\rho)]. \qquad (C.15)$$

At the poles of the exponential term, at $\sigma = -\infty$, the real singularities of $z(s)/p(s)$ each contribute π radians. Complex conjugate pairs contribute 2π and may therefore be disregarded. Designating the number of real singularities as q, the frequencies at which the asymptotes of the poles due to the delay cross the $j\omega$ axis are, from (C.15),

$$\omega_\rho = [(m+1)\pi \pm 2n\pi + q\pi]/\tau_D. \qquad (\text{C.16})$$

At the zeros, at $+\infty$, each singularity of $z(s)/p(s)$ contributes $0°$ so that q in (C.15) equals zero. Hence, the frequencies at which the asymptotes of the zeros due to the delay cross the $j\omega$ axis are not a function of q:

$$\omega_\rho = [(m+1)\pi \pm 2n\pi]/\tau_D. \qquad (\text{C.17})$$

Example

Presence of delay makes an amplifier exhibiting a single pole potentially unstable. Consider

$$\text{LT}(s) = -\exp(-s\tau_D)/s\tau_0$$

for which $\sigma_b = -1/\tau_D$ from (C.13) and $\tau_0 \geq e\tau_D$ is the condition obtained from substituting σ_b in (C.14) for the principal path of the root locus to lie on the real axis. Furthermore, $m = 0$ and $q = 1$. Hence, the poles due to the delay factor are at $[-\infty \pm j2n\pi]/\tau_D$, and the zeros at $[\infty \pm j(2n+1)\pi]/\tau_D$.

The principal part of the resulting root locus is shown in Fig. C.3. Recalling the effect of the zero in (C.10), drawing the root locus towards the RHS of the complex frequency plane, we expect a similar behavior due to delay, which introduces an infinite number of zeros in the RHS. This is confirmed by Fig. C.3, which shows the trajectory of the pole at the origin and of the one due to the delay at $-\infty$, whose angle equals zero. The complete root locus contains an infinite number of poles, which migrate towards the corresponding zeros as k_0 increases from zero to

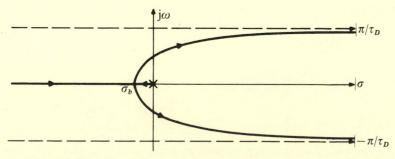

Fig. C.3. Part of root locus due to single pole, in presence of delay.

infinity, but their effect on the response can be shown to be negligible if $k_0 \leqslant k_0(\sigma_b)$.

Problem
C.1. For
$$LT = -\exp{(-s\tau_D)}/s\tau_0(1 + s\tau_1),$$

approximate the effect of the delay, replacing the exponential term by the first three terms of its series expansion, and draw the root locus for this case.

C.3 The gain-bandwidth theorem

In the transfer function of an amplifier, poles at the origin do not occur in practice, since they imply infinite gain at dc (positive feedback may produce such a pole of questionable stability). Instead, we are dealing with 'dominant poles', whose time constants are typically by two or more orders of magnitude longer than τ_0, the unity-gain time constant.

The LT of such an amplifier, including a single parasitic pole at the frequency $-1/\tau_1$, is given by

$$LT(s) = k_0/[s + 1/A(0)\tau_0](s + 1/\tau_1), \qquad k_0 = -1/\tau_0\tau_1, \quad \text{(C.18)}$$

with $A(0)\tau_0$ being the dominant time constant.

But the root locus of (C.18) differs only slightly from that of (C.8) shown in Fig. C.1(a), provided that

$$A(0)\tau_0 \gg \tau_1. \tag{C.19}$$

Moreover, evaluation of σ_b for (C.18) yields

$$\sigma_b = \frac{-1}{2}\left[\frac{1}{A(0)\tau_0} + \frac{1}{\tau_1}\right]$$

which again is very close to the corresponding value obtained for Fig. C.1(a). Hence, if condition (C.19) is satisfied, we may substitute (C.8) for (C.18), since the root locus for the two LTs is for all practical purposes identical within the region for which

$$|s| \gg 1/A(0)\tau_0. \tag{C.20}$$

This substitution is desirable, since it simplifies the computation of σ_b.

For the general case, let

$$LT_1(s) = -A(0)B'(0)/\exp{(s\tau_D)}[1 + sA(0)\tau_0]\prod^{n}(1 + s\tau_i), \quad \text{(C.21)}$$

$$LT_1'(s) = -B'(0)/\exp{(s\tau_D)}s\tau_0\prod^{n}(1 + s\tau_i). \tag{C.22}$$

Provided that $LT_1(s)$ is a positive real function, and the conditions

$$A(0)B'(0) \gg 1, \tag{C.23}$$

$$\tau_C - B'(0)\tau_D \geqslant B'(0) \sum_{i}^{n} \tau_i \tag{C.24}$$

are satisfied, it can be shown that the zeros of $[1\text{-}LT_1(s)]$ are practically equal to those of $[1\text{-}LT_1'(s)]$ within the region defined by (C.20). The inclusion of τ_D in condition (C.24) is valid only within the region of s, in which $\exp(s\tau_D)$ may be replaced by the first two terms $(1+s\tau_D)$ of its series expansion.

By similar reasoning let

$$LT_2(s) = -A_1(0)A_2(0)B'(0)s\tau_{03}/[1+A_1(0)\tau_{01}][1+sA_2(0)\tau_{02}]\prod^{n}(1+s\tau_i),$$
$$\tag{C.25}$$

$$LT_2'(s) = -B'(0)\tau_{03}/s\tau_{01}\tau_{02}\prod^{n}(1+s\tau_i). \tag{C.26}$$

The appropriate conditions for substituting $LT_2'(s)$ for $LT_2(s)$ are then that the span between the frequencies $1/A_1(0)\tau_{01}$ and $1/A_2(0)\tau_{02}$ and between the gain-bandwidth product $\omega_0 = B'(0)\tau_{03}/\tau_{01}\tau_{02}$ be large, and that (C.24) relating to the span between ω_0 and the parasitic poles be satisfied:

$$\left.\begin{array}{l} A_1(0)B'(0) \gg \tau_{02}/\tau_{03}, \\[2mm] A_2(0)B'(0) \gg \tau_{01}/\tau_{03}, \\[2mm] \tau_{01}\tau_{02} \geqslant B'(0)\tau_{03} \sum_{i}^{n} \tau_i. \end{array}\right\} \tag{C.27}$$

If the parasitic poles may be subdivided into k significant and $(n-k)$ insignificant poles, whose relationship is given by

$$\sum_{k+1}^{n} \tau_i \ll \sum_{1}^{k} \tau_i, \tag{C.27a}$$

then the effect of the *insignificant poles* on stability conditions is negligible and may be safely disregarded.

Equations (C.23), (C.24) and (C.27) define the gain-bandwidth theorem,[4] which replaces the dominant poles of $LT(s)$ by poles at $s = 0$ in $LT'(s)$. The asymptotic frequency response of $LT(s)$ is preserved in $LT'(s)$ at frequencies beyond $1/A(0)\tau_0$, and $LT'(0)$ at dc becomes infinitely large. Stability conditions are practically the same for LT(s) and LT'(s), but computation of σ_b is greatly simplified.

C.4 Design examples for LT yielding real closed-loop poles

The condition of 'real singularities only', whilst not necessarily mono-tonic, is well suited to supply initial design data for a model which will be subsequently optimized by trial and error or by computer simulation.

Throughout this section we shall employ LTs to which the gain-band-width theorem has been applied, i.e. whose dominant poles have been moved to the origin. This kind of modified LT is then described by its gain-bandwidth product ω_0 and by the parasitic poles plus possible doublets introduced by the amplifier and the feedback network. Among these parameters, the parasitic poles cannot be modified by the designer, since they are a function of the speed of the gain elements used in the amplifier and therefore constitute the ultimate design limit for the speed of the closed-loop response. But ω_0 and most zeros or doublets are 'designable' parameters, since they can be controlled by passive components.

As an example, consider the pattern for

$$LT(s) = -1/s\tau_0(1 + s\tau_1)(1 + s\tau_1/a) \qquad (C.28)$$

shown in Fig. C.4, with $1 \le a \le \infty$. Evaluation of σ_b from the derivative of (C.7) and its substitution in (C.28) yields, for

$$\infty \ge a \ge 1,$$

$$s_1/2 \ge \sigma_b \ge s_1/3,$$

$$4\tau_1 \le \tau_0 \le 27\tau_1/4.$$

This is an instructive result, since it reveals the necessary span between $\omega_0 = 1/\tau$ and a single or double parasitic pole in LT(s), for real poles of the closed-loop response.

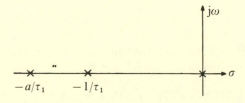

Fig. C.4. Pole-zero pattern for (C.28).

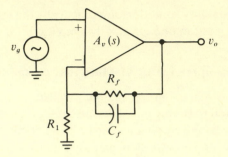

Fig. C.5. Feedback amplifier accommodating double pole at origin.

Next consider the amplifier shown in Fig. C.5, whose gain function (after application of the gain-bandwidth theorem) we assume to be

$$A_v(s) = -1/s^2 \tau_{01} \tau_{02} \prod_2^n (1 + s\tau_i), \qquad (C.29)$$

and for which

$$B(s) = B'(s) = [1 + sA_f(0)\tau_1]/A_f(0)(1 + s\tau_1), \qquad (C.30)$$

with $A_f(0) = (R_1 + R_f)/R_1$, $A_f(0)\tau_1 = R_f C_f$, and $\tau_1 = (R_f \| R_1)C_f$. Equation (C.29) indicates inclusion of two gain stages with unity-gain time constants of τ_{01} and τ_{02}, respectively. This is an interesting configuration, since the zero due to $A_f(0)\tau_1$ in $B'(s)$ allows two dominant poles to be accommodated in the gain function. But we also note that the zero introduced by $B'(s)$ is necessarily accompanied by the parasitic pole at $-1/\tau_1$, which is not due to the amplifier and whose presence sets in this case a design limit to the highest value for ω_0 compatible with real closed-loop poles.

For

$$\sum_2^n \tau_i \ll \tau_1 < \tau_{01}\tau_{02}/\tau_1 = \tau_0 = 1/\omega_0,$$

with τ_i denoting the parasitic poles due to the amplifier, only the pole at frequency $1/\tau_1$ due to the feedback network will be significant and those due to the amplifier will be insignificant. For this condition we shall consider the root locus of

$$LT(s) = A(s)B'(s) = -[1 + sA_f(0)\tau_1]/s^2 A_f(0)\tau_{01}\tau_{02}(1 + s\tau_1). \qquad (C.31)$$

Equation (C.31) yields for $k_0(\sigma_p)$ as defined by (C.7) a third order equation, which, after differentiation, yields the following solution for σ_b:

$$|\sigma_b| = \frac{1}{4A_f(0)\tau_1}\{3 + A_f(0) \pm [9 - 10A_f(0) + A_f^2(0)]^{1/2}. \quad (C.32)$$

Hence, we obtain two real solutions for σ_b, provided that either $0 \leq A_f(0) \leq 1$, or $A_f(0) > 9$. Discarding $A_f(0) \leq 1$ as undesirable, the root locus for $A_f(0) > 9$ is shown qualitatively in Fig. C.6(a), and the corresponding values of k_0 are shown in Fig. C.6(b).

For $A_f(0) = 9$ we obtain a single solution for σ_b, i.e. a triple pole at $-1/3\tau_1$ as shown in Fig. C.6(c). In this case the closed-loop gain-bandwidth product occurs at the same frequency, i.e. $\omega_0 = 1/3\tau_1 = \tau_1/\tau_{01}\tau_{02}$, yielding $\tau_{01}\tau_{02} = 3\tau_1^2$ as the design condition for τ_{01} and τ_{02}.

If $A_f(0) > 9$, the $|\sigma_b|$ yielding the highest k_0 for real poles and hence lowest sensitivity is the larger one between the two defined by (C.32). For

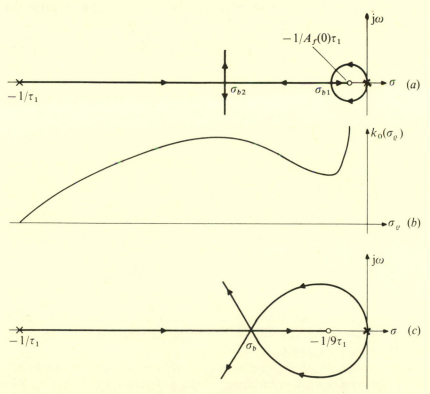

Fig. C.6. Root locus for (C.31). (a) $A_f(0) > 9$, $|\sigma_{b1}| < 1/3\tau_1 < |\sigma_{b2}|$. (b) $k_0(\sigma_p)$ for case (a). (c) $A_f(0) = 9$, $\sigma_b = -1/3\tau_1$.

high values of $A_f(0)$, i.e. $A_f(0) \geq 100$, the larger σ_b becomes $|\sigma_b| \simeq [3 + 2A_f(0)]/4A_f(0)\tau_1 \simeq 1/2\tau_1$, requiring a span of four between ω_0 and the single significant parasitic pole at $|1/\tau_1|$: $(1/\tau_1)/(\tau_1/\tau_{01}\tau_{02}) \simeq 4$ or

$$\tau_{01}\tau_{02} \simeq 4\tau_1^2. \tag{C.33}$$

The corresponding root locus becomes practically identical with that of Fig. C.1.

In the closed-loop transfer function $H(s)$, the zero at $-1/A_f(0)\tau_1$ cancels against the pole of the same value, due to $1/B(s)$, which in turn introduces a zero at the frequency $-1/\tau_1$. Hence, for $A_f(0) = 9$ and $\tau_{01}\tau_{02} = 3\tau_1^2$,

$$H(s) = 9(1 + s\tau_1)/(1 + 3s\tau_1)^3. \tag{C.34}$$

Designing LT from H(s)

An elegant technique for evaluating the singularities of LT from a desired closed-loop response has been employed by Waldhauer[5]

$$H(s) = [-\mathrm{LT}(s)]/B(s)[1 - \mathrm{LT}(s)] \tag{C.35}$$

yields

$$\mathrm{LT}(s) = -H(s)B(s)/[1 - H(s)B(s)]. \tag{C.36}$$

Assume, as an example, the desired closed-loop response to be

$$H(s) = \frac{A_f(0)}{(1 + s\tau)^3} = \frac{1}{B(0)} \frac{-\mathrm{LT}(s)}{[1 - \mathrm{LT}(s)]},$$

with

$$A_f(0) = 1/B(0) \quad \text{and} \quad B'(0) = B(0),$$

yielding

$$\mathrm{LT}(s) = A(s)B'(0) = -B'(0)/s\tau_0(3 + 3s\tau + s^2\tau^2)$$

$$= -1/s\tau(3 + 3s\tau + s^2\tau^2). \tag{C.37}$$

Hence, $A(s)$ requires a unity-gain time constant of $\tau_0 = B'(0)\tau$ and a complex conjugate pole pair at $-(1.5 \pm j3^{1/2}/2)/\tau$. The root locus for (C.37) is shown in Fig. C.7.

This is an interesting configuration, yielding a truly monotonic closed-loop response (a third order negative real pole) from an open-loop gain function exhibiting a dominant pole and two complex conjugate ones. Note, however, that here we are dealing with the case in which both an

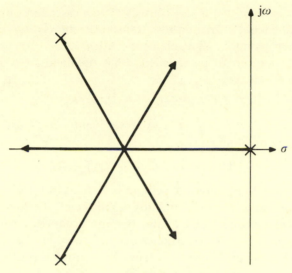

Fig. 6.7. Root locus for (C.37).

increase and a decrease in the critical value of $k_0(\sigma)$ causes the migrating poles to branch out from the real axis. In the practical realization of this response,[5] the conjugate pole pair is created by the secondary feedback loop of the output stage consisting of two transistors.

Mosquito on the tank

A closed-loop response may contain complex poles but still be monotonic, if the oscillatory response is damped out by a 'sufficiently dominant' pole. Fig. C.8 shows a suitable circuit with

$$Z_T(s) = -R_T A(0)/(1+sR_T C_T)(1+sA(0)\tau_0) \prod^{n} (1+s\tau_i), \quad (C.38)$$

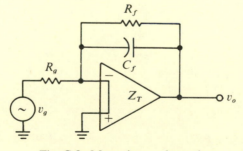

Fig. C.8. Mosquito on the tank.

Here, we assume that $Z_T(s)$ consists of two cascaded gain elements, one transimpedance element whose transfer function is $R_T/(1 + sR_TC_T)$, and another voltage gain element $A(0)/[1 + sA(0)\tau_0]$. Assuming mismatch conditions to be satisfied, B' is obtained from Fig. C.8 as $(1 + sR_fC_f)/R_f$. Hence

$$LT(s) = Z_T(s)(1 + sR_fC_f)/R_f. \tag{C.39}$$

For $R_fC_f \geq R_TC_T \gg \tau_0$, $C_f \geq C_T$, $A(0) \gg 1$, and $\omega \gg 1/R_TC_T$, we may replace LT by LT', yielding

$$LT'(s) = -(C_f/C_T)/s\tau_0 \prod_{}^{n} (1 + s\tau_i). \tag{C.40}$$

The remarkable thing about this circuit is its behavior if C_f is gradually increased (without violating stability conditions). This increases the gain-bandwidth product of LT' and thereby produces an increasingly non-monotonic closed-loop response due to a complex pole pair migrating into the complex frequency plane. But, simultaneously, $1/B(s) = -R_f/R_g\ (1 + sR_fC_f)$ introduces a pole which becomes progressively more dominant with increasing C_f, and therefore attenuates the frequencies of the oscillatory response. For a sufficiently large ratio C_f/C_T, the damping due to R_fC_f will make the overall response monotonic. A student nicknamed this configuration 'the mosquito on the tank'; the mosquito is obviously incapable of imposing his high frequency of vibration on the tank.

The relationship between a single significant pole at $-1/\tau_1$, ω_0 of LT and σ_b for the various configuration discussed is summarized in Table C.1:

Table C.1

$-LT(s)$	$-\sigma_b$	Order of σ_b	Condition	ω_0
$\dfrac{B'(0)}{s\tau_0(1 + s\tau_1)}$	$1/2\tau_1$	2	$\tau_0/B'(0) = 4\tau_1$	$1/4\tau_1$
$\dfrac{B'(0)}{s\tau_0(1 + s\tau_1)^2}$	$1/3\tau_1$	2	$\tau_0/B'(0) = 27\tau_1/4$	$4/27\tau_1$
$\dfrac{1 + sA_f(0)\tau_1}{s^2 A_f(0)\tau_{01}\tau_{02}(1 + s\tau_1)}$	$1/3\tau_1$	3	$A_f(0) = 9$ $\tau_{01}\tau_{02} = 3\tau_1^2$	$1/3\tau_1$
$\dfrac{B'(0)}{s\tau_0(3 + 3s\tau + s^2\tau^2)}$	$1/\tau$	3	$\tau_0/B'(0) = \tau$	$1/\tau$

Appendix D

Characterization of a transfer function in the time and frequency domains by time moments

Emphasis is laid throughout this book on system behavior which asymptotically approaches a response open to simple computational evaluation. Application of the gain-bandwidth theorem to a feedback system is one example, and the limiting condition for real poles of the closed-loop response is another. This appendix deals with a different aspect of the same approach, namely with the characterization of a response by time moments.

Time moments can be interpreted in terms of ω_{45} and ω_h in the frequency domain, and in terms of the delay time τ_D and rise time τ_R in the time domain, if they are related to the response of an infinite number of identical cascaded integrators (Gaussian response). This interpretation can, however, also be applied to a small number of non-identical stages and approaches the accurately computed values for ω_{45}, ω_h, τ_D and τ_R very closely, provided that their transfer function exhibits a monotonic magnitude versus frequency plot and step response.

Definition of time moments

Let $h(t)$ be the normalized impulse response of the system to be characterized. Then we define, following Emore,[1]

$$T_1 = \int_{-\infty}^{\infty} th(t)\, dt \qquad (D.1)$$

as the *first time moment* of $h(t)$ with respect to $t = 0$, or its center of gravity. Similarly,

$$T_2^2 = \int_{-\infty}^{\infty} (t - T_1)^2 h(t)\, dt, \qquad (D.2)$$

which is defined as the *second time moment* of $h(t)$ with respect to T_1, or the radius of gyration with respect to its center of gravity.

The normalization means that $\int_{-\infty}^{\infty} h(t)\,dt = h_{-1}(\infty) = 1$, i.e. the final value of the step response equals unity.

Relationship between the time moments and the frequency response of a system

A normalized impulse response of general nature is given in the frequency domain by

$$H(j\omega) = \prod^m (1 + j\omega\tau_i) \Big/ \prod^n (1 + j\omega\tau_j) = \exp(A + jB). \qquad \text{(D.3)}$$

For this function it can be shown[1,2] that

$$A = \ln |H(j\omega)| = -\tfrac{1}{2}(\omega T_2)^2 + \tfrac{1}{4}(\omega T_4)^4 - \tfrac{1}{6}(\omega T_6)^6 + \cdots$$

and

$$B = \measuredangle H(j\omega) = -\omega T_1 + \tfrac{1}{3}(\omega T_3)^3 - \tfrac{1}{5}(\omega T_5)^5 + \cdots, \qquad \text{(D.4)}$$

with

$$(T_k)^k = \sum^n (\tau_j)^k - \sum^m (\tau_i)^k. \qquad \text{(D.5)}$$

The term T_k is defined as the kth order time moment of $h(t)$, with the moments of first and second order being given by (D.1) and (D.2), respectively.[1,2]

The response of a system involving feedback is not directly obtained in factorized form, as in (D.3). Hence, it is of special interest to obtain the time moments from the following relationship:

$$\prod^m (1 + s\tau_i) = 1 + sa_1 + s^2 a_2 + \cdots, \qquad \text{(D.6)}$$

where

$$a_1 = \tau_1 + \tau_2 + \cdots + \tau_m = \sum^m \tau_i, \qquad \text{(D.7)}$$

$$a_2 = \tau_1\tau_2 + \tau_1\tau_3 + \cdots + \tau_2\tau_3 + \cdots, + \tau_{m-1}\tau_m$$

$$= \tfrac{1}{2}\left[\left(\sum^m \tau_i \right)^2 - \sum^m \tau_i^2 \right]. \qquad \text{(D.8)}$$

In accordance with our previous definition, the time moments of first and second order are obtained from (D.7) and (D.8) in terms of a_1 and a_2

as

$$T_1 = a_1, \tag{D.9}$$

and

$$a_2 = (T_1^2 - T_2^2)/2 \quad \text{or} \quad T_2^2 = a_1^2 - 2a_2. \tag{D.10}$$

Higher order time moments can be correspondingly obtained, but have no bearing on our present consideration.

Finally, if

$$H(s) = (1 + sa_1 + s^2 a_2 + \cdots)/(1 + sb_1 + s^2 b_2 + \cdots), \tag{D.11}$$

then its first and second order moments, in accordance with (D.9), (D.10) and (D.5), are

$$T_1 = b_1 - a_1, \tag{D.12}$$

$$T_2^2 = b_1^2 - a_1^2 + 2(a_2 - b_2). \tag{D.13}$$

Equations (D.12) and (D.13) establish simple relationships between the time moments and the frequency response. What remains to be done is to find their physical significance in the respective domains.

Relationships between the Gaussian response and the time moments[2]

The Gaussian response is defined in the frequency domain as

$$G(j\omega) = \lim_{n \to \infty} (1 + j\omega\tau)^{-n} = \exp\left[-\tfrac{1}{2}(\omega T_2)^2 - j\omega T_1\right]. \tag{D.14}$$

Hence, phase and amplitude of the Gaussian response are only functions of T_1 or T_2, respectively, with

$$T_1 = n\tau \quad \text{and} \quad T_2^2 = n\tau^2. \tag{D.15}$$

From (D.14), the frequency response of $G(j\omega)$ is characterized in terms of the time moments as

$$\omega_{45} = \pi/4T_1 \quad \text{and} \quad \omega_h = (\ln 2)^{1/2}/T_2. \tag{D.16}$$

In the time domain, the impulse response of $G(j\omega)$ is

$$g(t) = \frac{1}{(2\pi)^{1/2} T_2} \exp\left\{-\frac{1}{2}\left[\frac{t - T_1}{T_2}\right]^2\right\}, \tag{D.17}$$

which is again a Gaussian function. Since in the time domain the time

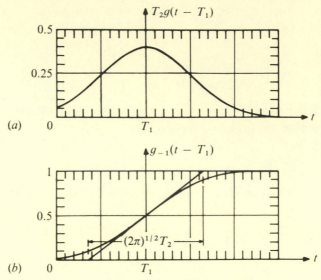

Fig. D.1. Gaussian response. Adapted from [2]. (a) Impulse response. (b) Step response.

moments are related to the step response $g_{-1}(t)$, we shall also define

$$g_{-1}(t) = \int_{\infty}^{t} g(t)\,dt = 0.5\left[1 + \mathrm{erf}\left(\frac{t - T_1}{2^{1/2}T_2}\right)\right]. \qquad (\mathrm{D}.18)$$

Equation (D.17) and (D.18) are plotted in Fig. D.1. Note that the step response is dimensionless, whereas its derivative $g(t)$ has the dimension $(\mathrm{time})^{-1}$.

It is common engineering practice to characterize a monotonic step response by its delay time τ_D and its rise time τ_R. The delay time τ_D is customarily defined as the elapsed time between the application of the input step and the instant at which the output reaches 50% of its final value, and τ_R as the time required for the output to increase from 10% to 90% of its final value.

An inspection of Fig. D.1 yields

$$\tau_D = T_1 \qquad (\mathrm{D}.19)$$

according to the above definition. A mathematical interpretation of T_1 may be obtained, drawing a parallel with the theory of probability. Considering T_1 as the expected time of arrival of the impulse response, it is defined as the first moment of $h(t)$ with respect to $t = 0$, as given by (D.1). That $h(t)$ is a non-causal response does not pose any mathematical problem, it merely brings out the fact that the Gaussian response is not physically realizable, and only serves as a useful mathematical tool.

The rise-time τ_R is defined as the inverse derivative of $g_{-1}(t)$ at $t = T_1$, which is obtained as

$$1/g(T_1) = (2\pi)^{1/2} T_2 = \tau_R. \tag{D.20}$$

Note the excellent agreement between the above definition of τ_R and its 10%–90% value, since τ_R $(10\%-90\%) = 2.563\ T_2$ for $g_{-1}(t)$, whereas $(2\pi)^{1/2} = 2.507$.

The mathematical interpretation of T_2, again taken from the theory of probability, is the second moment with respect to T_1 or the variance of $g(t)$, as given by (D.2).

Applicability of characterization by the time moments to monotonic transfer functions

Comparing (D.14) and (D.4) shows that, for transfer functions of general nature, higher than first or second order time moments contribute to ω_{45} and ω_h, respectively. It may be shown, however, that for responses with a monotonic magnitude versus frequency plot, these higher order time moments do not contribute significantly. Hence, ω_{45} and ω_h may for this kind of response still be characterized by T_1 or T_2, in accordance with (D.16), without making a significant error. Similarly, the delay time τ_D and rise time τ_R of a monotonic step response are to a good approximation given by (D.19) and (D.20), respectively.

Appendix E

The sampling theorem

Consider a linear gate as shown in Fig. E.1, and the corresponding waveforms in Fig. E.2: (a) an analog function $f(t)$; (b) a train of sampling pulses $\delta_T(t)$ of repetition rate $1/T$ and unit intensity; and (c) the sampled waveform described mathematically by the product $f(t) \times \delta_T(t)$. The absolute values of the Fourier transforms of (a), (b), and (c) are shown in (d), (e) and (f), respectively. The signal spectrum is assumed not to contain any frequencies beyond f_m. Since multiplication in the time domain transforms into convolution in the frequency domain, the transform of $f(t) \times \delta_T(t)$ yields $F_s(j\omega)$, the Fourier transform of $f_s(t)$ repeating itself every $1/T$ Hz.

Now assume that the sampled waveform $F_s(j\omega)$ is processed by an ideal low-pass filter of rectangular shape and a cut-off frequency $f_0 = 1/2T$, as shown in Fig. E.2(f). Then, provided that $f_m < f_0$, the original signal will be faithfully recovered from the low-pass filter output.

If, however, $f_m > f_0$, then the two nearest convolved frequency spectra of $F_s(j\omega)$ overlap with the center lobe as shown in Fig. E.3. Hence, the output from the low-pass filter of bandwidth $1/2T$ not only lacks part of the frequencies contained in the original spectrum, but also contains components (called *aliases*) due to the adjacent folded portion of the spectrum which interfere with the recovered signal. These latter frequencies cannot be removed from the sampled data, because the system has no way of knowing which frequency components really belong to the sampled data representing the actual input signal and which are aliases.

It follows that a signal having highest frequency components up to f_m can be faithfully recovered from its samples if and only if the sampling

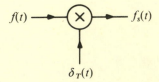

Fig. E.1. Sampling of $f(t)$ by train of impulses $\delta_T(t)$.

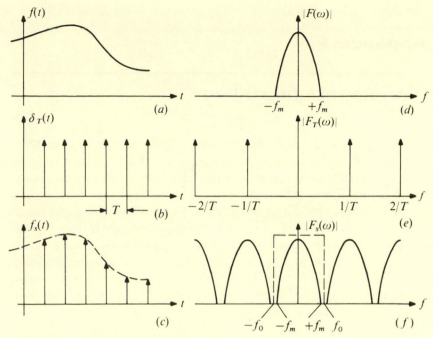

Fig. E.2. Waveforms for Fig. E.1 and corresponding Fourier transforms.[1]

frequency

$$1/T > 2f_m. \tag{E.1}$$

The effect of (E.1) not being satisfied can be visualized by a motion-picture sequence in which a wagon wheel is observed to stand still ($f_m = n/2T$) or rotate backwards ($f_m > n/2T$), with $n = 1, 2, \ldots$.

Equation (E.1) defines the *sampling theorem* due to Nyquist. Since ideal filters are non-existent and undesirable signal frequencies can only be attenuated, it is common practice to sample at 2.5 to 5 times the Nyquist rate given by (E.1).

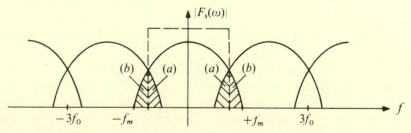

Fig. E.3. Sampled spectrum for $f_m > f_0$. Adapted from [1]. (a) Interfering portion of spectrum. (b) Missing portion of spectrum.

Appendix F

Digital averaging modes[1]

Averaging is fundamental to all random signal analysis, and to SNR improvement by repetitive measurements. Most digital statistical analysers provide two averaging modes. One is straightforward normalized summation, in which the average is computed by summing m sample products and then dividing the total by m. Summation averaging is here a process of pre-established length; it is therefore particularly useful for the analysis of a limited quantity of data obtained from a stationary process.

The second averaging mode, exponential, is a continuous updating process which forgets old information and thus follows changing phenomena. If A_{m-1} is the current value of the running average and a new input I_m arrives, the new average is

$$A_m = A_{m-1} + (I_m - A_{m-1})/m, \tag{F.1}$$

$m = 1, 2, \ldots$. This running-average algorithm, although digital, acts like an RC averaging circuit: the averager output responds exponentially to a step input, with a time constant approximately equal to m times the interval Δt between samples.

Because it is difficult to divide by an arbitrary integer m in a high-speed system, the exponential averaging mode is practically implemented by dividing by the power of 2 nearest to m. The division can then be achieved simply by shifting, and the algorithm becomes

$$A_m = A_{m-1} + (I_m - A_{m-1})/2^n. \tag{F.2}$$

No error is introduced into the results by this substitution of 2^n for m. The averages simply take a little longer to approach to within a given percentage of their final values (assuming a stationary process, in which the final value is not changing).

An important application of the exponential mode is the measurement of non-stationary signals, i.e. signals whose statistical properties change with time. If the signal may be considered as quasi-stationary during a time interval corresponding to the averaging time constant, then the function stored by the analyser will follow the statistical property of the signal.

A powerful feature of the exponential averaging algorithm is the effect of the increasing value of m during the initial phase of measurement. Initially, m is one; then, as the measurement proceeds, the value of m is incremented. This has the effect of averaging with a capacitor whose value increases during the measurement. High-frequency noise is averaged out immediately, leaving the lower frequencies for later, thus giving a rapid preliminary estimate of the final average. After the first M samples have been taken, the weighting factor of the new sample remains M and hence no further change of time constant takes place.

The average is now

$$A_m = \frac{1}{M} \sum_{i=1}^{m} \left(\frac{M-1}{M}\right)^{m-i} I_i \qquad \text{(F.3)}$$

where i is the running index of input I. The weight of the $(m-M)$th sample in (F.3) is e^{-1}, which shows that the weighting factor of the samples taken reduces exponentially with a 'time constant' of M samples.

Digital exponential averaging has two important advantages – stability and flexibility – over the analog RC averaging method. It is more stable because it is not subject to the drift normally associated with analog circuits, and more flexible because changing the time constant entails not a change of physical components, but simply a changed divisor in the averaging algorithm. Digital averaging has the added advantage that it makes possible very long time constants, and hence low-frequency analysis without physically large components. Time constants from a few milliseconds to many days represent no problem for digital integrators, but are quite impossible with RC averagers.

Problem

F.1. (a) Prove (F.1), using $A_m = \sum_{1}^{m} I_i / m$.

(b) Assuming that the standard deviation of the individual samples I_i from their expected value is σ, find the standard deviation of A_m in (F.3) for $m \to \infty$.

(c) Show that (F.1) also holds for (F.3).

Appendix G

Selected answers to problems

Chapter 1

1.1. Equation (1.21) is obtained by substituting (1.19b) for F_{min}.

In the computation of $\overline{v_{no}^2}$ for Fig. 1.14, we note that the noise spectral density at the input of the amplifiers consists of a common term $4k \mathcal{T} n G_p R_g^2$ which is fully correlated, and a second term $4k \mathcal{T} R_s$ which is uncorrelated. Hence, the total noise spectral density at the output is

$$S_{vo} = 4k \mathcal{T} n^2 |A_v^2|(R_g + R_s/n + n G_p R_g^2),$$

from which (1.23) follows directly.

1.2. (a) $\quad F = 1 + (R_1 + R_s + G_p R_1^2)/R_g + R_g G_p + 2R_1 G_p$
$$+ r_0^2(1 + \omega^2 \tau^2)/A^2(0) R_L R_g.$$

   (b) $\qquad\qquad r_0^2 \ll A(0)(R_L R_g)/(1 + \omega^2 \tau^2).$

Inspection of the circuit diagram shows that R_L does not contribute noise if $r_0 = 0$, since it is short-circuited by the active source A_v.

   (c) $\qquad \overline{v_{no}^2} = 4k \mathcal{T}[R_g + R_1 + R_s + G_p(R_1 + R_g)^2]/4\tau$

Note that for $r_0 \neq 0$, and disregarding physical limitations, R_L would contribute white noise.

1.4. (a) Series series feedback.

   (b) $\qquad\qquad \dfrac{1}{B} = \lim_{|LT| \to \infty} i_o/v_g = (V_S - v_g)/\alpha_Q R_f,$

$LT = -\alpha_Q R_f A_v/r_e \ (1 + \alpha_Q R_L/r_e)$ (obtained by source splitting). R_L is the load resistance (equals zero in Fig. 1.35).

   (c) Assuming the circuit to employ a basic transadmittance amplifier, Q should be considered as a transadmittance element. Further, assuming mismatch to be satisfied between A_v and Q_1, the input impedance of Q is disregarded. Hence, the output admittance is computed from Fig. G.1 as

$$-i_o/v_o = -i_o/v_E = gm/(1 + A_v \ gm \ R_f), \ gm = \alpha_Q/r_e.$$

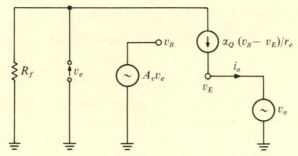

Fig. G.1. Problem 1.4(c). Circuit for computing output admittance.

1.5.

$$i/v = G_1 + G_2\{1 - A_v/[1 + A_v G_2/(G_1 + G_2)]\}. \tag{G.1}$$

From (G.1), which may be written by inspection (G_2 is reduced by the Miller effect), the results given in (a) and (b) can be derived.

(c) Source splitting yields, for $R_1 = R_2 = R_3 = R_4 = R$,

$$LT = -A_v R/2(R + 2Z_L). \tag{G.2}$$

1.6. (a) $$R_{\text{out}} \approx R[1 + A_1 A_2/(2A_1 + A_2)]. \tag{G.3}$$

(b) Since we are dealing with a linear system, we compute first R_{out} from (G.3) with $A_i = 10^5$ and $\delta_i = 0$, then with infinite A_i and $\delta_i = 10^{-3}$, and finally compare the two results.

(c) The solution is straightforward, if we substitute $A_i(s)$ in (G.3).

(d) If $\delta_i \neq 0$, then

$$\frac{1}{B_2} = \frac{R_2}{(R_1 + R_2)} + \frac{(R_3 + R_4)}{R_3} \neq 1.$$

If we assume that $1 - 1/B_2 = \varepsilon \gg \zeta$, where $\zeta = [2/A_2(0) + 1/A_1(0)]$, then

$$Z_{\text{out}} = \frac{R}{\varepsilon} \frac{[1 + s(\tau_{01} + 2\tau_{02}) + 2s^2 \tau_{01}\tau_{02}]}{[1 + s(\tau_{01} + 2\tau_{02})/\varepsilon + 2s^2 \tau_{01}\tau_{02}/\varepsilon]}. \tag{G.4}$$

Equation (G.4) indicates that $\varepsilon > 0$ is essential for stability, and that an increase in ε reduces R_{out} but simultaneously increases the frequency range $0 \leqslant \omega \ll \varepsilon/(\tau_{01} + 2\tau_{02})$ over which Z_{out} remains resistive. The computation also shows that, for $0 \leqslant \varepsilon \ll \zeta$, Z_{out} approaches asymptotically the value

$$Z_{\text{out}} = \frac{R}{\zeta} \frac{[1 - s(\tau_{01} + 2\tau_{02}) + 2s^2 \tau_{01}\tau_{02}]}{[1 + s(\tau_{01} + 2\tau_{02})/\zeta + 2s^2 \tau_{01}\tau_{02}/\zeta]}. \tag{G.5}$$

1.7. (a) We assume $I^- = I^+$. The dc conditions are, for $I^+ \ll I_{CQ}$,

$$V_E = I_{CQ}R_3 = V_{R3} = V_{R2} = (I^-)R_2.$$

Hence,

$$V_C = V_{BE} + (R_4 + R_5)I_{CQ}R_3/R_2.$$

(b) The capacitor C_3 prevents negative feedback in the signal path from v_C to the non-inverting input. The time constants due to C_2 and C_3, in this loop, should be widely different in order to prevent non-monotonicity at low frequencies.

(c) Over the frequency range, for which the impedance of C_1 and C_2 is negligible,

$$1/B = R_2/R_1 R_3. \tag{G.5}$$

Assuming mismatch conditions to be satisfied,

$$LT = -Z_T/R_2. \tag{G.6}$$

(d) This effect is not reduced by feedback and therefore enters the ideal response $1/B$ as $di_{out} = dv_{BE}/R_3$.

(e) The current gain α is also excluded from the loop, hence the answer to (d) is also correct with regard to α.

1.8. (a) $V_E = V_S R_2/R_4$ and $I_{CQ} \simeq (V_S - V_E)/R_5 - V_E/R_2$.

(b)
$$\frac{1}{B} = \frac{\alpha_0(1 + R_2/R_5)\{1 + s[R_3 + (R_2\|R_5)]C_2\}}{R_1(1 + sR_3 C_2)(1 + s\tau_\alpha)}.$$

(c) Drawing the diagram is straightforward.

1.9. Proofs of (a), (b) and (c) are straightforward.

(d)
$$S^H_{k_0} = \frac{\partial H}{\partial k_0}\frac{k_0}{H} = \frac{b_{21}b_{43}}{F^2}\frac{k_0}{H} = \frac{H - H_0}{FH}.$$

1.10. (a) For the non-inverting amplifier, $B' = B \triangleq B_1$,

$$A_a/A_b = (CMRR - 1)/(CMRR + 1)|_{CMRR \gg 1} \simeq 1 - 2/CMRR,$$

and

$$A_b B_1 \simeq LT.$$

For the inverting amplifier, $v_b = 0$ and $v_{o2} \neq f(A_b)$.

(b) In the non-inverting amplifier, v_g appears as common mode voltage at the input terminals of the amplifier. In the inverting amplifier, v_a is at virtual ground and no common mode appears at the amplifier input.

1.11. (a) $\mathbf{R} = \begin{bmatrix} v_{gb} \\ v_{ga} \end{bmatrix}$, $\mathbf{E} = \begin{bmatrix} (v_{a1} - v_{b1}) \\ v_{a2} \end{bmatrix}$, $\mathbf{Y} = \begin{bmatrix} v_{o1} \\ v_{o2} \end{bmatrix}$,

$$\mathbf{T}_{11} = \begin{bmatrix} -3/2 & 2 \\ (3/2 + G_2/2G_1) & (2 + G_2/G_1) \end{bmatrix},$$

$$\mathbf{B} = \begin{bmatrix} 0 & -1 \\ -G_2/G_1 & -1 \end{bmatrix}, \quad \mathbf{A} = \begin{bmatrix} A_1 & 0 \\ 0 & A_2 \end{bmatrix}, \quad \mathbf{I}_{22} = \begin{bmatrix} 1 & 0 \\ 0 & 1 \end{bmatrix}.$$

(b) For $A_1 = \infty$, $A_2 = \infty$, the two local LTs are infinite, and

$$\mathbf{Y} = \mathbf{B}^{-1}\mathbf{R}, \quad \mathbf{B}^{-1} = \begin{bmatrix} G_1/G_2 & -G_1/G_2 \\ -1 & 0 \end{bmatrix}.$$

Hence, $v_{o1} = G_1(v_{gb} - v_{ga})/G_2$ and the CMRR is infinite.

(c) Inspection of Fig. 1.44 shows, for $A_1 = \infty$, $A_2 = \infty$, that $v_{a2} = 0$ and $v_{a1} = v_{b1}$. The latter condition also requires that $v_{a1} = 0$, $v_{b1} = 0$, since any increase in v_{a1} or v_{b1} is immediately compensated by a corresponding reduction in v_{o2}. Thus, no current flows through the resistors connecting v_{a1} and v_{b1} to v_{a2}, and both v_{a1} and v_{b1} are at virtual ground. After these relationships have been established, analysis is straightforward.

For $v_{ga} = 0$, $v_{o2}/v_{gb} = -1$ and $v_{o1}/v_{o2} = -R_2/R_1$. Hence, $v_{o1}/v_{gb} = R_2/R_1$. Also, for $v_{gb} = 0$, v_{a1} is still acting as virtual ground, $v_{o2} = 0$, and finally $v_{o1}/v_{ga} = -R_2/R_1$. Hence, $v_{o1} = (R_2/R_1)(v_{gb} - v_{ga})$. For a common mode signal, $v_{gb} = v_{ga}$ and $v_{o1} = 0$, QED.

(d) Elimination of $\mathbf{E}$ in the flow matrix yields

$$A_{\mathrm{CM}} = C_1/(C_3 + C_2A_2),$$

with

$$C_1 = A_1G_2/G_1; \quad C_2 = [3 + G_2/2G_1 + A_1G_2/G_1]A_2;$$
$$C_3 = (6 + 5G_2/2G_1) + 2A_1G_2/G_1.$$

So

$$S^{A_{\mathrm{CM}}}_{A_2} = \frac{\partial(A_{\mathrm{CM}})}{\partial(A_2)} \frac{A_2}{A_{\mathrm{CM}}}\bigg|_{A_1 \gg 3G_1/G_2 + 5/4} \approx \frac{-A_2}{(1 + A_2)}.$$

1.12. $v_o = [v_{gb}(A_b/A_a) - v_{ga}]\dfrac{R_2}{R_1}\dfrac{(-LT)}{(1 - LT)} = v_{gb}\mathscr{A}_b - v_{ga}\mathscr{A}_a,$

$$\mathrm{CMRR} = (\mathscr{A}_b + \mathscr{A}_a)/(\mathscr{A}_b - \mathscr{A}_a) = (A_b + A_a)/(A_b - A_a).$$

Hence, the CMRR of the feedback amplifier equals that of the basic amplifier, which shows that even an ideal feedback network cannot improve the latter.

1.13. (a) Derivation is straightforward.

(b) The two sensitivities are the same. This may be concluded directly from the relationship $\omega_0 = A_v(0)\omega_h$, since $A_v(0)$ is constant for the above sensitivity computation. More specifically, equation (1.74) shows that sensitivity increases for both from very low values at low frequencies up to unity, beyond $\omega_0 B$.

1.14. Derivation of (a) and (b) is straightforward.

(c) We assume, for simplicity, that the output impedance of the amplifier is zero. Hence,

$$LT_1 = -A_v r_e / (r_e + R_f), \tag{G.7}$$

$$LT_2 = -R_1 / (r_e + R_f), \tag{G.8}$$

$$A_v \triangleq R_1 / r_e.$$

(d)
$$g_{in} = (g_e + G_f)(1 - LT_1) = g_e + G_f(1 + A_v), \tag{G.9}$$

$$g_{in} = (g_e + G_f)(1 - LT_2) = G_f + g_e(1 + R_1 / R_f). \tag{G.10}$$

The dominant terms in (G.9) and (G.10) are $G_f A_v$ and $g_e R_1 / R_f$, respectively, which are identical. However, in (G.9) we are dealing with voltage gain and the term $G_f A_v$ describes the reduction of R_f due to the Miller effect, whereas the term $g_e R_1 / R_f$ in (G.10) describes the reduction of r_e by the current LT. Note the duality in these relationships.

1.15.
$$F = 1 + \frac{2R_s}{R_1} + R_g \left\{ \frac{1}{R_p} + \frac{1}{R_1} \left[1 + \frac{2(R_s + R_2)}{R_1} \right] \right\} + \frac{R_s}{R_g}. \tag{G.11}$$

1.16. This is a straightforward problem. Transmission of noise to the output is preferably computed first for small-signal sources. As a sample computation we shall find the transmission to the output of series noise S_{va1} due to the amplifier fed by v_{ia1}. Omitting indexes a and b of nominally equal resistors,

$$\frac{d(\overline{v_{no}^2})}{df} = S_{va}(1 + 2R_2/R_1)^2 R_4^2 / R_3^2.$$

Note that the two noise voltages at the differential output of the first stage, due to s_{va}, are fully correlated.

1.17. (a) $S_{vo} = S_g(R_1 + R_2)^2 + S_i(R_1^2 + R_2^2)$

$$+ S_v(R_1 + R_2 + R_g)^2 / R_g^2 + S_{vR1} + S_{vR2}. \tag{G.12}$$

where $S_v = S_{va} + S_{vb}$; $S_g = 4k\mathcal{T}/R_g$; $S_{vRi} = 4k\mathcal{T}R_i$.

$$S_{ia} = 4k\mathcal{T}G_{pa} = S_{ib} = S_i.$$

(b)
$$F = 1 + R_g \left\{ \frac{(1 + RG_p)}{2R} + R_s \left(\frac{1}{2R} + \frac{1}{R_g} \right)^2 \right\}.$$

(c)
$$R \ll 1/G_p.$$

1.19. The transimpedance of the two circuits is

$$R_{Ta} = R_1;$$

$$R_{Tb} = R_2 + R_3(1 + R_2/R_4) \simeq R_2(1 + R_3/R_4)$$

since $R_2 \gg R_4$.

$$R_{Ta} = R_{Tb} = R_T,$$

$$F_a = 1 + R_g/R_T,$$

$$k = R_3/R_2 + R_3/R_4 + 1 > 1.$$

Hence $F_b > F_a$, i.e. the circuit shown in Fig. 1.68(a) is preferable regarding its noise performance.

1.20. For the basic amplifier,

$$D_2 = 2 \times 10^{-2} = 2\%.$$

For the feedback amplifier,

$$D_2/|1 - LT(j\omega_2)| \simeq 8 \times 10^{-6} = 8 \times 10^{-4}\%.$$

$f_2 = 100$ Hz.

1.21. (a)
$$\tau_{s1} = (3 \ln 10)\tau_0 = 6.9\tau_0.$$

(b) $\tau_{s2} = \{A(0)/[1 + A(0)]\}\{3 \ln 10 - \ln [1 + 1/A(0)]\}\tau_0 < \tau_{s1}.$

Hence, reduction of $A(0)$ improves the settling time slightly.

(c) $\varepsilon(t) = (10/11) \exp[-t/(10/11)\tau_0] + (1/11) \exp(-t/110\tau_0),$

$$\tau_{s3} \simeq 110\tau_0 \ln (10^3/11) = 496\tau_0.$$

(d)
$$\varepsilon(t) = (10/9) \exp(-t/\tau_0) - (1/9) \exp(-t/10\tau_0),$$

which indicates that response is not monotonic.

$$\tau_{s4} \simeq 10\tau_0 \ln (10^3/9) \simeq 47\tau_0.$$

(e) Comparison of the characteristic equation

$$s^2 + s(1/\tau_z + 1/\tau_0)/m + 1/m\tau_0\tau_z$$

with (3.43) yields the desired result.

1.22. For this amplifier, $B' = B$.

(a) $$B_{min} = \tau_{0\,min}/4\tau_1.$$

(b) $$\tau_{R\,min} = 4(\pi)^{1/2}\tau_1.$$

(c) $$1/B\tau_{R\,min} = \omega_{0\,max}/(\pi)^{1/2}$$

1.23. The ratio between the dominant time constant of the closed-loop response and τ_z is

$$[\tau_p + LT(0)\tau_z]/[1 + LT(0)]\tau_z \simeq m/LT(0) + 1.$$

1.24. The root loci for $\tau_2/\tau_1 = 0.8$ and $\tau_2/\tau_1 = 1.4$ are shown in Figs. G.2 and G.3, respectively. Endpoints of solid lines correspond to $\omega_0\tau_1 = 0.28$.

Chapter 2

2.1. $$V_o = V_c C_g/C_f \propto x.$$

$$C_f = A\varepsilon/x.$$

The beauty of this lies in its simplicity.

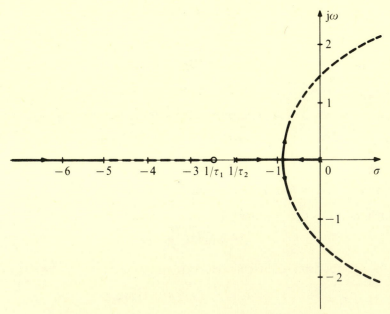

Fig. G.2. Problem 1.24. Root locus for $\tau_2/\tau_1 = 0.8$.

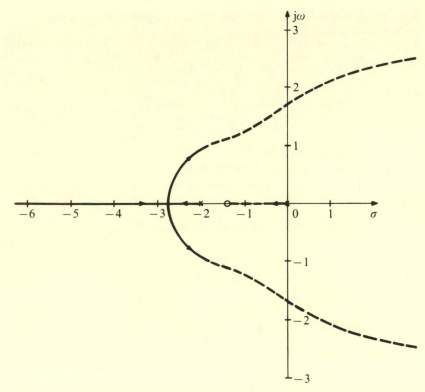

Fig. G.3. Problem 1.24. Root locus for $\tau_2/\tau_1 = 1.4$.

Chapter 3

3.2. The angular frequency $1/A_f(0)\tau_0$ is the gain-bandwidth product of the LT.

3.3. For this amplifier, $B = B' = 1/A_f(0)$, and for $\tau_1 = \tau_2 = \tau$, we obtain

$$\frac{v_o}{v_g} = \frac{A_f(0)}{1 + sA_f(0)\tau_0(1 + s\tau)^2}. \qquad (G.13)$$

If migration of the parasitic poles in the root locus is insignificant, we expect to be able to factorize the denominator of (G.13) as indicated in (3.18) which can be easily shown. A comparison between the factors associated with s and s^2 in the denominator of (G.13) and those in the denominator of (3.18) shows indeed that they are approximately equal, if we assume that $A_f(0)\tau_0 \gg 2\tau$ and disregard differences of second order.

3.4. We define the local LT of the voltage amplifier as LT_1, with

$$LT_1(s) = -A_vC_f/C_T + C_f) = -(1/\tau_0'), \qquad (G.14)$$

$$\tau_0' = \tau_0(C_T + C_f)/C_f.$$

Hence,

$$\frac{v_o}{i_e} = \frac{1}{B}\frac{(-LT_1)}{(1-LT_1)} = \frac{-1}{j\omega C_f}\frac{1}{(1+j\omega\tau_0')}. \qquad (G.15)$$

3.5. We index the noise parameters of the input transistor Q and of the voltage amplifier by 1 and 2, respectively. Also $R_{p1}' = (R_{p1}\|R_1)$, $R_{p2}' = (R_{p2}\|R_2)$, and $A = R_f/R$. Then the mean squared output noise voltage of the differentiator, filtered by an integrator of time constant $\tau = RC$, is

$$\overline{v_{no}^2} = 4k\mathcal{T}(A^2/20\tau)(R_{s1} + R'),$$

with

$$R' = R\{1 + (r_e + r_x)G_{p1}[1 + (r_e + r_x)G]$$
$$+ 5[1 + R_f(G_{p1}' + G_{p2}' + R_{s2}(G_2^2 + 0.25G_f^2(1 + C_T/C_f)^2))]/A\}.$$

Normalization of the differentiator gain with respect to T_2 is not applicable here, since the response is non-monotonic. We therefore compute the midband gain at the frequency of maximum response (see Appendix A.4), which yields $[H(f_d)]^2 \approx 0.24A^2$. Finally, $\overline{v_{ni}^2} = \overline{v_{no}^2}/0.24A^2 = 4k\mathcal{T}\Delta f(R_{s1} + R')$, $\Delta f = \text{ENB} \approx 1/4.8\tau$.

3.6. Computation of noise referred to input, of Fig. 3.16. First, $\overline{v_{no}^2}$ is obtained from a small-signal diagram corresponding to that of Fig. 3.14(b). Noting that $RC \gg A\tau_0$; $R_fC_f = RC = \tau$ and $A = R_f/R$, the various transmissions are:

$$\frac{v_{no}}{v_{nR}} = H(j\omega) = \frac{-j\omega A\tau}{(1 + j\omega\tau)^2(1 + j\omega\tau_0)}, \qquad (G.16)$$

$$\frac{v_{no}}{(i_{nf} + i_{np})} = \frac{H(j\omega)}{Y_g} = \frac{-j\omega A\tau(R + 1/j\omega C)}{(1 + j\omega\tau)^2(1 + j\omega\tau_0)}, \qquad (G.17)$$

$$\frac{v_{no}}{v_{ns}} = H(j\omega)(Y_g + Y_f)/Y_g = H(j\omega)B/B'$$

$$= \frac{(1 + j\omega A\tau)(1 + j\omega\tau/A)}{(1 + j\omega\tau)^2(1 + j\omega\tau_0)}. \qquad (G.18)$$

The final result is

$$\overline{v_{ni}^2} = (2)^{1/2}k\mathcal{T}/T_2. \qquad (G.19)$$

3.7. The transfer function of each amplifier is, with $R_f = R$,

$$H(s) \simeq -[(1+sRC_f)(1+s\tau_0)]^{-1}. \tag{G.20}$$

The input current to the second amplifier is, with $RC_f \gg \tau_0$,

$$\frac{v_g s(RC_f + \tau_0)[1+s(1/\tau_0+1/RC_f)^{-1}]}{R(1+sRC_f)(1+s\tau_0)} \simeq \frac{v_g s C_f}{(1+sRC_f)}. \tag{G.21}$$

Hence,

$$v_o/v_g \simeq -sRC_f/(1+sRC_f)^2(1+s\tau_0). \tag{G.22}$$

3:8. If $LT = \infty$, $v_3 = 0$ and $i_3 = 0$. Hence, from the third row of the matrix, we obtain

$$v_2/v_1 = -y_{31}/y_{32}.$$

The LT is obtained (also from the third row) as a function of the transmission v_3/v_2, if $v_1 = 0$ and $i_3 = 0$:

$$LT = -A_v(v_3/v_2) = -A_v y_{32}/y_{33}.$$

3.9. For interconnections as shown in (a),

$$\frac{1}{B} = -\frac{(z_{22}-z_{21})}{(z_{11}-z_{12})} \frac{(1-XY)}{(1+X)},$$

where

$$X = (z_{33}-z_{34})/(z_{44}-z_{43})$$

and

$$Y = (z_{11}-z_{12})/(z_{22}-z_{21}).$$

For interconnections as shown in (b),

$$\frac{1}{B} = -\frac{(z_{22}-z_{21})}{(z_{11}-z_{12})} \frac{(1+X)}{(1-X/Y)}.$$

3.10. (a) $1/B = [1+2sRC/4Q_i^2 + s^2 R^2 C_i^2/4Q_i^2]^{-1}.$

(b) For this pole pattern, σ_i is constant for all i:

$$\sigma_i = -\omega_{ni}/2Q_i = \sigma. \tag{G.23}$$

Furthermore,

$$\omega_{di} = i\omega_{d1}. \tag{G.24}$$

Hence, by definition,

$$D = \omega_{ni}/2Q_i\omega_{d1} = i\omega_{ni}/2Q_i\omega_{di}. \tag{G.25}$$

Substituting (3.44) in (G.25)

$$D^2 = i^2/(4Q_i^2 - 1),$$
$$Q_i = (1 + i^2/D^2)^{1/2}/2. \tag{G.26}$$

(c) At high values of Q, $\omega_d \simeq \omega_n$. Hence, all transfer functions may be computed at ω_n, which greatly simplifies the result.

$$1/B = Q,$$

$$S_{1/B}^Q = \frac{\partial Q}{\partial(1/B)} \frac{1/B}{Q} = 1,$$

$$S_A^H \simeq \frac{1}{A_v B'} = \frac{(1 + 1/4\,Q_i^2)^{1/2}}{A_v(1 + 1/2\,Q_i^2)^{1/2}} \simeq \frac{1}{A_v},$$

$$S_A^Q \simeq 1/A_v.$$

3.11. (a) The flow graph is shown in Fig. G.4.

(b) For high local LTs, the feedback-stabilized gain of the corresponding branches in the flow graph are $b_1 = -Z_1/R_{g1}$, $b_2 = -Z_2/R_2$, $b_3 = -R_3/R_5$, $b_4 = -R_4/R_6$, $b_5 = -R_4/R_{g2}$, $b_6 = -R_4/R_7$, $b_7 = -Z_1/R_8$, $b_8 = -Z_2$ and $b_9 = -R_3$, with $Z_1 = R_1/(1 + sR_1C_1)$ and $Z_2 = 1/sC_2$.

The multivariable feedback circuit is defined by the three amplifiers A_{v1}, A_{v2} and A_{v3}. The amplifier A_{v4} is connected externally and is not part of the multivariable loop. Hence, the rank of the desired matrix is

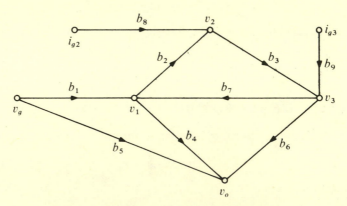

Fig. G.4. Problem 3.11. Flow graph for Fig. 3.5.

three, and v_o is obtained *after* solving for v_1, v_2 and v_3.

$$\begin{bmatrix} v_1 \\ v_2 \\ v_3 \end{bmatrix} = \frac{1}{F} \begin{bmatrix} b_1 & (b_3b_7b_8) & (b_7b_9) \\ (b_1b_2) & b_8 & (b_2b_7b_9) \\ (b_1b_2b_3) & (b_3b_8) & b_9 \end{bmatrix} \begin{bmatrix} v_g \\ ig_2 \\ ig_3 \end{bmatrix}, \qquad (G.27)$$

$$F = 1 - b_2b_3b_7. \qquad (G.28)$$

(c) $$H = v_o/v_g = b_5 + b_1(b_4 + b_2b_3b_6)/F. \qquad (G.29)$$

Equation (G.29) shows that the flow graph of Fig. G.4 is already in the bilinear form, since the gain element a_1 appears in the single branch b_1 only. Hence,

$$S_{b_1}^H = \frac{1}{F}\left(1 - \frac{H_0}{H}\right), \qquad (G.30)$$

$$H_0 = b_5,$$

where H_0 is the transmission between v_g and v_o with the gain element $a_1 = 0$. Finally,

$$S_{a_1}^{b_1} = 1/(1 - LT_{11}), \qquad LT_{11} = -a_1Z_{11}/Z_1 \quad \text{and} \quad (Z_{11})^{-1}$$
$$= Y_1 + G_{g1} + G_8.$$

3.12. $A_{DM} = A_v(1 + s\tau_a)$ and $A_{CM} = A_v s\delta\tau/(1 + s\tau_a)(1 + s\tau_b)$,

so

$$A_{DM}/A_{CM} = (1 + s\tau_b)/s\delta\tau. \qquad (G.31)$$

(b) Substitution of $(v_a + v_b)/2$ for v_G yields

$$\begin{bmatrix} v_g \\ v_{GL} \end{bmatrix} = \begin{bmatrix} [1 + s(\tau_a + \tau_b)/2] & -[1 + s(\tau_a + \tau_b)/2] \\ -s\tau_b/2 & (1 + s\tau_b/2) \end{bmatrix} \begin{bmatrix} v_a \\ v_b \end{bmatrix}. \qquad (G.32)$$

Solving for v_a and v_b yields

$$v_a - v_b = v_g/[1 + s(\tau_a + \tau_b)/2] \neq f(v_{GL}).$$

Hence, with $CMRR_A = \infty$, $v_o(v_{GL}) = 0$.

(c) Substituting the solution of (G.32) in $v_o = -A_v[(v_a - v_b) + (v_a + v_b)/CMRR_A]$ yields

$$v_o = -A_v[v_g(1 + 1/CMRR_A + s\tau/CMRR_A)$$
$$+ 2v_{GL}(1 + s\tau)/CMRR_A]/(1 + s\tau) \qquad (G.33)$$

with $\tau_a = \tau_b = \tau$.   Hence, $CMRR = A_{DM}/A_{CM}|_{\omega \ll CMRR/\tau} \approx$
$2(1 + s\tau)$.

3.13. With a step of amplitude V applied to the input, and $A_v = \infty$, the voltage across R_1 equals

$$v_{R1}(t) = -Vt/R_gC_T \qquad (G.34)$$

and the output voltage

$$v_o(t) = \left\{v_{R1}(t) - \frac{1}{C_T} \int_0^t \left[\frac{V}{R_g} - \frac{v_{R1}(t)}{R_1}\right] dt\right\} = \frac{-Vt}{R_gC_T}\left(2 + \frac{t}{2R_1C_T}\right).$$
$$(G.35)$$

Chapter 4

4.1. (a) According to the characterization of a filter response by time moments, the rise time τ_{RF} of the filter response to a step input and its half-power frequency are related as

$$\tau_{RF} = (2\pi)^{1/2}T_2 = (2\pi \ln 2)^{1/2}/\omega_h. \qquad (G.36)$$

The values of τ_{Ro}^2 and $\overline{v_{no}^2}$ are given as

$$\tau_{Ro}^2 = \tau_{RS}^2 + \tau_{RF}^2, \; \overline{v_{no}^2} \propto \Delta f = \omega_h/4 \propto 1/\tau_{RF}.$$

Hence,

$$\overline{v_{no}^2}\tau_{Ro}^2 \propto \tau_{RS}(\tau_{RS}/\tau_{RF} + \tau_{RF}/\tau_{RS}). \qquad (G.37)$$

For a given τ_{RS}, equation (G.37) becomes minimum if $\tau_{RF} = \tau_{RS}$, or

$$\omega_h = (2\pi \ln 2)^{1/2}/\tau_{RS}. \qquad (G.38)$$

(b) If we approximate the signal at the output of the filter by a ramp of rise time τ_{Ro}, then $\overline{v_{no}^2}\tau_{Ro}^2$ is the variance of the timing signal derived from a suitable level crossing.

(c) We consider the vertical amplifier as the filter referred to above, whose ENB equals Δf. If the midband gain of the amplifier is $A_v(0)$ and its noise is dominant, we may consider $\overline{v_{no}^2}$ as being due to white noise $\overline{v_n^2}$ at its input:

$$\overline{v_{no}^2} = \overline{v_n^2}A_v^2(0)\,\Delta f. \qquad (G.39)$$

In order to observe the signal on the scope with maximum accuracy, a compromise should be struck between a desirable reduction of $\overline{v_{no}^2}$ by reducing Δf, and the resulting deterioriation in the observed signal rise time.

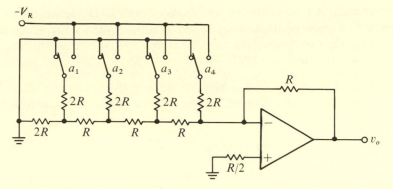

Fig. G.5. Problem 5.2. Voltage switching employed with $R-2R$ network.

Chapter 5

5.1. If the value of R_c equals the resistance seen by the inverting input terminal of the amplifier, and if ΔI_B of the amplifier equals zero, then no output offset voltage is created by the offset currents of the amplifier. In the circuit shown in Fig. 5.4, that resistance is a function of the switch positions.

5.2. The circuit is shown in Fig. G.5.

5.3. The non-inverting input terminal of the voltage follower sees a constant resistance R. Connection of a resistor valued R between the inverting input terminal and the output of the voltage follower provides the desired solution.

5.4. An analytic solution is rather involved, whereas a computerized evaluation of the error due to the different resistors, one at a time, is relatively simple.

Chapter 6

6.1. An error of 7% in a four-bit converter exceeds 1 LSB and causes the D/A converter to be non-monotonic at the transition from 0111 to 1000, for the given conditions. For a rising ramp at the input, the output of a monotonic digital ramp A/D converter stops at that digital number, whose converted analog value is closest above the analog input. A total error of +7% for the second and third bit causes the code 0111 to correspond to an analog input of 8.12/16 of FS, and the code 1000 to $\frac{1}{2}$ of FS, which prevents the converter from stopping at 1000 and thus creates a missing code. A total error of 6% is smaller than 1 LSB. Hence, the D/A converter is monotonic, but the range of input voltages for which the converter changes from 0111 to 1000 is much smaller than the analog value corresponding to 1 LSB.

6.2. Both offset voltage and current of the integrator cause an error in the integrated analog voltage, which is proportional to the time T of integration:

$$\Delta v_o = (\Delta V_{BE}/RC + \Delta I/C)T \qquad\qquad (G.40)$$

Polarity has been disregarded in (G.40), but it should be noted that the individual contributions of ΔV_{BE} and ΔI will either add or subtract, depending on their relative polarity.

The offset due to the comparator has no effect on the result N, since a single comparator with a fixed level of comparison is employed in sensing the start and the end of the dual slope conversion process. Hence, the absolute level of comparison is irrelevant. The difference in the start and the stop level due to hysteresis is also unaffected by the offset.

6.3. Advantage: The switches are at ground potential, which makes for a simpler design of their control voltages.

Disadvantage: Any mismatch between the two resistors introduces an additional error. However, closely matched resistors are available, making this error relatively insignificant.

6.4. An R–S flip flop is connected as follows: S to a_1; R to a_2; $\bar{Q}$ to S_1 and to S_2.

Chapter 10

10.1 The mean square noise at the filter output is proportional to Δf. Hence, after averaging,

$$(\text{SNR})^2 \propto m/\Delta f, \qquad\qquad (G.41)$$

or, for constant SNR, m is proportional to Δf.

10.2. The total mean square noise in the sum of all samples equals

$$\overline{v_{nT}^2} = E[v_n(T) + v_n(2T) + \cdots + v_n(iT) + \cdots + v_n(mT)]^2. \quad (G.42)$$

For stationary noise,

$$E[v_n(aT)v_n(bT)] = R_n[(a-b)T]. \qquad\qquad (G.43)$$

Hence,

$$\overline{v_{nT}^2} = mR_n(0) + 2 \sum_{i=1}^{m-1} (m-i)R_n(iT), \qquad (G.44)$$

and the averaged mean squared noise equals

$$\overline{v_n^2} = \overline{v_{nT}^2}/m^2 = [R_n(0)/m]\left[1 + \frac{2}{m}\sum_{i=1}^{m-1} (m-i)\rho_n(iT)\right] \qquad (G.45)$$

The signal is correlated, hence its value after averaging remains unchanged.

Since $R_n(0)$ corresponds to the mean squared noise without averaging the SNR after averaging improves by the factor

$$(m)^{1/2}\left[1 + \frac{2}{m}\sum_{i=1}^{m-1} (m-i)\rho_n(iT)\right]^{-1/2}.$$

Chapter 11

11.1 and 11.2.

$$\overline{v_{noD}^2} = 4\mathcal{k}\mathcal{T}R'_{eq}\int_0^\infty [(1+\omega^2\tau_1^2)(1+\omega^2\omega^2\tau_2^2)]^{-1}\,dr = 4\mathcal{k}\mathcal{T}R'_{eq}/4(\tau_1+\tau_2),$$

$$f_{c1} = 1/\tau_1, \quad f_{c2} = 1/\tau_2, \quad \Delta f_i = \pi f_{ci}/2, \quad i = 1, 2.$$

For $\Delta f_2 \ll f_{c1}$ we have $\tau_1 \ll \tau_2$ and $\overline{v_{noD}^2} \approx 4\mathcal{k}\mathcal{T}R'_{eq}/4\tau_2$.

For $\Delta f_2 \gg f_{c1}$ we have $\tau_2 \ll \tau_1$ and $\overline{v_{noD}^2} \approx 4\mathcal{k}\mathcal{T}R'_{eq}/4\tau_1$.

If $S_{v1}(f)$ and $|H_2(j\omega)|^2$ are plotted, it can be observed that in the first case the value of $S_{v1}(f) \approx S_{v1}(0)$ within the frequency range of $H_2(j\omega)$, and vice versa in the second case.

11.3. The noise spectral density at the input of the preamplifier is

$$d(\overline{v_{ni}^2})/df = \mathcal{k}\mathcal{T}R_D/(1+\omega^2R_D^2C_I^2/4). \qquad (G.46)$$

Hence, $R_{eq}(0) = R_D/4$.

11.4. $Z_{in} = Z_a + Z_c = Z_o \cosh(u)/\sinh(u)$

$$= (1 + j\omega R_D C_D/2 - \omega^2 R_D^2 C_D^2/4 + \cdots)/(j\omega C_D - \omega^2 R_D^2 C_D^2 + \cdots).$$

For $\omega \to 0$, the terms containing ω^2 are negligibly small. Hence,

$$\lim_{\omega \to 0} Z_{in} = R_D/2 + 1/j\omega C_D.$$

Appendix A

A.1. Note the relationships

$$E[x(k)\mu_y + y(k)\mu_x] = 2\mu_x\mu_y, \quad \text{and} \quad E[\mu_x\mu_y] = \mu_x\mu_y = E[x(k)]E[y(k)].$$

A.2.

$$R_{yx}(\tau) = \lim_{T \to \infty} \frac{1}{T} \int_{-T/2}^{T/2} x(t-\tau)y(t)\, dt$$

$$= \lim_{T \to \infty} \frac{1}{T} \int_{-T/2}^{T/2} x(t-\tau) \int_{-\infty}^{\infty} h(u)x(t-u)\, du\, dt$$

$$= \int_{-\infty}^{\infty} h(u) \left\{ \lim_{T \to \infty} \frac{1}{T} \int_{-T/2}^{T/2} x(t-\tau)x(t-u)\, dt \right\} du$$

$$= \int_{-\infty}^{\infty} h(u)R_x(\tau-u)\, du.$$

If the noise applied to the input is white, its autocorrelation R_x is an impulse for $\tau = u$ and zero otherwise. In this case, the value of $h(u)$ is constant over the infinitesimally small range $u = \tau$ for which $R_x(\tau - u)$ exists. Hence we may write

$$R_{yx}(\tau) = h(\tau) \int_{-\infty}^{\infty} R_x(\tau - u)\, du = h(\tau)W_x(0).$$

A.3. (a) Substituting $H(j\omega)S_x(f) = S_{yx}(f)$ and $S_y(f) = S_x(f)|H(j\omega)|^2$ in (A.56) provides the desired proof.
   (b) follows directly from (a).

A.4. (a) We introduce $H_3(j\omega) = H_1(j\omega) \exp(-j\omega\tau_D)$. Hence,

$$S_{vo}(f) = S_v|H_2(j\omega) + H_3(j\omega)|^2$$
$$= S_v\{|H_2(j\omega)|^2 + 2\text{Re}[H_2(j\omega)H_3^*(j\omega)] + |H_3(j\omega)|^2\},$$
$$v_{no}^2 = R_2(0) + 2\rho_{23}[R_2(0)R_3(0)]^{1/2} + R_3(0),$$

$$R_i(0) = S_v \int_0^{\infty} |H_i(j\omega)|^2\, df.$$

(b)

$$\rho_{23} = \left\{ \int_0^{\infty} \text{Re}[H_1(j\omega)H_2^*(j\omega)] \cos \omega\tau_D\, df \right.$$

$$\left. - \int_0^{\infty} \text{Im}[H_2(j\omega)H_1^*(j\omega)] \sin \omega\tau_D\, df \right\} [|H_1(j\omega)|^2|H_2(j\omega)|^2]^{-1/2}$$

A.5. (a) $\quad S_a = 4k(\mathcal{T}_1/R_1 + \mathcal{T}_2/R_2)\,\text{Re}\,[Y_3(j\omega)]/|Y(j\omega)|^2]$,

$\qquad Y_3(j\omega) = (R_3 + 1/j\omega C_3)^{-1}, \; Y = G_1 + G_2 + j\omega C_1 + Y_3(j\omega)$.

(b) $\qquad S_b = 4k\mathcal{T}_3\,\text{Re}\,[Y_3(j\omega)](G_1 + G_2)/|Y(j\omega)|^2$

(c) $\qquad \dfrac{S_a}{S_b} = \dfrac{\mathcal{T}_1 G_1 + \mathcal{T}_2 G_2}{\mathcal{T}_3(G_1 + G_2)} > \dfrac{\mathcal{T}_2 G_1 + \mathcal{T}_2 G_2}{\mathcal{T}_3(G_1 + G_2)} = \dfrac{\mathcal{T}_2}{\mathcal{T}_3} > 1.$

(d) $\qquad W_{ab} = \displaystyle\int_0^\infty (S_a - S_b)\,\mathrm{d}f.$

A.6. The exact expression for the noise spectral density referred to the input is

$$S_i = [S_1 + S_2/A^2(0)]\{1 + S_2\omega^2 A^2(0)\tau_0^2/[A^2(0)S_1 + S_2]\}.$$

Appendix B

B.1. $\qquad y_{ia} \gg Y_1 + Y_f, \; y_{oa} \gg Y_2 + Y_f, \; LT = -Z_T/Z_f.$

B.2. With the exception of the amplifier shown in Fig. B.2(a), all amplifiers employ enhancing combinations in which the input and output immittances of the basic amplifiers are enhanced by the feedback applied. The input immittance of the amplifier shown in Fig. B.2(a) employing a basic voltage amplifier is

$$I_{in} = \{(Y_1 + Y_f)[1 + A_v Y_f/(Y_1 + Y_f)]\}^{-1} = [Y_1 + Y_f(1 + A_v)]^{-1}.$$

B.3. The Kirchhoff matrix equation is

$$\begin{bmatrix} v_{g1} - v_{g2}/(1 - G_2/A_v y_{oa}) \\ 0 \end{bmatrix} = \begin{bmatrix} Y_{11}/G_1 & -G_2/G_1 \\ (A_v y_{oa} - G_2) & Y_{22} \end{bmatrix} \begin{bmatrix} v_e \\ v_2 \end{bmatrix},$$

$$Y_{11} = G_1 + G_2 + y_{ia} \qquad Y_{22} = y_{oa} + G_2 + Y_2.$$

(b) $\qquad y_{ia} \ll G_1 + G_2, \qquad y_{oa} \gg G_2 + Y_2,$

$$LT = -A_v R_1/(R_1 + R_2)$$

B.4. (a) The gain of the active device will be preferably chosen as a function of the error parameter i_1, i.e. as current gain: $A_c = -Y_T z_i$.
Hence,

$$\begin{bmatrix} v_1 \\ v_2 \\ i_3 \end{bmatrix} = \begin{bmatrix} z_i & 0 & 1 \\ -Y_T z_i z_o & z_o & 1 \\ -1 & -1 & Y_f \end{bmatrix} \begin{bmatrix} i_1 \\ i_2 \\ v_3 \end{bmatrix}.$$

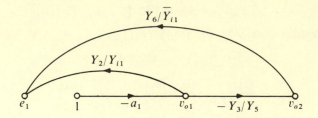

Fig. G.6. Problem B.5. Flow graph for return difference for a_1 in Fig. B.12.

(b) and (c)

$$\begin{bmatrix} v_1 \\ v_2 \end{bmatrix} = \begin{bmatrix} (z_i + Z_f) & Z_f \\ -(Y_T z_i z_o - Z_1) & (z_o + Z_f) \end{bmatrix} \begin{bmatrix} i_1 \\ i_2 \end{bmatrix},$$

$$z_i \gg Z_{f1} \qquad z_o \gg Z_{f1} \qquad 1/B = 1/Z_f.$$

$$\begin{bmatrix} v_1 \\ i_3 \end{bmatrix} = \begin{bmatrix} z_i & 1 \\ -(Y_T z_i + 1) & (Y_f + y_0) \end{bmatrix} \begin{bmatrix} i_1 \\ v_3 \end{bmatrix},$$

$$y_o \ll Y_f, \qquad 1/B = 1.$$

$$\begin{bmatrix} i_3 \\ v_2 \end{bmatrix} = \begin{bmatrix} (Y_f + y_i) & -1 \\ (1 + Y_T z_o) & z_o \end{bmatrix} \begin{bmatrix} v_3 \\ i_2 \end{bmatrix},$$

$$y_i \ll Y_f, \qquad 1/B = -1.$$

For all configurations, $LT = -Y_T Z_f$.

B.5. Proofs of (a) and (b) are straightforward.

(c) The return difference for a_1, for high local LTs is obtained from the flow graph shown in Fig. B.13, as shown in Fig. G.6.

Appendix C

C.1. The expansion yields two complex zeros at $[1 \pm j(2)^{1/2}]/\tau_D$, which draw the two migrating poles into the RHS of the complex frequency plane, at high values of LT. The resulting root locus approaches closely the accurate one, in which the two pole paths are confined within the assymptotes at $\pm \pi/\tau_D$ as shown in Fig. C.3 for a single pole.

Appendix F

F.1. (a) Substitution of

$$A_{m-1} = \frac{1}{(m-1)} \sum_1^{m-1} I_i, \quad \text{in} \quad A_m = \frac{1}{m} \left(\sum_1^{m-1} I_i + I_m \right)$$

yields the desired result.

(b) Expansion of (F.3) and summation of the resulting series of signal and noise samples yields the SNR after averaging:

$$\lim_{m \to \infty} \frac{S}{N} \frac{[1 - (1 - 1/M)^m]}{\left[\dfrac{1 - (1 - 1/M)^{2m}}{2M - 1} \right]^{1/2}} = \frac{S}{N} (2M - 1)^{1/2}$$

where S/N is the ratio between signal and noise of a single sample.

(c) The proof is obtained by substituting for $A_{m-1} = I_{m-1}(M-1)/M$.

Appendix H

Abbreviations, symbols and notation

Abbreviations

A/D	analog-to-digital
B	buffer
BCD	binary-coded decimal
CAD	computer aided design
CB	common base
CCCS	current controlled current source
CCVS	current controlled voltage source
CE	common emitter
CM	common mode
CMRR	common mode rejection ratio
CRT	cathode ray tube
D/A	digital-to-analog
DDL	double delay line
DDP	digital data processor
DM	differential mode
EEG	electroencephalogram
EKG	electrocardiogram
EMI	electromagnetic interference
ENB	equivalent noise bandwidth
FDNR	frequency-dependent negative resistance
FET	field effect transistor
FS	full scale
F/V	frequency-to-voltage
FWHM	full-width half-maximum
IC	integrated circuit
LHS	left hand side of complex frequency plane
LPF	low-pass filter
LSB	least significant bit
LSI	large-scale integration
LT	loop transmission
MOS	metal oxide silicon

MSB	most significant bit
NMR	nuclear magnetic resonance
NPD	noise power density
NTM	normalized time moment
PDF	probability density function
ppm	parts per million
PROM	programmable read-only memory
P/S	parallel-to-series
PTRC	precision timing reference circuit
RHS	right hand side of complex frequency plane
ROM	read-only memory
RSS	root sum squared (error)
SAW	surface acoustic wave
SCA	single channel analyser
SNR	signal-to-noise ratio
TAC	time-to-amplitude converter
VCCS	voltage controlled current source
VCO	voltage controlled oscillator
VCVS	voltage controlled voltage source
V/F	voltage-to-frequency
VTC	voltage to time converter

Symbols

A_f	gain of feedback amplifier
A_i	current gain
A_v	voltage gain
B'	attenuation of feedback loop
$1/B$	transfer function of feedback amplifier for $\lim |LT| \to \infty$
C_g	source capacitance
C_i	input capacitance
C_L	load capacitance
d	duty ratio of periodic signal
D	drain (of FET)
f_c	clock frequency; corner frequency (used in noise computations)
F	noise figure
$\mathscr{F}$	return difference
gm	transconductance of active element
$g(t)$	Gaussian impulse response
$g_{-1}(t)$	Gaussian step response

G	conductance $(= 1/R)$; gate (of FET)
G_I	instrument ground
G_S	ground at signal source
G_T	transconductance
$G(s)$	Gaussian frequency response
h	Planck's constant
$h(t)$	impulse response
$h_{-1}(t)$	step response
$H(s)$	transfer function
i	small-signal current variable
i_i	small-signal input current
i_e	small-signal error current
i_o	small-signal output current
i_L	small-signal load current
I	dc current
I_{in}	input immittance
I_{jj}	self-immittance of loop or node j
k_0	gain factor of LT (used in constructing root locus)
$k_0(\sigma_b)$	breakaway factor
$\pounds$	Boltzmann's constant
LT	loop transmission
LT'	LT obtained after applying the gain-bandwidth theorem
LT_s	LT obtained by source splitting
$p(x)$	probability density function
P_a	power gain of amplifier
$P(x)$	probability distribution function
q	electron charge
Q	quality factor
r_o	output resistance of gain element or amplifier
R_c	compensating resistance
R_C	cable resistance
R_{eq}	equivalent series noise resistance (represents total noise)
R_g	generator or source resistance
R_L	load resistance
R_p	equivalent parallel noise resistance $(= 1/G_p)$
R_s	equivalent series noise resistance
$R_x(\tau)$	autocorrelation function
$R_{xy}(\tau)$	crosscorrelation function
s	complex variable
S	slewing rate; source (of FET)
$S_i(f)$	mean square current spectral density

$S_v(f)$	mean square voltage spectral density
$S_x(f)$	physical power spectral density function
$S_{xy}(f)$	physical cross-spectral density function
t	time variable
t_D	delay time
t_m	time of measurement
T	time interval; period of repetitive signal
T_r	response time of comparator
T_1, T_2	first and second time moments
$\mathscr{T}$	temperature
$\mathscr{T}_{eq}$	equivalent noise temperature
$\mathscr{T}_0$	reference temperature for specifications
v	small-signal voltage variable
v_A	analog voltage
v_B	base voltage
v_C	collector voltage
v_D	distortion voltage
v_e	error voltage
v_E	emitter voltage
v_g	source voltage
v_G	guard voltage
v_{GL}	ground-loop voltage
v_H	hysteresis voltage
v_i	input voltage
v_o	output voltage
V	dc voltage
V_R	reference voltage
V_s	sense voltage
V_S	supply voltage
V_T	voltage equivalent of temperature
$w(t)$	weighting function
W	power
$\bar{W}$	normalized power (measured across resistance of 1 Ω)
W_{na}	available noise power of amplifier referred to input
W_{ng}	available noise power from source
W_{sg}	available signal power from source
$W_x(f)$	mathematical power spectral density function
$\bar{W}_x$	normalized noise power of v_x
Y_T	transadmittance
z_i	amplifier input impedance ($= 1/y_i$)
z_o	amplifier output impedance ($= 1/y_o$)

Z_g	source impedance
Z_L	load impedance
$\gamma^2(f)$	coherence function
Δf	equivalent noise bandwidth
ΔI	offset current
ΔV	offset voltage
μ	mean value
ρ	correlation coefficient
$\rho(\tau)$	correlation function coefficient
σ	standard deviation
σ^2	variance
τ	time displacement; time constant
τ_a	aperture time
τ_{c0}	unity-gain time constant
τ_D	delay
τ_R	rise time
τ_s	settling time
τ_{0LT}	unity-gain time constant of LT $(=1/\omega_{0LT})$
ω_{c0}	unity-gain frequency
ω_d	damped natural frequency
ω_h	half-power frequency
ω_{hH}	half-power frequency of $H(j\omega)$
ω_{hLT}	half-power frequency of LT
ω_{hOL}	half-power frequency of open loop
ω_n	undamped natural frequency
ω_0	extrapolated gain-bandwidth product
ω_{0H}	gain-bandwidth product of $H(j\omega)$
ω_{0LT}	gain-bandwidth product of LT
ω_{45}	frequency at which phase angle equals $\pi/4$

Notation

$X \| Y$	X in parallel with Y
X^*	complex conjugate of X
$\hat{X}$	estimator of X
$\bar{X}$	ensemble average of X
$X \triangleq Y$	X corresponds to Y

References

Chapter 1

[1] *Selection guide and catalog handbook to operational amplifiers*, copyright © 1969, Analog Devices, Inc., Norwood, Massachusetts. Adapted by permission.

[2] J. E. Solomon, 'The monolithic op amp: a tutorial study', *IEEE journal of solid-state circuits* **SC-9**, 6, December 1974, pp. 314–32.

[3] See, for example, M. S. Ghausi, *Electonic circuits*, Van Nostrand Reinhold, New York, 1971, Section 2.1.

[4] 'Representation of noise in linear two-ports', tutorial paper by IRE Sub-committee on Noise, *Proceedings of the IRE*, January 1960, pp. 69–74.

[5] *Noise bibliography*

   (*a*) C. D. Motchenbacher and F. C. Fitchen, *Low noise electronic design*, John Wiley, New York, 1973.

   (*b*) Mahdu S. Gupta (ed.), *Electrical noise: fundamentals and sources*, IEEE Press, John Wiley, New York, 1977.

   (*c*) S. Lester and N. Webster, 'Noise in amplifiers', *IEEE spectrum* **7**, August 1970, pp. 67–75.

   (*d*) S. Hsu, 'Bistable noise in operational amplifiers', *IEEE journal of solid-state circuits* **SC-6**, 6, December 1971.

   (*e*) Donn Soderquist, *Minimization of noise in operational amplifier applications*, Application Note **AN-15**, Precision Monolithics, Inc., April 1975.

[6] L. W. Read, *Computer optimization of small signal amplifier design*, Cornell University Conference Book, 1969, pp. 227–38.

[7] P. E. Allen, 'Slew induced distortion in operational amplifiers', *IEEE journal of solid-state circuits* **SC-12**, 1, February 1977, pp. 39–44.

[8] R. J. Widlar, 'Some circuit design techniques for linear integrated circuits', *IEEE transactions on circuit theory* **CT-12**, 6, December 1965, pp. 586–90.

[9] *Application notes on the LM3900 quad amplifier*, National Semiconductor Corp., Inc.

[10] J. I. Smith, *Modern operational circuit design*, Wiley Interscience, New York, 1971, p. 129.

[11] B. M. Oliver and J. N. Cage, *Electronic measurements and instrumentation*, McGraw-Hill, New York, 1971, p. 202.

[12] J. K. Roberge, *Operational amplifiers, theory and practice*, John Wiley, New York, 1975.

[13] J. G. Graeme, 'Applications of operational amplifiers', *Third generation techniques*, McGraw-Hill, New York, pp. 91–2.

[14] A. P. Brokaw *et al.*, 'An improved monolithic instrumentation amplifier', *IEEE journal of solid-state circuits* **SC-10**, 6, December 1975, pp. 417–23.

[15] R. J. Van de Plassche, 'A wide-band monolithic instrumentation amplifier', *IEEE journal of solid-state circuits* **SC-10**, 6, December 1975, pp. 424–31.

[16] H. W. Bode, *Network analysis and feedback amplifier design*, Van Nostrand, New York, 1945.

[17] S. K. Mitra, *Analysis and synthesis of linear active networks*, John Wiley, New York, 1969, Sections 4.8 and 5.2.

[18] T. Furukawa and J. Arisawa, 'Common-mode rejection of the differential operational amplifier with shunt common-mode feedback', *IEEE journal of solid-state circuits* **SC-10**, 6, December 1975, pp. 539–40.

[19] See, for example, J. G. Truxal, *Automatic feedback control system synthesis*, McGraw-Hill, New York, 1955.

[20] See, for example, P. B. Lathi, *Signals, systems and communications*, John Wiley, New York, 1965, pp. 303–10.

[21] K. Hatch, 'High stability nuclear pulse amplifier analysis', *IEEE transactions on nuclear science* **NS-12**, 1, February 1965, p. 314.

[22] V. Radeka, 'Signal, noise and resolution in position-sensitive detectors', *IEEE transactions on nuclear science* **NS-21**, 1, February 1974, pp. 51–64.

[23] T. J. Aprille, 'Wideband amplifier design using major multiloop feedback techniques', *Bell system technical journal* **54**, 7, September 1975.

[24] F. S. Goulding *et al.*, 'An opto-electronic feedback preamplifier for high-resolution nuclear spectroscopy', *Nuclear instruments and methods* **71**, 1969, pp. 273–9.

[25] D. A. Landis *et al.*, 'Pulsed feedback techniques for semiconductor detector radiation spectrometers', *IEEE transactions on nuclear science* **NS-18**, 1, February 1971, pp. 115–24.

[26] R. I. Demrow, 'Settling time of operational amplifiers' *Analog dialogue* **4**, 1, June 1970, pp. 1–11. Adapted by permission.

[27] J. J. Stefano III, A. R. Stubberud and I. J. Williams, *Schaum's outline of theory and problems of feedback and control systems*, McGraw-Hill, New York, 1967. Adapted by permission.

[28] F. D. Waldhauer, 'Analog circuits of large bandwidth', *IRE conv. rec.* 1963, Part 2, pp. 200–7.

[29] B. Y. Kamath, R. Meyer and P. R. Gray, 'Relationship between frequency response and settling time of operational amplifiers', *IEEE journal of solid-state circuits* **SC-9**, 6, December 1974, pp. 347–52.

[30] R. J. Apfel and P. R. Gray, 'A fast settling monolithic feedforward op-amp using doublet compression techniques', *IEEE journal of solid-state circuits* **SC-9**, 6, December 1974, pp. 332–40.

[31] W. C. Elmore, 'Fast pulse amplifiers for nuclear research', *Nucleonics* **3**, September 1949, pp. 48–55.

[32] W. C. Elmore, 'The transient response of damped linear networks with particular regard to wideband amplifiers', *Journal of applied physics* **19**, January 1948, pp. 55–63.

[33] D. J. Comer and J. M. Griffith, 'Calculating the minimum number of interactive stages in a wideband amplifier', *IEEE transactions on circuit theory* **CT-20**, September 1968, pp. 280–1.

[34] U. Weiser, 'A logarithmic preamplifier for laser signal detecting', M. Sc. thesis, Technion, Israel Institute of Technology, Haifa, March 1975.

[35] U. Weiser, A. Adin and A. Arbel, 'Speed limitations of feedback amplifiers due to signal delay', *International journal of electronics* **41**, 3, 1976, pp. 249–55.

[36] *Data sheet of type 1430 200 ns to 0.01% settling FET operational amplifier*, Teledyne Philbrick, November 1976.

[37] G. J. Murphy, *Basic automatic control theory*, Van Nostrand Reinhold, New York, 1957, pp. 230–1, 244–5.

[38] Black, US Patent 1 686 792, October 1929.

[39] M. O. Deighton, E. H. Cooke Yarborough and G. L. Miller, 'A method of enhancing stability and linearity of transistor amplifier systems', *Proceedings of the Northeast Regional Electrical Manufacturers Conference, Boston 1964*.

[40] R. K. Jurgen, 'Feedforward correction: a late-blooming design', *IEEE spectrum* **9**, April 1972, pp. 41–3.

[41] P. J. Walker, 'Current dumping audio amplifier', *Wireless world*, December 1975, pp. 560–2.

Chapter 2

[1] K. L. Lion, 'Transducers: problems and prospects', *IEEE transactions on industrial electronic and control instrumentation* **IECI-16**, 1, July 1969.

[2] Kenneth Arthur, *Transducer measurements*, Measuring Concepts Series, Tektronix, Inc., Beaverton, Oregon, 1970.

[3] J. D. Lenk, *Handbook of electronic test equipment*, Prentice-Hall, Englewood Cliffs, Chapter 9.

[4] B. A. Gregory, *An introduction to electrical instrumentation*, Macmillan Press, London, 1975.

[5] L. A. Geddes, 'Interface design for bioelectric systems', *IEEE spectrum* **9**, October 1972, pp. 41–63.

[6] Peter Strong, *Biophysical measurements*, Measuring Concepts Series, Tektronix, Inc., Beaverton, Oregon, 1970.

[7] John Webster (ed.), *Medical instrumentation, application and design*, Houghton Mifflin Co., Boston, 1978.

[8] Fred Alt (ed.). *Advances in biomedical engineering*, Plenum Press, New York, 1966, pp. 93–102, 169–70.

[9] J. Miyara, 'Measuring air flow using a selfbalancing bridge', *Analog dialogue* **5**, 1, January 1971, pp. 13–14. Adapted by permission.

Chapter 3

[1] G. E. Tobey, J. G. Graeme and L. P. Huelsman, *Operational amplifiers: design and applications*, McGraw-Hill, New York, 1971.

[2] Graeme, Chapter 1, [13].

[3] Roberge, Chapter 1, [12].

[4] Smith, Chapter 1, [10].

[5] Larry L. Schick, 'Linear circuit applications of operational amplifiers', *IEEE spectrum* **8**, April 1971, pp. 36–50.

[6] G. B. Clayton, *Operational amplifiers*, Newnes-Butterworth, London, 1971.

[7] J. V. Wait, L. P. Huelsman and G. A. Korn, *Introduction to operational amplifier theory and applications*, McGraw-Hill, New York, 1975.

[8] J. G. Graeme, *Designing with operational amplifiers*, McGraw-Hill, New York, 1977.

[9] D. H. Sheingold, *Analog digital conversion handbook*, Analog Devices, Inc., Norwood, Massachusetts, 1972.

[10] Solomon, Chapter 1, [2].

[11] P. R. Gray and R. G. Meyer, 'Recent advances in monolithic operational amplifiers design', *IEEE transactions on circuits and systems* **CAS-21**, 3, May 1974, pp. 317–27.

[12] G. Wilson, 'Compensation of some operational amplifier based *RC* active networks', *IEEE transactions on circuits and systems* **CAS-23**, 7, July 1976, pp. 443–6.

[13] E. H. Goldberg, 'Stabilization of wide-band DC amplifiers for zero and gain', *RCA review* **2**, 1950, pp. 296–300.

[14] R. L. Young, 'Lift IC op amp performance', *Electronic design*, February 15, 1973, pp. 66–9.

[15] A. Arbel and E. Berg, unpublished notes.

[16] L. Su. Kendall, 'Active filters', *IEEE transactions on circuits and systems* **CAS-10**, 5, October 1976, pp. 2–8 (includes extensive bibliography).

[17] G. C. Temes and S. K. Mitra, *Modern filter theory and design*, John Wiley, New York, 1973.

[18] Claude S. Lindquist, *Active network design with signal filtering applications*, Steward & Sons, Long Beach, California, 1977.

[19] L. Storch, 'Synthesis of constant time-delay networks using Bessel polynomials', *Proceedings of the IRE* **42**, November 1954, pp. 1666–75.

[20] Y. J. Wong and W. E. Ott, *Function circuits, design and applications*, Burr Brown Electronic Series, McGraw-Hill, New York, 1976.

[21] E. A. Guillemin, *Synthesis of passive networks*, John Wiley, New York, 1957, Chapter 9.

[22] H. Babić, 'A linear phase network for pulse shaping', *Inst. Ruder Bošković Report* **1**, Zagreb, Yugoslavia, 1970.

[23] C. H. Mosher, 'Pseudo-Gaussian transfer functions with superlative baseline recovery', *IEEE transactions on nuclear science* **NS-23**, 1, February 1976, pp. 226–8.

[24] H. Rudin, 'Automatic equalization using transversal filters', *IEEE spectrum* **4**, January 1967, pp. 53–9. Copyright © 1967 by the Institute of Electrical and Electronics Engineers, Inc. Adapted by permission.

[25] H. E. Kallman, 'Transversal filters', *Proceedings of the IRE* **28**, July 1940, pp. 302–10.

[26] N. Wiener and Y. W. Lee, US Patent 2 124 599, July 1938.

[27] See, for instance, *Proceedings of the IEEE* **64**, 8, August 1976 (issue on adaptive systems).

[28] C. S. Hartman, D. T. Bell Jr. and R. C. Rosenfeld, 'Impulse model design of acoustic surface-wave filters'. *IEEE transactions on sonics and ultrasonics* **SU-20**, 2, February 1973, pp. 80–93.

[29] R. Melen and D. Buss (eds.), *Charge-coupled devices: technology and applications*, IEEE Press, John Wiley, New York, 1977.

[30] Séquin, 'Antialiasing inputs for charge-coupled devices', *IEEE journal of solid-state circuits* **SC-12**, 6, December 1977, pp. 609–16.

[31] J. L. Berger and J. L. Coutures, 'Cancellation of aliasing in CCD low-pass filters', *IEEE journal of solid-state circuits* **SC–12**, 6, December 1977, pp. 617–25.

[32] G. L. Miller, 'Transversal filters for pulse spectroscopy', *Proceedings of the Second Ispra Nuclear Electronics Symposium, Stresa, 1975*, pp. 9–19.

[33] J. F. Watson, 'Matched filter lumped approximations for pulse applications', *IEEE transactions on circuits and systems* **CAS-21**, 3, May 1974, pp. 364–8.

[34] L. A. Zadeh and J. R. Raggazini, 'Optimum filters for the detection of signals and noise', *Proceedings of the IRE*, October 1952, pp. 1223–31.

[35] W. B. Davenport and W. L. Root, *Random signals and noise*, McGraw-Hill, New York, 1958, p. 246.

[36] B. M. Dwork, 'Detection of a pulse superimposed on fluctuation noise', *Proceedings of the IRE* **38**, July 1971, pp. 771–4.

[37] H. Den Hartog and F. A. Muller, 'Optimum instrument response for discrimination against spontaneous fluctuations', *Physica* **13**, 9, November 1947, pp. 571–80.

[38] A. S. Buchman, 'Noise control in low level data systems', *Electromechanical design*, September 1962, pp. 64–81.

[39] C. Engle, *Techniques to analyze and optimize noise rejection ratio of low level differential data systems*, technical paper **521**, Dana Laboratories, Inc., Irvine, California, December 1965.

[40] H. M. Schlicke and O. J. Struger, 'Getting noise immunity in industrial controls', *IEEE spectrum* **10**, 1973, pp. 30–5.

[41] *Floating measurements and guarding*, application note **123**, Hewlett-Packard Co., Palo Alto, California.

[42] Jack Riedel, *Vibration measurements handbook*, Endevco, San Juan Capistrano, California, Section III on isolation and shielding. Adapted by permission.

[43] R. R. Fullwood, 'Transmission line transformers for ground loop noise suppression', *Nuclear instruments and methods* **75**, 1969, pp. 349–50.

[44] P. Vettiger, 'Linear signal transmission with optocouplers', *IEEE journal of solid-state circuits* **SC-12**, 3, June 1977, pp. 298–302.

[45] B. Widrow, 'Adaptive noise cancelling: principles and applications', *Proceedings of the IEEE*, **63**, 12, December 1975, pp. 1692–716.

[46] E. A. Gere and G. L. Miller, 'The design of a linear gating multiplexer and its use in nuclear spectroscopy', *IEEE transactions on nuclear science* **NS-16**, 1, p. 436.

[47] Van de Plassche, Chapter 1, [15].

[48] W. D. Little and A. C. Capel, 'Digital multiplexing analog signals', *IEEE transactions on computers*, August 1972, p. 920.

[49] D. H. Sheingold (ed.), *Nonlinear circuits handbook*, Analog Devices, Inc., Norwood, Massachusetts, 1974.

[50] J. Giles and A. Seales, *A new high-speed comparator, the AM 685*, application note, June 1972. Copyright © 1978 Advanced Micro Devices, Inc., Sunnyvale, California. Adapted with permission of the copyright owner.

[51] Richard Brunner, *A high-speed dual differential comparator, the MC1514*, application note **AN-547**, Motorola Semiconductor Products, Inc., Phoenix, Arizona, 1971.

[52] L. B. Robinson, 'Reduction of baseline shift in pulse amplitude measurements', *Review of scientific instruments* **36**, 1965, pp. 1830–9.

[53] R. L. Chase and L. R. Paulo, 'A high precision DC restorer', *IEEE transactions on nuclear science* **NS-14**, 1, February 1967, pp. 83 ff.

[54] D. Hary, 'A fetal monitoring system during labor and delivery', M.Sc. thesis, Technion, Israel Institute of Technology, Haifa, 1976.

[55] V. Radeka, 'The effect of baseline restoration on signal-to-noise ratio in pulse amplitude measurements', *Proceedings of the Conference on Semiconductor Nuclear-Particle Detectors and Circuits, Gatlinburg 1967*. Publication **1593**, National Academy of Sciences, 1969.

Chapter 4

[1] A. Sanchez, 'Understanding sample-hold modules', *Analog dialogue* **5**, 6, 1971, pp. 6–9. Adapted by permission.

[2] R. Goldstone and D. Hertz, 'Signal processing A/D converters', *N. E. Electr. Res. & Eng. Meeting*, 1971.

[3] B. M. Gordon and W. H. Seaver, 'Designing sampled-data system', *Control engineering*, April 1961, pp. 127–32.

[4] G. Erdi and P. Henneuse, 'A precision FET-less sample-and-hold with high charge to droop ratio', *IEEE journal of solid-state circuits* **SC-13**, 6, December 1978, pp. 864–73.

[5] P. C. Dow Jr., 'An analysis of certain errors in electronic differential analyzers II – Dielectric absorption', *IRE transactions on electronic computers*, March 1958, pp. 17–22.

[6] A. Arbel, 'Transistorized analog to digital converter for pulse-height analysis', *Proceedings of Conference on Nuclear Electronics, Belgrade, May 1961*, IAEA, Vienna, 1962, Part 2, pp. 3–9.

[7] Graeme, Chapter 3, [8].

[8] D. A. Gedcke and C. W. Williams, *High resolution time spectroscopy*, Ortec, Inc., Oak Ridge, Tennessee, August 1968. Adapted by permission.

[9] See, for example, P. W. Nicholson, *Nuclear electronics*, John Wiley, London, 1974.

[10] B. Sabbah and A. Suhami, 'An accurate pulse-shape discrimator for a wide range of energies', *Nuclear instruments and methods* **58**, 1968, pp. 102–10.

[11] See, for instance, R. Nutt, D. A. Gedecke and C. W. Williams, 'A comparison of constant fraction and leading edge timing with Na I (TI)

scintillators', Nuclear Science Symposium 1969, *IEEE transactions on nuclear science* **NS-17**, 1, February 1970, pp. 299–306.

Chapter 5

[1] D. H. Sheingold and R. A. Ferrero, 'Understanding A/D and D/A converters', *IEEE spectrum* **9**, September 1972, pp. 47–56. Copyright © 1972 by the Institute of Electrical and Electronics Engineers, Inc. Adapted by permission.

[2] For a more detailed treatment see, for instance, E. R. Hnatek, *A user's handbook of D/A and A/D converters*, John Wiley, New York, 1976.

[3] Bruce K. Smith, 'Digital to analog converters and their performance specifications', EEE **12**, 2, November 1970, pp. 57–9.

[4] W. B. Barbour, 'Simplified PCM A/D converter using capacitive charge transfer', *Proceedings of Telemetering Conference*, **1961**, pp. 4.1–4.11.

[5] J. L. McCreary and P. Gray, 'All-MOS charge redistribution A/D conversion technique, Part I', *IEEE journal of solid-state circuits* **SC-10**, 6, December 1975, pp. 370–9.

[6] R. Suárez, P. Gray and D. A. Hodges, 'All-MOS charge redistribution A/D conversion techniques, Part III', *IEEE journal of solid-state circuits* **SC-10**, 6, December 1975, pp. 379–85.

[7] R. J. Van de Plassche, 'Dynamic element matching for high accuracy monolithic D/A converters', *IEEE journal of solid-state circuits* **SC-11**, 6, December 1976, pp. 795–800.

[8] A. J. Monroe, *Digital processes for sampled data systems*. John Wiley, New York, 1962.

Chapter 6

[1] E. Renschler, *Analog to digital conversion techniques*, application note **AN-471**, Motorola Semiconductor Products, Inc., Phoenix, Arizona. Adapted by permission.

[2] Sheingold and Ferrero, Chapter 5, [1].

[3] H. Schmid, *Electronic analog/digital conversions*, Van Nostrand Reinhold, New York, 1970.

[4] E. R. Hnatek, Chapter 5, [2].

[5] De Lotto, P. F. Manfredi, P. Maranesi and R. Vecchio, 'Very precise method for A/D conversion of continuous waveform samples or random pulses'. *International Symposium on Nuclear Electronics, Versailles 1968*, paper 94, La Documentation Française, Volume 2.

[6] D. H. Wilkinson, 'A stable ninety-nine channel pulse amplitude analyser for slow counting', *Proceedings of the Cambridge Philosophical Society* **46**, pp. 508–18.

[7] G. Kelly, 'Pulse amplitude analyzers for spectrometry', *Nucleonics* **10**, 1952, p. 34.

[8] P. N. Nordstrom, 'High speed integrated A/D converter', *International Solid State Circuit Conference, Philadelphia, 1976, digest of technical papers*, p. 151.

[9] F. D. Waldhauer, US Patent 3 187 325, 1965.

[10] Patent 42002072, West Germany, 28 June 1978; 1497806, Britain, 10 May 1978; others pending.

[11] A. Arbel and R. Kurz, 'Fast ADC', *IEEE transactions on nuclear science* **NS-22**, 1975, pp. 440–51.

[12] G. Bekey and W. Karplus, *Hybrid computation*, John Wiley, New York, 1968.

[13] 'B. Gordon of Analogic speaks out on what's wrong with A/D converter specs', *EEE* **17**, 2, February 1969, pp. 54–61. Adapted by permission.

[14] C. Cottini, E. Gatti and V. Svelto, 'A new method for analog to digital conversion', *Nuclear instruments and methods* **24**, 1963, pp. 241–2.

[15] Patent 36694, Italy, 1 July 1963.

[16] T. G. Stockham Jr., 'A–D and D–A converters: their effect on digital audio fidelity', *Digital signal processing*, IEEE Press, John Wiley, New York, 1972, pp. 484–96.

[17] J. Bruce, 'Evaluation of A/D performance', *Raytheon memo* **BAU-502**, 20 April 1971.

[18] B. Dunbridge, 'Analysis of A/D converter aperture time error', *Proceedings of the National Telemetering Conference* **CA-1-1**, 1966, pp. 262–5.

[19] Goldstone and Hertz, Chapter 4, [2]. Adapted by permission.

[20] Marce Eleccion, 'A/D and D/A converters', *IEEE spectrum* **8**, July 1972, pp. 63–6. Copyright © 1972 by the Institute of Electrical and Electronics Engineers, Inc. Adapted by permission.

Chapter 7

[1] Eugene L. Zuch, 'Voltage to frequency converters. Versatility now at low cost', *Electronics*, 15 May 1975. Copyright © McGraw-Hill, Inc., 1975. Adapted by permission.

[2] D. Arnold, 'A logarithmic counter', *Hewlett-Packard journal* **27**, 9, May 1976, pp. 7–8. Adapted by permission.

[3] See, for example, Sheingold, Chapter 3, [9].

[4] See, for example, Application Notes **AN-17**, September 1975, and **AN-19**, December 1976, Precision Monolithics, Inc.

[5] D. Degryse and B. Guerin, 'A logarithmic transcoder', *IEEE transactions on computers* **C-21**, 11, November 1972, pp. 1165–8.

[6] W. A. Fischer and W. B. Risley, 'Improved accuracy and convenience in oscilloscope timing and voltage measurements', *Hewlett-Packard journal* **26**, 4, December 1974, pp. 2–9.

Chapter 8

[1] T. O. Anderson, 'New approaches to data-acquisition system design', *Analog dialogue* **5**, 1, January 1971, pp. 6–8. Adapted by permission.

Chapter 9

[1] *Fast Fourier transform processing*, Spectra Data, Inc., Northridge, California. Adapted by permission.

[2] A. F. Heers, *Statistical analysis of waveforms and digital time-waveform measurements*, Application Note **93**, February 1969, Hewlett-Packard Co., Palo Alto, California. Adapted by permission.

[3] *Field training manual Model 3721 correlator*, May 1970, Hewlett-Packard Co., South Queensferry, West Lothian, Scotland, Adapted by permission.

[4] I. M. Langenthal, 'Correlation and probability analysis', *Saicor signals* **TB 14**, 1970. Signal Analysis Industrial Corp., Hauppauge, New York. Adapted by permission.

[5] See, for example, Nicholson, Chapter 4, [9].

[6] Radeka, Chapter 3, [55].

[7] *Signal correlators and fourier analyzer Models 100A, 101A, 102*, November 1969, Princeton Applied Research Corp., Princeton, New Jersey. Adapted by permission.

[8] H. Wolf, 'Signal identification by power density spectra', *Electro-technology*, January 1966, pp. 44–5. Adapted by permission.

[9] Y. W. Lee, *Statistical theory of communication*, John Wiley, New York, 1960, Section 13.9.

Chapter 10

[1] Charles R. Trimble, 'What is signal averaging'. *Hewlett-Packard journal* **19**, 8, April 1968, pp. 2–7. Adapted by permission.

[2] T. Coor, 'Signal averaging computers', *Industrial research*, May 1972, pp. 52–6. Adapted by permission.

[3] Langenthal, Chapter 9, [4].

[4] *Signal averagers Models TDH-9, CW-1, 160*, Princeton Applied Research Corp., Princeton, New Jersey. Adapted by permission.

[5] *Time interval averaging*, application note **162–1**, Hewlett-Packard Co., Palo Alto, California. Adapted by permission.

[6] See also Arbel, Chapter 4, [6].

Chapter 11

General reference: John J. Allan (ed.), *CAD systems*, North Holland Elsevier 1977.

[1] R. A. Siders, 'Computer aided design', *IEEE spectrum* **4**, November 1967, pp. 84–92. Copyright © 1967 by the Institute of Electrical and Electronics Engineers, Inc. Adapted by permission.

[2] C. Machover, 'Graphic displays I and II', *IEEE spectrum* **14**, August 1977, pp. 24–32; October 1977, pp. 22–7.

[3] C. W. Beardsley, 'Computer aids for IC design, artwork, and mask generation', *IEEE spectrum* **8**, September 1971, pp. 64–79.

[4] R. B. Owen and M. L. Awcock, 'One and two-dimensional position sensing semiconductor detectors', *IEEE transactions on nuclear science* **NS-15**, 1968, pp. 290–303.

[5] J. L. Alberi and V. Radeka, 'Position sensing by charge division', *IEEE transactions on nuclear science* **NS-23**, 1, 1976, pp. 251–63.

[6] J. Borkowski and M. K. Kopp, 'Design and properties of position sensitive proportional counters using resistance-capacitance position encoding', *Review of scientific instruments* **46**, 1975, pp. 951–62 (includes bibliography).

[7] J. Borkowski and M. K. Kopp, 'Some applications of one and two-dimensional position sensitive proportional counters', *IEEE transactions on nuclear science* **NS-17**, 3, 1970, p. 340.

[8] Frank H. Speckhart and Walter L. Green, *A guide to use CSMP*, Prentice Hall, Englewood Cliffs, 1976.

[9] See, for instance, S. V. Pollak and T. D. Sterling, *A guide to PL/I*, Holt, Rinehart and Winston, New York, 1969.

[10] N. Shenhav, 'Optimization of position resolution for a position sensitive detector, employing the rise time method', M. Sc. thesis, Technion, Israel Institute of Technology, Haifa, January 1978.

[11] D. H. Trelaven *et al.*, 'Modelling operational amplifiers for computer-aided circuit analysis', *IEEE transactions on circuit theory* **CT-18**, 1971, pp. 205–7.

[12] G. R. Boyle *et al.*, 'Macromodeling of integrated circuit operational amplifiers', *IEEE journal of solid-state circuits* **SC-9**, 6, 1974, pp. 353–64.

[13] N. B. Rabbat *et al.*, 'A review of macromodeling techniques', *IEEE transactions on circuits and systems* **CAS-9**, 4, 1975, pp. 3–9.

[14] R. Ranfft *et al.*, 'A simple but efficient analog computer for simulation of high speed integrated circuits', *IEEE journal of solid-state circuits* **SC-12**, 1, 1977, pp. 51–8.

[15] I. E. Getreu *et al.*, 'An IC comparator macromodel', *IEEE journal of solid-state circuits* **SC-11**, 6, 1976, pp. 826–33.

Appendix A

[1] J. S. Bendat and A. G. Piersol, *Measurement and analysis of random data*, John Wiley, New York, 1966.

[2] W. R. Bennet, 'Methods of solving noise problems', *Proceedings of the IRE* **44**, May 1956, pp. 609–38.

[3] R. D. Thornton *et al.*, *Characteristics and limitations of transistors*, SEEC Vol. 4, John Wiley, New York, 1966.

[4] Chapter 9, [3].

[5] D. G. Lampard, 'Response of linear networks to random noise', *IRE transactions on circuit theory* **CT-2**, 1, February 1955, pp. 49–57.

[6] F. S. Goulding, 'Pulse shaping in low noise nuclear amplifiers, a physical approach to noise analysis', *Nuclear instruments and methods* **100**, 1972, pp. 493–504.

[7] V. Radeka, 'Optimum signal processing for pulse amplitude spectrometry in the presence of high rate effects and noise', *IEEE transactions on nuclear science* **NS-15**, 3, June 1968, pp. 455–70.

[8] H. Nyquist, 'Thermal agitation of charge in conductors', *Physics Review* **32**, 1928, p. 110.

[9] E. F. Vandivere, 'Noise output of a multiple filter relative to that of the ideal square-response filter', *Proceedings of the IEEE* **61**, 12, December 1963, p. 1771.

[10] See, for example, Davenport and Root, Chapter 3, [35].

Appendix B

[1] P. E. Gray and C. L. Searle, *Electronic principles*, John Wiley, New York, 1969, Chapter 18.

[2] J. Millman and C. C. Halkias, *Integrated electronics*, McGraw Hill, Kogakusha Ltd., Tokyo, 1972, Section 13.7.

[3] A. Arbel, 'Teaching EE undergraduates negative feedback', *EE publication* **283**, July 1976, Faculty of Electrical Engineering, Technion – Israel Institute of Technology, Haifa, Israel.

[4] A. Arbel, 'Identification of the return difference matrix of a multivariable linear control system, in terms of its inverse closed-loop transfer matrix', *Journal of circuit theory and applications* **1**, 1973, pp. 187–90.

[5] See, for example, Ghausi, Chapter 1, [3], Chapter 2.

[6] A. Arbel, 'Mismatch oriented design of feedback circuits and its application to nuclear electronics', *IEEE transactions on nuclear science* **16**, 4, August 1969, pp. 3–15.

[7] W. F. Lovering, 'Analog computer simulation of transfer functions', *Proceedings of the IEEE* **53**, March 1965, pp. 306–7.

[8] Bode, Chapter 1, [16], p. 49.

Appendix C

[1] A. H. Zemanian, 'Properties of pole and zero locations for non-decreasing step responses', *AIEE transaction communications and electronics*, September 1960, pp. 421–6.

[2] See, for example, Truxal, Chapter 1, [19].

[3] Roberge, Chapter 1, [12].

[4] A. Arbel, 'Multistage transistorized current modules', *IEEE transactions on circuit theory*, **CT-13**, 2, September 1966, pp. 302–10.

[5] Waldhauer, Chapter 1, [28].

Appendix D

[1] Elmore, Chapter 1, [32].

[2] R. D. Thorton *et al.*, *Multistage transistor circuits*, SEEC Vol. 5, John Wiley, New York, 1965, Section 8.2.

Appendix E

[1] Adapted from Chapter 1, [20].

Appendix F

[1] Adapted in part from Chapter 9, [7].

Index

Page numbers in *italics* indicate definitions of the indexed terms.